Sustainable Transportation in Natural and Protected Areas

Protected areas are at the centre of nature-based tourism, which is increasingly popular across the world. As visitor numbers increase, so does awareness of the harmful effects that large crowds may have on both natural resources and individuals' recreational experience. This volume considers the challenge of transportation to and within natural and protected areas, the improvement of which has already been recognised as having great potential for mitigating the environmental impacts of tourism.

While several books have focused considerable attention on the management of protected areas in general, little has been said about the specific issue of sustainable transport, an emerging trend that is already reshaping visitation patterns in natural settings. This book provides current knowledge on issues associated with the transportation of visitors in natural and protected areas, and a comprehensive overview of the technical and strategic options available to tackle these issues.

It approaches the subject via three main topics: preferences, or the visitors' attitudes towards transportation; practices, where current approaches are assessed through examples and case studies of successful experiences and methodologies from around the world; and policies, where suggestions and recommendations are put forward for both local scale strategies and broad-scale regulatory action with global relevance. Contributors include academics in the field of natural resource management and tourism, with extensive experience in protected area management and active partnerships with natural park administrations.

Francesco Orsi is Assistant Professor in the Department of Geography at Kansas State University, USA. Previously he was a post-doctoral researcher in the Department of Civil, Environmental and Mechanical Engineering at the University of Trento, Italy.

Sustainable Transportation in Natural and Protected Areas

Edited by Francesco Orsi

LONDON AND NEW YORK

First published 2015
by Routledge

2 Park Square, Milton Park, Abingdon, Oxfordshire OX14 4RN
711 Third Avenue, New York, NY 10017

Routledge is an imprint of the Taylor & Francis Group, an informa business

First issued in paperback 2017

© 2015 Francesco Orsi, selection and editorial material; individual chapters, the contributors

The right of the editor to be identified as the author of the editorial material, and of the authors for their individual chapters, has been asserted in accordance with sections 77 and 78 of the Copyright, Designs and Patents Act 1988.

All rights reserved. No part of this book may be reprinted or reproduced or utilised in any form or by any electronic, mechanical, or other means, now known or hereafter invented, including photocopying and recording, or in any information storage or retrieval system, without permission in writing from the publishers.

Trademark notice: Product or corporate names may be trademarks or registered trademarks, and are used only for identification and explanation without intent to infringe.

British Library Cataloguing-in-Publication Data
A catalogue record for this book is available from the British Library

Library of Congress Cataloging in Publication Data
Sustainable transportation in natural and protected areas / edited by Francesco Orsi.
pages cm
Includes bibliographical references and index.
1. National parks and reserves—Transportation. 2. Protected areas—Public use. 3. Natural areas—Public use. 4. Transportation—Environmental aspects. 5. Sustainable tourism. I. Orsi, Francesco, 1979-
SB486.P83S87 2015
333.78'3—dc23

2014048888

ISBN: 978-1-138-78857-2 (hbk)
ISBN: 978-0-8153-9535-5 (pbk)

Typeset in Bembo
by diacriTech, Chennai

Contents

List of contributors ix
Acknowledgements xv

1 Introduction 1
FRANCESCO ORSI

PART I
Concepts and definitions **9**

2 Sustainability requisites of transportation in natural and protected areas 11
FRANCESCO ORSI

3 Sustainability potential of various transport modes in natural settings 28
FRANCESCO ORSI

PART II
Preferences – Attitudes towards sustainable transportation **43**

4 Visitor preferences toward scheduled bus use in natural and protected areas 45
JO GUIVER, NICK DAVIES AND RICHARD WESTON

5 The shift from automobiles to alternatives and the role of intelligent transportation systems 57
KOURTNEY K. COLLUM AND JOHN J. DAIGLE

6 Park visitor and gateway community perceptions of mandatory shuttle buses 70
JOSHUA D. MARQUIT AND BRITTON L. MACE

7 Sustainable mobility within natural areas from the perspectives of persons with disabilities 82
BRENT LOVELOCK

PART III
Practices – Experiences around the world **97**

8 Managing sustainable mobility in natural areas: the case of South Tyrol (Italy) 99
ANNA SCUTTARI AND MARIA DELLA LUCIA

9 On-demand transport systems to remote natural areas: the Swiss case of Bus Alpin and AlpenTaxi 115
ROGER SONDEREGGER AND WIDAR VON ARX

10 Reducing visitor car use while securing economic benefits in protected areas: application of a market segmentation approach in the Lake District National Park (UK) 127
DAVINA STANFORD

11 Cycle tourism development in parks: the experience of the Peak District National Park (UK) 140
RICHARD WESTON, NICK DAVIES AND JO GUIVER

12 Estimating the effects of 'carrot and stick' measures on travel mode choices: results of a survey conducted in the Dolomites (Italy) 150
FRANCESCO ORSI AND DAVIDE GENELETTI

13 Exploring future opportunities and challenges of alternative transportation practice: a systematic-wide transit inventory across US national parks 168
JOHN J. DAIGLE

14 Participatory planning for the definition of sustainable mobility strategies in small islands: a case study in São Miguel Island (Azores, Portugal) 181
ARTUR GIL, CATARINA FONSECA AND HELENA CALADO

15 Achieving a balance between trail conservation and road development: the case of Bhutan 193
TAIICHI ITO

16 Case studies: lessons learned 208
FRANCESCO ORSI

PART IV
Policies – Strategies and policies for sustainable transportation **213**

17 From conventional to sustainable transportation management in national parks 215
ROBERT MANNING, CHRISTOPHER MONZ, JEFFREY HALLO, STEVEN LAWSON AND PETER NEWMAN

18 Identifying key factors for the successful provision of public transport for tourism 228
ANDREAS KAGERMEIER AND WERNER GRONAU

19 Increasing the economic feasibility of public transport supply in natural areas 239
WERNER GRONAU AND ANDREAS KAGERMEIER

20 Elements that encourage bicycling and walking to and within natural areas 249
NATALIE VILLWOCK-WITTE

21 Helping gateway communities embrace alternative transportation 261
ANNE DUNNING

PART V
Conclusion **275**

22 A glimpse into future research on sustainable transportation in natural settings 277
FRANCESCO ORSI

Index 282

List of contributors

Helena Calado
Assistant Professor
CIBIO – Research Centre in Biodiversity and Genetic Resources (Azores Unit)
Department of Biology
University of the Azores
9501-801 Ponta Delgada
Portugal
calado@uac.pt

Kourtney K. Collum
Doctoral Candidate
Department of Anthropology
University of Maine
5773 South Stevens Hall Room 234B
Orono, ME 04469
USA
kourtney.collum@maine.edu

John J. Daigle
Professor
Program Leader of the Parks, Recreation and Tourism program
School of Forest Resources
University of Maine
221 Nutting Hall
Orono, ME 04469-5755
USA
jdaigle@maine.edu

Nick Davies
Research Assistant
Institute of Transport and Tourism
School of Sport, Tourism and The Outdoors

University of Central Lancashire
UK
njdavies@uclan.ac.uk

Maria Della Lucia
Aggregate Professor in Economics and Business Management
Department of Economics and Management
University of Trento
Via Inama 5
38122 Trento
Italy
maria.dellalucia@unitn.it

Anne Dunning
Associate Professor
Department of Urban Planning
University of Kansas
311 Marvin Hall
1465 Jayhawk Blvd
Lawrence, KS 66045
USA
dunning@ku.edu

Catarina Fonseca
PhD Student
CIBIO – Research Centre in Biodiversity and Genetic Resources (Azores Unit)
Department of Biology
University of the Azores
9501-801 Ponta Delgada
Portugal
catarinafonseca@uac.pt

Davide Geneletti
Associate Professor
Department of Civil, Environmental and Mechanical Engineering
University of Trento
Via Mesiano 77
38123 Trento
Italy
davide.geneletti@unitn.it

Artur Gil
Postdoctoral Researcher

Ce3C - Centre for Ecology, Evolution and Environmental Changes
Azorean Biodiversity Group
Department of Biology
University of the Azores
9501-801 Ponta Delgada
Portugal
arturgil@uac.pt

Werner Gronau
Professor
Tourism, Travel and Transport
School of Business
Stralsund University of Applied Sciences
Germany
werner.gronau@fh-stralsund.de

Jo Guiver
Researcher
Institute of Transport and Tourism
School of Sport, Tourism and The Outdoors
University of Central Lancashire
UK
jwguiver@uclan.ac.uk

Jeffrey Hallo
Associate Professor
Department of Parks, Recreation, and Tourism Management
280B Lehotsky Hall
Clemson, SC 29634-0735
USA
jhallo@clemson.edu

Taiichi Ito
Professor of Wildland Planning and Protected Area Management
Faculty of Life and Environmental Sciences
University of Tsukuba
1-1-1 Tennodai
Tsukuba, Ibaraki 305-8572
Japan
ito.taiichi.ft@u.tsukuba.ac.jp

Andreas Kagermeier
Full Professor
Freizeit- und Tourismusgeographie / Leisure and Tourism Geography

Universität Trier / University of Trier
Campus II - Behringstr.
D-54286 Trier
Germany
andreas.kagermeier@uni-trier.de

Steven Lawson
Director
Resource Systems Group, Inc.
55 Railroad Row
White River Junction, VT 05001
USA
steve.lawson@rsginc.com

Brent Lovelock
Associate Professor
Department of Tourism
University of Otago
Dunedin
New Zealand
brent.lovelock@otago.ac.nz

Britton L. Mace
Full Professor and Chair
Department of Psychology
Southern Utah University
Cedar City, UT 84720
USA
mace@suu.edu

Robert Manning
Steven Rubenstein Professor of Environment and Natural Resources
The Rubenstein School of Environment and Natural Resources
University of Vermont
Burlington, VT 05405
USA
robert.manning@uvm.edu

Joshua D. Marquit
Instructor in Psychology
Psychology Department
Penn State University
Brandywine, PA

USA
j.marquit@aggiemail.usu.edu

Christopher Monz
Associate Professor
Department of Environment and Society
Utah State University
5215 Old Main Hill
Logan, UT 84322-5205
USA
chris.monz@usu.edu

Peter Newman
Professor
Department of Recreation, Park, and Tourism Management
Penn State University
801 G Donald H. Ford Building
University Park, PA 16802
USA
pbn3@psu.edu

Francesco Orsi
Assistant Professor
Department of Geography
Kansas State University
118 Seaton Hall
Manhattan, KS 66506
USA
checco.orsi@libero.it

Anna Scuttari
Researcher
Institute for Regional Development and Location Management
EURAC research
Viale Druso 1
39100 Bolzano
Italy
anna.scuttari@eurac.edu

Roger Sonderegger
Lecturer
School of Business
Lucerne University of Applied Sciences and Arts

Switzerland
roger.sonderegger@hslu.ch

Davina Stanford
Senior Lecturer and Course Leader for the Responsible Tourism Management MSc
School of Tourism, Events and Hospitality
Leeds Beckett University
UK
D.J.Stanford@leedsbeckett.ac.uk

Natalie Villwock-Witte
Assistant Research Professor/Research Engineer
Western Transportation Institute
Montana State University
2327 University Way, Suite 6
Bozeman, MT 59715
USA
natalie.villwock-witte@coe.montana.edu

Widar von Arx
Professor
School of Business
Lucerne University of Applied Sciences and Arts
Switzerland
widar.vonarx@hslu.ch

Richard Weston
Senior Research Fellow
Institute of Transport and Tourism
School of Sport, Tourism and The Outdoors
University of Central Lancashire
UK
RWeston@uclan.ac.uk

Acknowledgements

This book was conceived and written in the context of the "AcceDo" research project, which has received funding from the Provincia Autonoma di Trento (Italy) through a Marie Curie action, European Union's 7th Framework Programme, COFUND-GA2008-226070, "Trentino project – The Trentino programme of research, training and mobility of post-doctoral researchers".

The editor wishes to thank Ashley Wright and other members of the editorial office at Taylor & Francis for constant support throughout the publication process.

Acknowledgements

1 Introduction

Francesco Orsi

In 1909, after several years of work, the so-called 'Große Dolomitenstraße' (the great road of the Dolomites) was finally completed. The road, which measured around 160 kilometres, linked Bolzano/Bozen, Cortina d'Ampezzo and Dobbiaco/ Toblach, crossing various mountain passes, and some of the finest dolomitic landscapes in what was then part of the Austro-Hungarian Empire and is today a portion of north-eastern Italy. The father of this endeavour was Theodor Christomannos, born in Vienna in 1854 from a family of Greek origin, who had very well understood the importance of roads to let people know those beautiful mountains and therefore kick-start tourism activities in the area (Christomannos, 1998; Faggioni, 2012). Until the late nineteenth century, in fact, only a few people adventured into the upper part of the dolomitic territory: these were mostly villagers or aristocrats who could hire local guides to climb some seemingly inaccessible peak. The perseverance of Christomannos, along with the support of the Alpine Club (Deutscher und Österreichischer Alpenverein), could eventually convince politicians to sign the approval for construction in 1897. In his writings, Christomannos speculated that, at a good pace, the entire road could be travelled in three days on foot or by coach (Christomannos, 1998). However, things went faster than he had anticipated and, right after completion, the first cars appeared on the road and proved the whole itinerary could be covered in just one day. Tourism development was not long in coming: annual overnight stays in the main villages passed from tens to thousands and, in a matter of few years, tourism had become the main economic activity in the region. Anyway, Christomannos (who died in 1911) could hardly imagine that a century later the thousands of visitors a year would become thousands a day and that the road during July and August would be systematically packed with vehicles of people aiming to pass from one valley to another, reach some popular trailhead or even just gain a nice viewpoint and take a picture. In fact, road traffic has become considerable in recent decades: noise and crowding are now major concerns, and administrators are afraid of the possible repercussions of traffic-related issues on tourism.

The story of Theodor Christomannos and the 'Große Dolomitenstraße' reminds us of the inextricable link between nature-based tourism and transportation, and warns about the possible detrimental effects of transportation on the environment and eventually tourism itself. Transportation and transportation-related infrastructures provide access to nature and allow people to enjoy it. In fact, there would be no hiking in the wilderness without a road and a car to reach a trailhead, there would be no whale watching without a harbour and a boat to sail in whale-inhabited waters, there would be no downhill skiing without cableways to quickly climb slopes. By enabling outdoor recreation, transportation greatly benefits society through the provision of unique experiences to natural areas' visitors and the support of tourism-based economies. Nevertheless, all of these benefits come at a cost as transportation may also negatively affect the areas it serves and the experiences it provides. Cars, buses, ferries, snow coaches and all the transportation modes that people rely on to enjoy natural places bring significant impacts on the environment by releasing pollutants such as carbon dioxide or particulate, which impair organisms and contribute to climate change, and generating noise, which causes considerable disturbance to humans and animals. Transportation infrastructures like roads, railways, harbours and parking lots deeply modify the naturalness of places contributing to harmful processes (e.g. excessive runoff, habitat fragmentation) and influencing visitors' and residents' perception of the environment. Further, by allowing people to reach places more easily or to discover new places, transportation contributes to an increase in human pressure on natural resources and favours overcrowding, which detracts from the quality of the recreational experience.

Nature-based tourism is an extremely popular activity: people search for natural experiences and are willing to travel long distances and spend considerable amounts of time and money to have them. Various studies tell us that nature-based tourism has been growing significantly over the years, with protected areas being the cornerstone of such trend (Buckley, 2000; Balmford *et al.*, 2009). Officially designated areas (e.g. national parks), however, are not the only destination of people seeking opportunities for outdoor recreation: in fact, many non-protected areas worldwide provide excellent opportunities too and receive millions of visits every year. These areas, which we will broadly refer to as 'natural areas' hereafter, are rather heterogeneous in terms of stable human presence (i.e. from semi-wild areas to rural areas) and are often reasonably accessible from urban centres, therefore offering city dwellers convenient getaways during weekends or short holidays. Due to increasing interest in nature-based recreation, many natural and protected areas, especially in affluent countries, have been experiencing significant transportation-related issues over the last 30 years or so. For example, the Peak District National Park, an easily accessible protected area between the cities of Manchester and Sheffield in the UK, has seen traffic levels on cross-park routes more than double in the period 1980-1999, this causing significant impact on the environment and frustration to both visitors and residents (PDNPA, 2010). The Shuswap Lake, a very popular recreation area in south-central British Columbia (Canada), has recently undergone a dramatic increase in boat traffic, which has resulted in noise, pollution and conflict between different users (e.g. kayakers vs. motorboat users) (Kramer, 2010). Visitors to the Shiretoko National Park, a popular tourist destination in the Hokkaido Island (Japan), experience traffic jams that can last for hours as they attempt to approach some of the most scenic spots with private vehicles (Ishikawa *et al.*, 2013).

The recognition of the seriousness of issues like those described in the examples above has raised global attention on the need for sustainable transportation systems in natural settings. Hence, managers of natural and protected areas worldwide have increasingly adopted rigorous measures to regulate the use of private motorized vehicles and foster a progressive shift to alternative forms of mobility. Among other initiatives, two can be cited for their extent and success: the Alternative Transportation Program (ATP) of the United States National Park Service (NPS) and the 'Alpine Pearls' project in Europe. The former, which was launched in 1998, is aimed at coordinating projects and policies for the implementation of alternative transportation systems (ATS) to and within units of the NPS (http://www.nps.gov.transportation/index.html). The latter, which is the outcome of two European Union's projects (Alps Mobility and Alps Mobility II – Alpine Pearls), started in 2006 as a cooperation between 29 municipalities in six Alpine countries (Austria, France, Germany, Italy, Slovenia, Switzerland) to promote soft mobility through improved public transportation systems (http://www.alpine-pearls.com). These and other experiences have shown that tackling some of the greatest issues commonly associated with transportation (e.g. traffic congestion, air pollution, noise) is actually possible and that visitors are willing to rethink their behaviour in order to address such issues. They have also shown that the use of alternative transportation makes it easier for managers to control visitor flows and use levels (e.g. a bus system lets managers know exactly how many people will get to a destination every hour), thus enhancing the protection of natural resources and recreational experiences. Nevertheless, designing and implementing truly sustainable transportation systems in natural settings is a hard challenge involving a wide array of intertwined environmental and socio-economic considerations that deserve great attention and in-depth technical knowledge.

This book explores the issues and opportunities associated with making transportation in natural and protected areas sustainable, and provides a set of concepts and strategic options for understanding the context and setting plans for action. The volume, which follows in the footsteps of that by Manning *et al.* (2014), adds new insights about the socio-economic implications of transportation in areas where people live and work, and provides an overview of sustainable transportation experiences in natural and protected areas outside the USA. The book hosts contributions from leading scholars working in the fields of transportation, outdoor recreation management and tourism. The structure of the book encompasses five parts, of which three – exploring preferences, practices and policies – constitute the core of the volume.

The first part of the book leads the reader through the concept of sustainable transportation, its complex space and time implications, its peculiar requisites in natural settings, and provides an overview of alternative transportation modes and how these should be used to assure sustainable transportation in natural areas.

Chapter 2, by Francesco Orsi, starts from the concept of sustainability to provide a tentative definition of sustainable transportation in natural settings. This is achieved by listing a set of nine sustainability requisites that transportation should fulfil. Such requisites pertain to the environmental externalities of transportation (e.g. pollution, noise), the role of transportation in managing visitor flows and guaranteeing mobility to as many visitor groups as possible, and eventually the relationship between transportation and the local economy. The chapter is introduced by an analysis of transportation in the light of sustainability's space and time dimensions, emphasizing the need to assess sustainability over adequate spatial and temporal scales.

Chapter 3, by Francesco Orsi, provides an overview of alternative transportation modes that are commonly adopted in natural settings, namely: buses, trains, boats, cableways and bicycles. The key message of the chapter is that sustainability does not come from the mere adoption of a 'green' technology, but from how that technology is designed for and managed in a specific context. Different modes are analysed considering three elements: their positive contribution to the sustainability of a transportation system, their negative contribution to the sustainability of a transportation system and the management and design conditions under which their use can be defined sustainable.

The second part of the book focuses on people's preferences and attitudes towards the characteristics of transportation in natural and protected areas. Knowing such preferences and attitudes is in fact key to defining transportation systems that maximize benefits to both nature and society. Preferences of both visitors and residents are considered, and special attention is paid to the perception of disabled people.

Chapter 4, by Jo Guiver, Nick Davies and Richard Weston, discusses the attitudes and preferences of visitors towards scheduled bus services in rural tourist areas. Special attention is paid to issues like service quality, timing, fares and ticketing, and recommendations on these same issues are eventually provided. The chapter, which is based on evidence from a research work conducted in the UK between 2005 and 2011, includes statements of people interviewed during various surveys and is introduced by a thorough discussion of the benefits of bus services in protected areas for visitors, residents and tourism providers.

Chapter 5, by Kourtney K. Collum and John J. Daigle, explores the shift from automobiles to alternative transportation in protected areas and, in particular, whether and how intelligent transportation systems (ITS) can foster such a shift. This is done through analysis of the results of extended research work conducted in Rocky Mountain National Park (USA). The chapter also includes a comprehensive historical introduction about the shift from automobiles to alternative forms of mobility in US national parks, along with an overview of research conducted to assess visitor perception towards alternative transportation.

Chapter 6, by Joshua D. Marquit and Britton L. Mace, analyses the perception of visitors and residents towards mandatory shuttle buses. The latter, which are used sometimes to exclude any level of private vehicle traffic, may in fact impair the freedom of visitors and the economic activities of communities living within or around a protected area. The chapter is based on the experience conducted in Zion National Park (USA) and sheds a light on the factors determining the success of a mandatory shuttle service and the fears and expectations of business owners.

Chapter 7, by Brent Lovelock, explores an often neglected issue, namely how disabled people perceive sustainable mobility in natural areas. Perspectives of persons with disabilities are explored considering the type and level of access to natural areas they think is appropriate. The chapter, which revisits a study conducted in New Zealand to better understand the perspective of disabled people towards motorized transport in natural settings, considers how sustainable transportation relates to disability and provides guidance for decisions in this field.

The third part of the book reviews current practices of sustainable transportation in natural settings by presenting eight case studies exploring alternative transportation technologies, sustainable mobility initiatives, opportunities for stakeholder involvement and economic issues associated with sustainable transportation. Chapters presenting an innovative method follow the standard structure of a scientific article (i.e. introduction, study area, method, results, discussion, conclusion), whereas chapters reviewing sustainable mobility experiences have a less strict outline. Nonetheless, all chapters provide a detailed description of the study area.

Chapter 8, by Anna Scuttari and Maria Della Lucia, presents a tourism traffic analysis aimed at estimating the environmental impact of inbound tourism, and an exploratory analysis of the effects of traffic management measures on tourism flows. Environmental impact was measured in terms of carbon emissions and energy consumption, whereas the effects of management measures (i.e. incentives to alternative transportation and disincentives to private vehicles) were estimated in terms of variations in tourist arrivals. The study was conducted in South Tyrol (Italy), a popular natural region in the Alps that is fostering sustainable tourism and mobility.

Chapter 9, by Roger Sonderegger and Widar von Arx, presents two innovative on-demand transportation options – Bus Alpin and AlpenTaxi - recently introduced in the Swiss Alps to reduce motorized traffic, while allowing visitors to reach places that are not accessible via traditional public transit. Bus Alpin is a network of small public transit lines serving tourist destinations, whereas AlpenTaxi is a taxi service based on existing private transportation options that is particularly devoted to alpinists and hikers. The chapter explores the success, efficiency and effects of these initiatives.

Chapter 10, by Davina Stanford, shows how market segmentation can be used to reduce the negative impacts of transportation in natural areas while ensuring economic benefits. The proposed method, which is based on Ajzen's theory of planned behaviour, seeks to identify market segments that might demonstrate a high propensity to a positive behavioural change and at the same time could strongly contribute to a destination in economic terms. The study was conducted in the Lake District National Park (UK).

Chapter 11, by Richard Weston, Nick Davies and Jo Guiver, explores how cycle tourism can contribute to improving access to and movement within protected areas, and to contrasting the social and environmental impacts associated with traditional mobility. The study presents the case of the Peak District National Park (UK), which has been historically known for cycle tourism and is implementing various initiatives to further strengthen its role as a cycling destination.

Chapter 12, by Francesco Orsi and Davide Geneletti, explores visitor preferences towards various travel modes to a popular hiking area in the Dolomites (Italy), and analyses the likely effects of various management strategies on travel mode choices. The chapter places an emphasis on the need to carefully design transport management strategies in areas where multiple transport modes serve multiple locations because modal shift in such contexts can generate largely unexpected, and potentially unintended, visitation patterns that harm natural resources and detract from the quality of a visitor's experience.

Chapter 13, by John J. Daigle, depicts the state of alternative transportation in US national parks through analysis of a national transit inventory conducted in 2012 by the National Park Service and the Volpe Center. The chapter provides a comprehensive overview of the funding sources, modal shares and business models of the many alternative transportation systems being run in US national parks. Findings of this and other studies are explored to identify opportunities and challenges for the future of alternative transportation in the national parks.

Chapter 14, by Artur Gil, Catarina Fonseca and Helena Calado, describes a participatory approach to involve stakeholders in the definition of sustainable mobility plans for small cities located in natural regions. The study, which was conducted in the city of Ponta Delgada in the Azores archipelago (Portugal), introduces a relevant but often neglected issue, namely how to make transportation sustainable in urban and peri-urban contexts that represent the gateway of natural areas. The study area – a small island state – presents characteristics of isolation and remoteness that make planning and management particularly complex.

Chapter 15, by Taiichi Ito, discusses the possibility to establish sustainable transportation systems in Bhutan, a country characterized by extended wilderness areas and centuries-old traditions that is experiencing rapid development, including the expansion of the road system. The chapter analyses the influence of road expansion on traditional trails and reviews management strategies and alternative modes of transport for implementation in Bhutan. The specific case of Jigme Dorji National Park is examined.

Chapter 16, by Francesco Orsi, further analyses the eight case studies to derive some key messages that should be retained for improving sustainable transportation in practice.

Table 1.1 summarizes the content of Chapters 8 to 15 by specifying whether they introduce an innovative methodology or review an experience of sustainable mobility; whether they refer to officially designated areas (protected areas) or non-designated areas (natural areas); and whether they explore transportation systems, visitor preferences, economic issues, policy design or policy assessment.

The fourth part of the book provides indications for the definition of policies and strategies that increase the ability of transportation systems in preserving natural resources, meeting visitor demand and safeguarding the quality of life of residents. Particular emphasis is placed on the financial sustainability of transportation systems, the identification of key factors to encourage bicycle and bus use, and recommendations to involve gateway communities in the planning of alternative transportation systems.

Chapter 17, by Robert Manning, Christopher Monz, Jeffrey Hallo, Steven Lawson and Peter Newman, explores the differences between conventional and sustainable transportation management in protected areas. The former is intended as

Table 1.1 Summary of case studies. Each chapter is classified based on its approach (i.e. methodological or review of experience), study area (i.e. designated or non-designated area) and topic (i.e. transportation systems, visitor preferences, economic issues, policy design, policy assessment).

Ch.	*Approach*		*Area*		*Topic*				
	Method	*Experience*	*Designated*	*Non-designated*	*Transport systems*	*Preferences*	*Economic issues*	*Policy design*	*Policy assess.*
8	●		●	●					●
9		●		●	●		●		
10	●		●			●	●	●	
11		●	●		●				●
12	●		●		●	●		●	
13		●	●				●		●
14	●			●				●	
15		●	●	●				●	

a demand-driven approach according to which facilities are designed and operated based on demand. The latter instead is an objective-driven approach according to which objectives describing the level of resource protection desired inform the design and operation of transport facilities. The strengths and weaknesses of the two approaches are described analysing evidences from research conducted in Rocky Mountain National Park (USA) and Denali National Park (USA).

Chapter 18, by Andreas Kagermeier and Werner Gronau, identifies key factors for the provision of efficient public transport in natural and protected areas. The chapter originates from the need to solve a transportation management dilemma: while a high quality transportation is needed to attract visitors, this is very expensive to run and the temporally volatile demand makes its costs hardly sustainable. Solutions to the dilemma may come from actions on both the demand side (i.e. target groups, catchment area) and the supply side (i.e. tourism- and transport-related stakeholders).

Chapter 19, by Werner Gronau and Andreas Kagermeier, explores ways to increase the economic feasibility of public transport supply in natural areas. Starting from considerations about the intrinsic inefficiency of public transport in natural areas (volatile demand, dispersed destinations), the chapter argues that tourists can represent the target to count on to increase demand and therefore interrupt the vicious circle of poor demand and supply. Intervening on a set of conditions (e.g. network design, accessibility, marketing plan), it is possible to attract the leisure demand that is needed to guarantee a better service to citizens while making tourism more sustainable. Evidence from a study in Germany is presented.

Chapter 20, by Natalie Villwock-Witte, explores elements that support bicycling and walking to and within natural areas. Through analysis of experiences in the United States and the Netherlands, four elements are eventually identified: presence of facilities (e.g. bicycle paths, bike rentals); connectivity between urban and natural areas (e.g. train with bike racks); convenience of bicycling/walking over other transport modes (e.g. fees for car users); promotion of alternative modes (e.g. websites, flyers).

Chapter 21, by Anne Dunning, provides guidance for involving gateway communities in the planning, support and use of alternative transportation systems with consideration of local resources and accountability for the performance of the adopted systems. The chapter explores the characteristics of gateway communities and the approaches to engage local stakeholders, with special reference to the identification of relevant groups, the teaching-learning process and communities' motivation to contribute to transportation planning. Finally, emphasis is placed on the importance of enabling local communities to evaluate the effects of alternative transportation systems.

The fifth part of the book comprises only one short chapter that concludes the volume by identifying and presenting areas for future research.

Chapter 22, by Francesco Orsi, explores research topics that should be accorded priority in the future to favour the sustainability of transportation in natural settings. These topics are: sustainable transportation to natural areas, sustainable transportation through ICT and ITS, sustainable transportation in developing countries and sustainable transportation and ecosystem services.

References

Balmford, A., Beresford, J., Green, J., Naidoo, R., Walpole, M., Manica, A. (2009) 'A global perspective on trends in nature-based tourism', *Plos Biology*, vol 7, no 6, pp. 1–6.

Buckley, (2000) 'Tourism in the most fragile environments', *Tourism Recreation Research*, vol 25, no 1, pp. 35–40.

Christomannos, T. (1998) *Die Dolomitenstrasse: Bozen-Cortina-Toblach*, Nordpress, Chiari, Italy.

Faggioni, S. (2012) *Theodor Christomannos: Geniale Pioniere del Turismo Nelle Dolomiti*, Reverdito, Trento, Italy.

Ishikawa, K., Hachiya, N., Aikoh, T., Shoji, Y., Nishinari, K., Satake, A. (2013) 'A decision support model for traffic congestion in protected areas: A case study of Shiretoko National Park', *Tourism Management Perspectives*, vol 8, pp. 18–27.

Kramer, B. (2010) 'Shuswap Lake Boat Traffic Report: A part of the Shuswap Lake Environmental Impact Study 2010/2011', Sicamous, BC, Canada, www.shuswaplakewatch.com/environment/pdffiles/BoatTrafficReport2010.pdf, accessed 7 October 2014.

Manning, R., Lawson, S., Newman, P., Hallo, J., Monz, C. (Eds.) (2014) *Sustainable Transportation in the National Parks: From Acadia to Zion*, University Press of New England.

PDNPA (2010) 'Traffic and Transport in the Peak District National Park, Fact Sheet 5', Peak District National Park Authority, UK.

Part I

Concepts and definitions

2 Sustainability requisites of transportation in natural and protected areas

Francesco Orsi

Introduction

The term sustainable transportation, or sustainable mobility, is used to broadly refer to modes of transport that minimize negative impacts on the environment while ensuring adequate mobility opportunities to individuals as well as economic viability. Over the last 40 years, sustainable transport measures have been increasingly invoked, conceived and eventually adopted in many urban contexts around the world to tackle transport-related environmental impacts (e.g. atmospheric pollution, noise, land use conversions) and socio-economic costs (e.g. high commuting times, accidents) (Newman and Kenworthy, 1989; Kemp and Rotmans, 2004; Schiller *et al.*, 2010). Only recently, however, a similar commitment to favouring more sustainable forms of mobility has been devoted to natural settings (e.g. national parks, rural areas) in response to declining recreational quality as a consequence of the many side effects of traditional mobility (Cullinane, 1997; Daigle and Zimmerman, 2004; Manning *et al.*, 2014). This commitment has resulted in several initiatives being implemented worldwide to reduce the reliance on private vehicles in natural areas by favouring modal shift to alternative forms of mobility (e.g. buses, trains) that guarantee better environmental performances. This is commonly achieved through a variety of measures that involve both incentives (e.g. shuttle buses, discounted bus tickets) to alternative transportation and limitations to the use of private vehicles (e.g. road pricing, restricted car access) (Holding and Kreutner, 1998; Scuttari *et al.*, 2013; Orsi and Geneletti, 2014). For example, in some UK national parks efficient bus services and cycling paths are a strong incentive to reduce car use (see Chapters 4 and 11). In the Alps, an increasing number of tourist destinations offer rapid connections to urban areas via public transit and offer a number of alternative transportation options therefore allowing real car-free vacations (see Chapter 8). In various US national parks, alternative transportation systems (ATS) like shuttle buses are available that enable visitors to reach some of the most interesting locations within a park (e.g. viewpoints, trailheads) without the need of a car (see Chapter 13). However, while these initiatives are commonly labelled 'sustainable' because they improve transit conditions or bring positive effects on the environment (e.g. reduction of pollutants), they might not be entirely so in real terms. In fact, these sustainability labels are often assigned based on a simplistic/partial interpretation of the sustainability concept that neglects the implications of transport-related measures across space and over time. For example, shifting visitors from private vehicles to another form of mobility (e.g. bus) may actually reduce pollutants, but it could concurrently rearrange visitation patterns in a way that is harmful for the environment (e.g. too many visitors simultaneously present at a sensitive location) and/or detrimental for the local economy (e.g. reduced revenues for businesses along the road). In other words, a seemingly sustainable transportation system may generate unintended consequences that can prove negative in the long run. Hence, which characteristics should transportation in natural settings have to be actually sustainable? This chapter aims to provide a definition of sustainable transportation in natural areas by introducing a set of requisites that reflect environmental, social and economic sustainability. The description of sustainability requisites is preceded by an analysis of the sustainability concept applied to transportation in the light of the concept's fundamental space and time dimensions.

Sustainable transportation: space and time

The idea of sustainable transportation derives from the broader concepts of sustainability and sustainable development, the latter of which denotes a kind of 'development that meets the needs of the present without compromising the ability of future generations to meet their own needs' (Brundtland, 1987). These concepts are generally criticized for their ambiguity and vagueness, which inevitably lead to the difficulty of establishing what is sustainable and what is not. A sustainable system is one which persists or survives. Some have argued, however, that this entails a series of hardly understandable questions regarding what should persist and for how long, and when we should assess whether the system has persisted (Costanza and Patten, 1995). Further, the actual assessment of the needs of future generations is not straightforward, as we do not know the technology that will be available in the future (Marshall and Toffel, 2005). This ambiguity has generally resulted in simplistic interpretations of the concept and subsequently the adoption of measures that do not necessarily guarantee long-term benefits to both the environment and society. In the case of transportation, for example, most considerations around sustainability place emphasis on transportation modes, looking at their direct environmental impacts (e.g. atmospheric pollution), but often neglecting the indirect large-scale effects (e.g. land use conversion) of overall transportation systems (Handy, 2005). Common to many interpretations of the sustainability concept is in fact the impossibility or inability to account for the space and time scales over which the concept must apply (Costanza and Patten, 1995). A technology (or action) adopted (taken) here today is actually sustainable if its impacts in the environmental, social and economic realms are acceptable elsewhere tomorrow. Hence, the sustainability of adopting a technology or taking an action can only be evaluated if we are able to identify the overall consequences across appropriate spatial and temporal scales.

The consideration of transport-related impacts across space and over time, though fundamental for a proper sustainability assessment, is inherently difficult as transportation has the power to affect human behaviour, thus generating hardly predictable (and often unintended) consequences that span across space and over time. The basic evidence of the influence of transportation on human behaviour is given by the observation that people go where transportation allows them to go. Hence, transportation establishes where human activity will take place. Further, the efficiency of transportation, especially in terms of speed, alters the perception of space and time. This process, also called time-space compression (Harvey, 2001), may lead to drastic changes in individual mobility (Gössling and Hall, 2006) and subsequently a new series of social interactions (Schäfer, 2000). Following on these considerations, we may assume that the transportation of people and goods from point A (whose spatial location is S_A) to B (whose spatial location is S_B), occurring between time t_A and time t_B, generates effects that systematically transcend the spatial and temporal frame proper of the specific transfer (highlighted by the dashed line), and eventually affect the overall sustainability of the process (Figure 2.1). Not accounting for the discrepancy between the spatial–temporal frame in which transportation takes place and the larger area of its effects inevitably leads to unsustainable transportation systems and policies.

Transportation-related impacts across space and over time may be analysed at three different scales: the transport network, the origin–destination pair and the destination (Figure 2.2). Transportation is based on networks made up of multiple

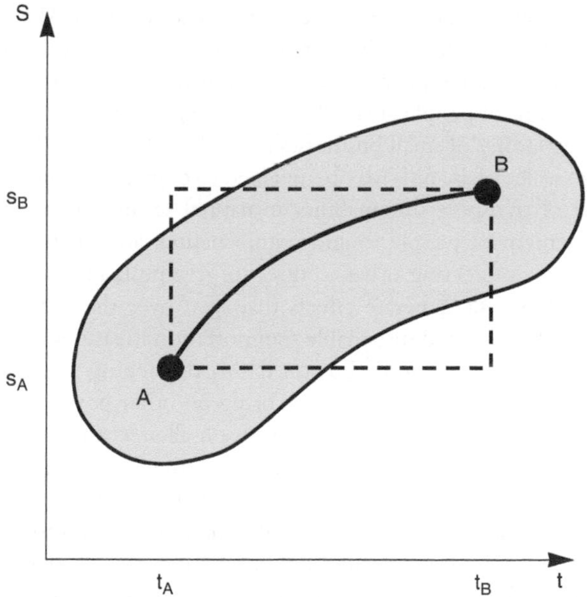

Figure 2.1 The transportation of people and goods between a point A (whose spatial location is S_A) and B (whose spatial location is S_B) occurring between time t_A and time t_B has environmental, social and economic effects that span well beyond the spatial–temporal frame in which transportation takes place.

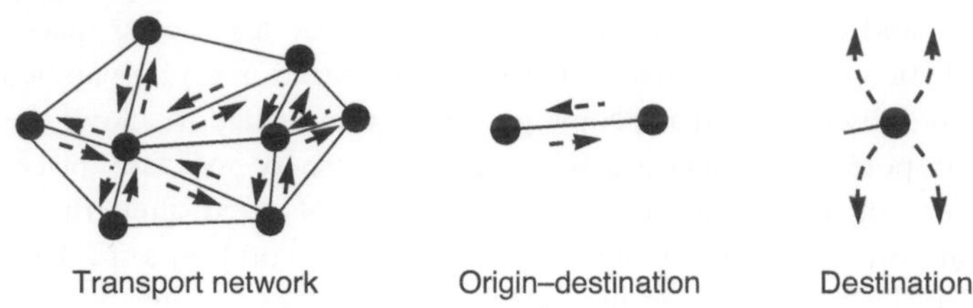

Figure 2.2 The impacts of transportation can be analysed at three scales: the transport network, the origin–destination pair and the destination.

arcs and nodes, and served by a variety of transportation modes. Each arc is (directly or indirectly) connected to the others in such a way that a transport-related measure applied on one arc (e.g. forbidding cars on a road) may bring consequences on the others (e.g. increased traffic on arcs adjacent to that road due to people changing their routes) and eventually a reconfiguration of global flows (Litman, 2001; McDonnell and Zellner, 2011). When considering a specific portion of the network, namely an origin–destination pair, the characteristics of the transport connection between the origin and the destination can have a major effect on the demand of transportation from the origin to the destination over time (e.g. the improvement of a railroad connection can stimulate more people to choose the train in the medium to long term, possibly exceeding the service's capacity) (Ben-Akiva and Lerman, 1985; Fröidh, 2005). Finally, the impact of transportation is not simply related to the actual movement of people between an origin and a destination but also to the presence and actions of people at the destination (e.g. people getting off a bus contribute to the economy of small businesses around the bus stop) (Kim *et al.*, 2004).

These three scales of analysis may be retained to depict the issues associated with the assessment of transport sustainability in natural settings across space and over time. The movement of people to and within natural areas is based on networks of arcs and nodes connecting urban centres to pristine lands. Intervening on any of these arcs is likely to generate effects that span over the entire network. The risk associated with so-called sustainable transportation measures focusing on single arcs (e.g. introduction of a shuttle bus on the road reaching a trailhead) is that of neglecting the impacts that such measures bring on other portions of the network and surrounding areas (e.g. car traffic increases on all arcs reaching the point from where the shuttle bus service is run). When restricting the perspective to a single origin–destination pair, it is important to consider how strongly transportation in natural settings may affect the choice of a destination (i.e. location to visit within a natural area) and therefore the likely transport demand to that destination. As opposed to urban dwellers, in fact, visitors of natural areas are generally free to choose their destination based on a variety of factors, including transport convenience (e.g. comfort, cost, travel time) (Khadaroo and Seetanah, 2008). Hence, by attempting to design a sustainable transportation link (e.g. hydrogen-powered shuttle bus service) to a place, park managers and administrators may end up with a service whose performance and comfort greatly stimulate demand over time, this entailing clearly unsustainable consequences (e.g. increased human presence in sensitive areas). Finally, transportation in natural areas may have enormous consequences at the destination, where visitors leave mechanized forms of mobility and enter pristine lands. The specific location where and the rate at which this occurs eventually determine the density of people at various locations across an area over the day. These outcomes contribute to the definition of the actual sustainability of a transportation system and must be accounted for by park managers and administrators in the planning phase. The issues presented here are summarized in Table 2.1, which also provides a series of questions that can drive the planning and sustainability assessment of transportation systems in natural settings. While Table 2.1 splits issues across the scales of analysis presented in this section (i.e. transport network,

Table 2.1 The spatial–temporal effects of transportation can be analysed at three scales: transport network, origin–destination pair and destination. For each level, the specific issue is reported along with an example from the real world and related questions that may support planning and sustainability assessment.

Scale	*Issue*	*Example*	*Questions*
Transport network	The adoption of a supposedly sustainable measure on one arc of the network may generate unsustainable consequences on other arcs.	Car use on one arc is forbidden and replaced by an electric rail transportation.	How does car traffic vary across the network as a consequence of the measure adopted? What is the environmental and visual impact of cars searching for parking spaces around the departure station? How does the volume of visitors vary in other portions of the area?
Origin–destination	The introduction of a supposedly sustainable link between an origin and a destination may affect the demand.	A shuttle bus service offers a faster and stress-free connection between a village and a lookout.	What are the environmental effects of an increasing number of buses along the road due to an increased demand? Will the transport conditions (e.g. crowding onboard) be acceptable by visitors in the future? Can the shuttle service technically and economically cope with the increased demand in the long term?

(*continued*)

Table 2.1 The spatial–temporal effects of transportation can be analysed at three scales: transport network, origin–destination pair and destination. For each level, the specific issue is reported along with an example from the real world and related questions that may support planning and sustainability assessment. (*continued*)

Scale	*Issue*	*Example*	*Questions*
Destination	The effects of transportation to a destination spread from the destination into the natural land.	A cableway delivers visitors to the heart of a pristine land.	What are the likely environmental effects of people being quickly delivered to the pristine land? At which rate should visitors be delivered to minimize overcrowding issues on the trail network departing from the arrival station? What are the economic impacts that businesses at the arrival station could experience?

origin–destination pair, destination), in the real world the three scales are widely overlapping and can generate a range of complex scenarios (e.g. a new transportation mode, increasing demand, stimulates traffic across the network and raises environmental pressure at the destination). Therefore, a thorough sustainability assessment is expected to embrace all levels simultaneously.

Sustainability requisites

Sustainable transportation systems are expected to bring positive impacts on a wide range of elements pertaining to the environmental (e.g. air pollution, noise, habitat loss), social (e.g. equity, human health, community cohesion) and economic (e.g. infrastructure costs, price for the user, accidents) components (Litman, 2007). Natural and protected areas aim to protect natural features of significant importance and provide visitors with valuable recreational opportunities. Transportation systems, by allowing the access and movement of visitors as well as by modifying the environment, can greatly contribute to or detract from the achievement of this dual goal. Hence, a transportation system operating in a natural context, in order to be sustainable, should preserve natural resources, enhance the recreational experience of visitors, safeguard the quality of life of local communities and be economically feasible. More precisely, a sustainable transportation system is one that

1 minimizes atmospheric pollution
2 minimizes noise
3 minimizes land use conversions
4 minimizes the direct impacts of visitation on the environment

5 minimizes the impacts of visitation on the recreational experience
6 safeguards the visual perception of naturalness
7 enables all visitor groups to move freely
8 ensures the protection of local communities' quality of life
9 is financially sustainable

These requisites, which comprehensively address the three components of sustainability (i.e. environment, society, economy), individually mostly refer to one or two of such components (Figure 2.3).

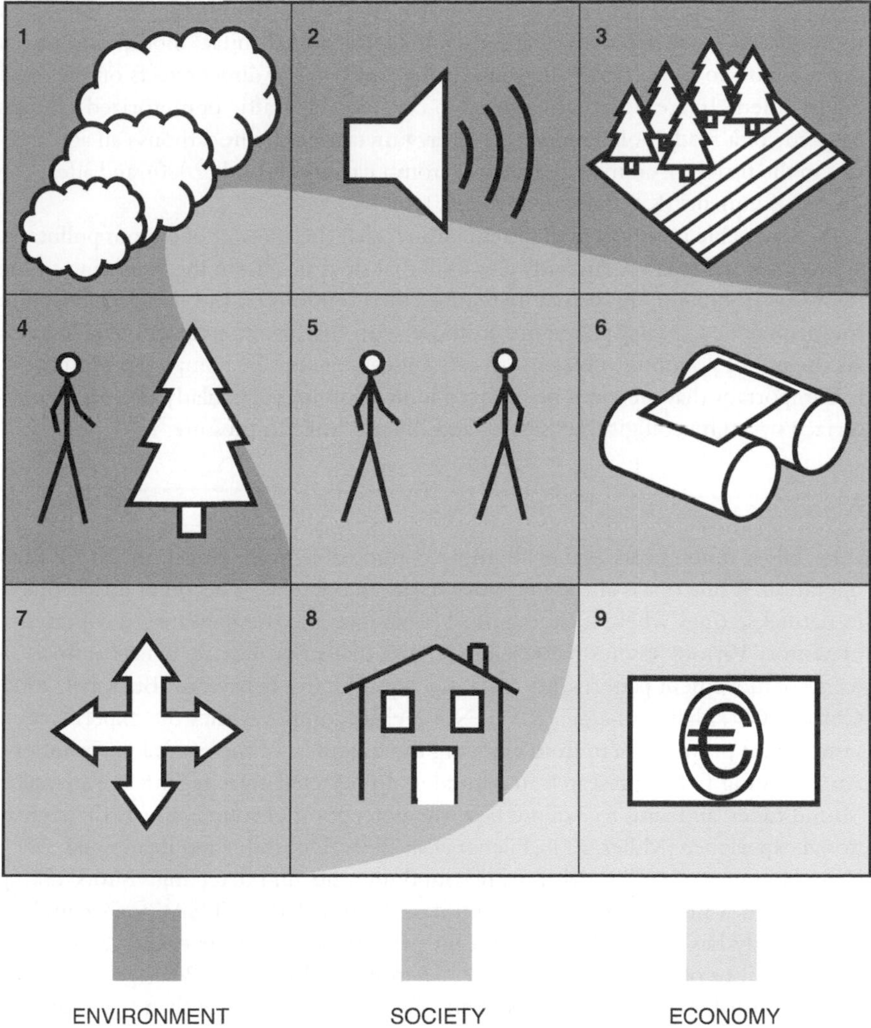

Figure 2.3 A sustainable transportation system should satisfy a number of requisites referring to the three components of sustainability (i.e. environment, society, economy). Each requisite individually may refer to one or even two of such components.

Atmospheric pollution

One of the main negative outcomes of transportation is the emission of various pollutants, including carbon dioxide (CO_2), benzene (C_6H_6), nitrogen compounds (NO_x), particulate, etc., which may affect the life of living organisms and alter the climate (Krzyzanowski *et al.*, 2005; Pope III and Dockery, 2006; Chapman, 2007). Carbon dioxide in particular is responsible for the well-known phenomenon of global warming, which in the next decades is expected to have dramatic consequences on the level of oceans, the availability of water and eventually the global economy (Lashof and Ahuja, 1990; Frankhauser, 1994; Barnett *et al.*, 2005; Meehl *et al.*, 2005; Nordhaus, 2006; Solomon *et al.*, 2009; Peters *et al.*, 2013). While transportation in natural settings can clearly contribute to climatic alterations and their global consequences (e.g. glacier meltdown, habitat modifications) in the medium to long term, it may also bring some major direct effects on the local environment. In areas that are exposed to considerable traffic of motorized vehicles (e.g. side of a road), pollutants such as heavy metals and hydrocarbons can accumulate in the tissues of animals and plants (Trombulak and Frissell, 2000), and affect the health of humans (Krzyzanowski *et al.*, 2005).

A sustainable transportation system ensures that the amount of various pollutants across a natural area is constantly at a level that does not harm the wellbeing of the living environment. This primarily requires interventions on the vehicles allowed in the natural area, giving preference to those with the lowest emission of pollutants. As the use of polluting vehicles (e.g. cars, buses) may not be completely eliminable, it is important that measures be taken to limit such use, particularly in areas characterized by high ecological sensitivity and intense tourism pressure.

Noise

Cars, buses, trains, boats, planes all imply considerable noise emissions during their operation. While this is already negative in urban contexts, it becomes unacceptable in natural settings where it alters animal behaviour and the recreational experience of visitors. Various studies have demonstrated that noise heavily affects animals in terms of movement patterns, life cycle and reproductive behaviour (Buckstaff, 2004; Coffin, 2007; Halfwerk *et al.*, 2011). Noise clearly impoverishes the experience of visitors as it prevents them from enjoying the quietness of the natural environment. An emerging field of research in natural and protected areas regards the so-called 'soundscapes' and aims to explore how the perception of sounds affects the recreational experience (Miller, 2008; Pilcher *et al.*, 2009; Dumyahn and Pijanowski, 2011; Weinzimmer *et al.*, 2014). Measuring sound intensity and detecting visitors' hiking patterns on a portion of Rocky Mountain National Park (USA), for example, a recent study has shown that a significant percentage of visitors cannot experience the conditions of natural quiet for over 15 minutes (Park *et al.*, 2010).

A sustainable transportation system ensures that noise levels are within the limits of acceptability of both animals and humans on the largest possible portion of a natural area. In particular, the noise generated by such a transport system must allow visitors to experience the natural quiet during most of their itineraries.

This is possible if preference is given to transportation modes that cause low noise emissions (e.g. electric vehicles), the volume of traffic is controlled and transport routes are carefully selected to avoid interference between vehicle passage and animal (e.g. breeding) and human (e.g. hiking) activities. Mitigation measures (e.g. sound barriers) are a further option available for minimizing interferences whenever simpler options are not feasible.

Land use conversions

A great impact of transportation is due to the conversion of natural land uses into artificial ones to make space for infrastructures. In fact, the creation of a road or a railway, the construction of a parking lot or a harbour may reduce the size of habitats, the ecological connectivity of a landscape and the space available for outdoor activities. This is especially evident in small and highly visited areas, where the space needed by transport-related facilities (e.g. parking lots) may be considerable as compared to the overall surface of such areas. Artificial surfaces (e.g. asphalt) are a reservoir of toxic contaminants that enter the environment via stormwater runoff (Coffin, 2007) and can be transported at great distances from the source (Forman *et al.*, 2003). Linear transport infrastructures (e.g. roads, railways), though not implying massive conversions of land use, constitute hardly surmountable barriers, which break habitats into pieces, thwarting the movement of animals and harming their survival (Alexander and Waters, 2000). As transport networks get more extended and complex, their effect on landscape changes and habitat fragmentation gets stronger, putting the life of some species at considerable risk (Forman and Alexander, 1998). In addition to that, linear infrastructures like roads or railways constitute a further risk for animals, which can be hit by vehicles moving on such infrastructures. Roadkill is in fact one of the main reasons of vertebrate mortality (Forman and Alexander, 1998).

A sustainable transportation system requires minimal land use conversions and ensures that these are made in the least sensitive portions of a natural area. The objective of minimizing land use conversions can be achieved if preference is given to transportation modes that need the smallest infrastructure per passengers transported in a unit of time. Particular attention should be paid to areas where visitors switch from one mode to another (e.g. car to shuttle bus) because, given visitation patterns and the difference in capacity between the two modes, considerable space may be needed for enabling modal shift (e.g. parking lots). In general, it is very important that the location, size and shape of transport infrastructures be conceived in a way that minimizes pressure on sensitive areas. In order to minimize the impact of linear infrastructures, ad hoc mitigation measures (e.g. under- or overpasses for fauna) can also be adopted.

Direct impacts of visitation on the environment

As described in this chapter, the sustainability of a transportation system should be assessed across space and over time, and consider impacts at the destination.

In natural settings, the mere presence of visitors may heavily affect natural resources. The daily passage of hundreds of people can have serious impacts on plant biodiversity and vegetation, including reduction in plant height and living biomass, damage to seedlings, soil loss and compaction, spread of weeds, among others (Pickering and Hill, 2007). The presence of visitors can also affect wildlife through habitat alteration and loss, modification of behaviour, reduction of health and fitness (Leung and Marion, 2000). All of these effects, which may also considerably reduce the quality of the visitor's experience (i.e. people get in contact with a degraded environment), clearly get more intense as the number of visitors per unit time increases. The transportation system plays a major role on the intensity of such impacts as it ultimately determines which locations people will visit and at what rate they will reach these locations.

A sustainable transportation system delivers visitors at locations and rates that keep disturbance to flora and fauna within acceptable limits. This is possible if the specific transportation modes, the routes they follow, the stops they make, the capacity they have are all defined based on the respect of the above-mentioned limits, rather than on some efficiency-oriented objective (e.g. maximize flow). Clearly, such a perfect design can only be obtained if administrators and park managers have in-depth information about the behaviour of visitors, because this will eventually allow them to understand the relationship between a given design and the subsequent distribution of visitors in space and time.

Impacts of visitation on the recreational experience

Driving on a congested panoramic road, travelling on an overcrowded sightseeing bus or encountering hundreds of hikers on a trail are experiences that do not really correspond to the image of quietness and naturalness that visitors commonly associate with natural areas. Starting from the concept of recreational carrying capacity (Wagar, 1964), a multitude of studies has shown that the presence of other people considerably affects the quality of an individual's recreational experience in natural contexts (Vaske *et al.*, 1980; Vaske and Donnelly, 2002; Arnberger and Brandenburg, 2007; Hallo and Manning, 2009). Transportation systems may contribute to the preservation of the quality of recreation, as they are both a way of directly experiencing natural areas (e.g. driving for pleasure in a national park) and a means to convey people to places where a recreational activity is performed (e.g. using a cableway to let people reach a trailhead).

A sustainable transportation system prevents visitors from experiencing conditions of overcrowding that may impair the quality of the recreation. This has two major implications for design. On the one hand, the system should accommodate the demand of visitors and allow these to travel under uncrowded conditions (e.g. empty road, bus with available seats). On the other hand, the system should deliver visitors at locations and rates that prevent visitors from experiencing crowding in their recreational activities. The design of such a system must rely on an analysis of how different groups of visitors (e.g. families, elders) are affected by encounters (both on the means of transportation and at destination). Knowing under which conditions different groups perceive crowding, which itineraries they prefer, and what behaviour they show, it will be possible for administrators and park managers to design a transportation system that maximally satisfies each group's expectations.

Visual perception of naturalness

Natural and protected areas should enable visitors to experience environments that are minimally influenced by human activities. The perception of human influence strongly depends on the visibility of human artefacts (Carver *et al.*, 2012; Orsi *et al.*, 2013). Transportation infrastructures like roads, railways, harbours and parking lots may be visible from very long distances, thus detracting much from the quality of the visitor experience on large portions of a natural area. For example, a remote and highly preserved place from which a road is visible can only offer limited wilderness experiences as compared to an equivalent place from which no road is seen.

A sustainable transportation system relies on facilities that are minimally visible from any point within a natural area. This requisite can be met by intervening at two levels: technology and design. While preference should be accorded to transportation modes that require minimal infrastructures, a careful design is fundamental to minimize those visual impacts that are inevitable. Linear infrastructures like roads or railways should follow paths that are minimally visible, especially from the most valuable portions of a natural area. Tunnels or other mitigation measures could be conceived to exclude impacts when these would be inevitable and particularly strong (e.g. in a narrow valley where the road or railway would be visible from all upper locations). The location and size of facilities like stations or harbours should also be carefully planned to make these poorly visible from elsewhere. Whenever possible, and if this does not bring other kinds of impact, actions should be taken to camouflage such facilities into the natural environment.

Freedom of movement

Outdoor recreation is intimately related to the possibility of moving freely across natural areas. In order to fully enjoy a park, visitors need to wander around it and reach locations from which to contemplate the scenery (e.g. lookouts) or perform recreational activities (e.g. trailheads). Nevertheless, this freedom of movement should not be the right of only specific visitor groups (e.g. young hikers), but a possibility for all. This is dependent on transportation systems that guarantee comfort, freedom and reliability to the largest possible public. In this respect, private vehicles (e.g. cars, motorbikes) are generally seen as the best option because they give visitors maximum freedom regarding where and when to go, and how much time to spend at a given location: something that can be hardly equalled by common public transport. However, a transportation system entirely based on private vehicles would not allow any recreation opportunity to, say, people without a car, thus contributing to a form of discrimination. A similar form of discrimination is due to transportation facilities tailored on the needs and expectations of able-bodied visitors, but totally neglecting those of disabled people (Lovelock, 2010).

A sustainable transportation system allows all visitor groups (e.g. families, elders, disabled people) to move freely across a natural area. Such a transportation system operates on routes that touch all key locations within a natural area. Schedules, intended as both hours of service (e.g. bus running from 8 am until 6 pm) and frequency of service (e.g. bus every 30 minutes), should guarantee a certain degree of freedom to visitors (e.g. a given location can be reached anytime throughout the day with acceptable wait times). The various transportation modes should be integrated in a unique schedule and fare system that minimizes losses of time (e.g. bus lines are coordinated) and money (e.g. only one ticket is needed for all legs of a complex itinerary) as well as constrained paths (e.g. visitors willing to move from A to B are not required to pass through C if this is off-the-track). Finally, vehicles and transportation infrastructures should be designed in a way that makes their use possible by all visitor groups. For example, a bus with no room for extra gear will be hardly used by a family with kids, just as a boat with a narrow and unsafe gangway will not allow a disabled person to get on board. The fact that a transportation system should give everyone the possibility to move does not mean, however, that all visitors should be put in the condition to go everywhere. Technical issues (e.g. maximum capacity of a bus) and management conditions (e.g. carrying capacity of a place) will ultimately determine the range of movement of each group.

Quality of life of the local communities

Transportation can deeply affect the quality of life of human communities living within or around a natural area for at least three reasons. First, transportation allows local residents to move and manage their daily activities. Second, transportation allows visitors to move, therefore sustaining outdoor activities, which contribute to the local economy. Third, transportation has direct negative impacts on the local population (e.g. noise, landscape alterations). Given the complex links between transportation and residents' quality of life, it is clear that the design of a transportation system can bring in a wide range of (often conflicting) consequences. A transportation system that is extremely convenient for visitors (e.g. rapid shuttle bus connecting the village and the hiking area), for example, could be detrimental for the local economy (e.g. lack of intermediate stops cuts revenues to local businesses along the road). While the issue of transportation and local communities in natural settings has been poorly investigated in the past, it is one of primary importance (Dunning, 2005).

A sustainable transportation system enhances the quality of life of local communities by guaranteeing freedom of movement, sustaining the local economy and minimizing negative externalities. The objective of guaranteeing visitors' mobility cannot be achieved at the expenses of residents' mobility (e.g. road restrictions should target visitors and residents differently) because this would have some heavy socio-economic consequences (e.g. impossibility to carry on normal businesses). Instead, visitors and residents could benefit from each other, as the former, by raising the demand of transportation, sustain the transit system that is also used by the latter. The design of a transportation system can do much for the local economy by driving visitor flows to the places where they would bring the greatest economic benefit (e.g. bus stop in front of a restaurant). Design is also fundamental to minimize the exposure of the local population to some of the negative outcomes of transportation, including pollution, noise, aesthetic impacts, that may be more easily hidden to visitors (e.g. big parking lot not visible from hotels but perfectly visible from a group of houses).

Financial sustainability

The transportation of goods and people has an economic cost. This is borne by both individuals, who pay for the expenses associated with the use of private vehicles, and administrations, which are responsible for public transit and transportation infrastructures. In natural areas, all attempts at shifting visitors from private vehicles to alternative forms of transportation are challenged by the cost of the latter and the fact that visitors may be reluctant to pay for it, especially if such forms of transportation do not guarantee an adequate degree of freedom (Orsi and Geneletti, 2014). This results in a vicious cycle where administrations, being constrained by the available funds, can only provide a limited transport service (e.g. few routes, low frequency), which in turn attracts few people and subsequently generates little revenues to be used for improvements (see Chapters 18 and 19). The solution to this problem is not straightforward, especially because the extra investment needed to make the service attractive may not always be paid back.

A sustainable transportation system has its costs entirely covered by direct or indirect revenues and is managed in a way that allows it to function efficiently over a long period. This requires both a detailed business plan, which thoroughly considers the revenues coming from public and private contributions, as well as the payment of tickets, and a sound design, which minimizes the cost of running the system. The business plan and the design should be conceived simultaneously and informed by detailed knowledge about how service quality and user fees affect the willingness of visitors to choose a given transportation mode. In an attempt to favour a shift from private to public transportation, managers may impose fees on the use of private vehicles (e.g. road tolls) through which to sustain public transit and improve its quality. Public-private partnerships are another valuable instrument to subsidize a transportation system. In this case, the private investor may not offer only the transport, but a wider package of services (e.g. panoramic buses, bars) that ensure higher ridership and adequate revenues. In natural areas with a significant resident population, the cost for transportation can be considerably limited by expanding the existent public transportation system (e.g. scheduled buses). This brings significant advantages to the local population and can invert the above-mentioned vicious cycle.

Conclusion

Providing a concise definition of sustainable transportation is a difficult task given the many effects of transportation on the environment, society and economy across space and over time. This is especially true in natural contexts where transportation guarantees the mobility of visitors while potentially harming the natural environment they came to enjoy. This chapter has tried to define sustainable transportation in natural areas by introducing nine basic requisites that transportation should satisfy. These reflect the contribution transportation can give to the maintenance of a natural area's main functions, namely the preservation of natural resources, the provision of high quality recreational opportunities and the enhancement of local communities' living conditions. As suggested by Figure 2.3, most of the requisites are transversal to the sustainability components (i.e. environment, society, economy) because the satisfaction of one requisite generally brings in cross-cutting benefits (e.g. noise reduction is good for animals and humans, and can improve the quality of tourism, which benefits local communities).

Consistent with frameworks developed for sustainable transportation (Litman, 2007) and protected area management (Manning, 2007), the actual sustainability of a transportation system can be assessed if, for each requisite, one or more measurable indicators are identified. Examples of indicators may be: the concentration of particulate in a 100-metre buffer around a parking lot, noise level in a 100-metre buffer around a road, number of people on a section of a trail, weekly revenues of local businesses. Each indicator should be assigned a standard or target, which expresses the level at which a requisite is satisfied (e.g. noise must be within 35 decibels during 80 per cent of the day). Standards should account for the environmental and socio-economic characteristics of an area (e.g. size, number of daily visitors). While the definition of environmental indicators may be simply based on data from the literature (e.g. noise beyond 50 decibels heavily affects the behaviour of a species), things are more complicated for indicators (e.g. number of cars per hour) measuring issues (e.g. traffic) that can be perceived differently by different people (e.g. 100 cars per hour may be acceptable for someone and absolutely inacceptable for someone else). In this case, in-depth studies are needed to understand the relationship between the levels of the indicator and the perception of people. A sustainable transportation system is ultimately one that ensures the respect of standards across space and over time. Monitoring campaigns, performed at a number of locations across the study area over a wide time horizon, are necessary to verify that standards are not exceeded. Simulation may be helpful to a priori assess the likely effects of a transportation system (e.g. visual impact of a cableway station) before it is actually put in operation, therefore informing its design.

While prior discussions on sustainable transportation have generally focused on the minimization of negative externalities, it is now time to consider the positive contribution of transportation to the environment, society and economy. In parks and protected areas, transportation is being increasingly considered as a powerful management tool, one that can be used by administrators to achieve/maintain conditions (e.g. crowding levels) that are beneficial to both natural resources and visitors (Manning, 2007; White *et al.*, 2011; Reigner *et al.*, 2012; Manning *et al.*, 2014). This new perspective, which is delineated by requisites referring to the direct impacts of visitation, suggests that the sustainability of a transportation system be evaluated in the light of how it contributes to the achievement of a natural area's management objectives.

References

Alexander, S. M. and Waters, N. M. (2000) 'The effects of highway transportation corridors on wildlife: a case study of Banff National Park', *Transportation Research Part C: Emerging Technologies*, vol 8, no 1–6, pp. 307–320.

Arnberger, A. and Brandenburg, C. (2007) 'Past on-site experience, crowding perceptions, and use displacement of visitor groups to a peri-urban national park', *Environmental Management*, vol 40, pp. 34–45.

Barnett, T. P., Adam, J. C. and Lettenmaier, D. P. (2005) 'Potential impacts of a warming climate on water availability in snow-dominated regions', *Nature*, vol 438, pp. 303–309.

Ben-Akiva, M. and Lerman, S. R. (1985) *Discrete Choice Analysis: Theory and Application to Travel Demand*, MIT Press, Cambridge, MA.

Brundtland, G. H. (1987) 'Our common future' in *Report of the World Commission on Environment and Development*, Oxford University Press, Oxford.

Buckstaff, K. C. (2004) 'Effects of watercraft noise on the acoustic behavior of bottlenose dolphins, Tursiops truncates, in Sarasota Bay, Florida', *Marine Mammal Science*, vol 20, no 4, pp. 709–725.

Carver, S., Comber, A., McMorran, R. and Nutter, S. (2012) A GIS model for mapping spatial patterns and distribution of wild land in Scotland, *Landscape and Urban Planning*, vol 104, no 3–4, pp. 395–409.

Cervero, R. and Kang, C. D. (2011) 'Bus rapid transit impacts on land uses and land values in Seoul, Korea', *Transport Policy*, vol 18, no 1, pp. 102–116.

Chapman, L. (2007) 'Transport and climate change: a review', *Journal of Transport Geography*, vol 15, no 5, pp. 354–367.

Coffin, A. W. (2007) 'From roadkill to road ecology: A review of the ecological effects of roads', *Journal of Transport Geography*, vol 15, no 5, pp. 396–406.

Costanza, R. and Patten, B. C. (1995) 'Defining and predicting sustainability', *Ecological Economics*, vol 15, no 3, pp. 193–196.

Cullinane, S. (1997) 'Traffic management in Britain's national parks', *Transport Reviews*, vol 17, no 3, pp. 267–279.

Daigle, J. J. (2008) 'Transportation research needs in national parks: a summary and exploration of future trends', *The George Wright Forum*, vol 25, no 1, pp. 57–64.

Dumyahn, S. L. and Pijanowski, B. C. (2011) 'Beyond noise mitigation: managing soundscapes as common-pool resources', *Landscape Ecology*, vol 26, pp. 1311–1326.

Dunning, A. E. (2005) 'Impacts of transit in national parks and gateway communities', *Transportation Research Record*, vol 1931, pp. 129–136.

Forman, R. T. T. and Alexander, L. E. (1998) 'Roads and their major ecological effects', *Annual Review of Ecology and Systematics*, vol 29, pp. 207–231.

Forman, R. T. T., Sperling, D., Bissonette, J. A., Clevenger, A. P., Cutshall, C. D., Dale, V. H., Fahrig, L., France, R., Goldman, C. R., Heanue, K., Jones, J. A., Swanson, F. J., Turrentine, T. and Winter, T. C. (2003) *Road Ecology: Science and Solutions*, Island Press, Washington, DC.

Frankhauser, S. (1994) 'The economic costs of global warming damage: a survey', *Global Environmental Change*, vol 4, no 4, pp. 301–309.

Fröidh, O. (2005) 'Market effects of regional high-speed trains on the Svealand line', *Journal of Transport Geography*, vol 13, no 4, pp. 352–361.

Gössling, S. and Hall, C. M. (2006) 'An introduction to tourism and global environmental change', in S. Gössling and C. M. Hall (eds) *Tourism and Global Environmental Change: Ecological, Social, Economic and Political Interrelationships*, Routledge, pp. 1–33.

Halfwerk, W., Holleman, L. J. M., Lessells, C. M. and Slabbekoorn, H. (2011) 'Negative impact of traffic noise on avian reproductive success', *Journal of Applied Ecology*, vol 48, no 1, pp. 210–219.

Hallo, J. C. and Manning, R. E. (2009) 'Transportation and recreation: a case study of visitors driving for pleasure at Acadia National Park', *Journal of Transport Geography*, vol 17, no 6, pp. 491–499.

Handy, S. (2005) 'Smart growth and the transportation-land use connection: what does the research tell us?', *International Regional Science Review*, vol 28, no 2, pp. 146–167.

Harvey, A. S. (2001) *Spaces of Capital: Towards a Critical Geography*, Routledge, New York.

Holding, D. M. and Kreutner, M. (1998) 'Achieving a balance between "carrots" and "sticks" for traffic in national parks: the Bayerischer Wald project', *Transport Policy*, vol 5, no 3, pp. 175–183.

Kemp, R. and Rotmans, J. (2004) 'Managing the transition to sustainable mobility', in B. Elzen, F. Geels and K. Green (eds) *System Innovation and the Transition to Sustainability: Theory, Evidence and Policy*, Edward Elgar, Cheltenham.

Khadaroo, J. and Seetanah, B. (2008) 'The role of transport infrastructure in international tourism development: A gravity model approach', *Tourism Management*, vol 29, no 5, pp. 831–840.

Kim, J. H., Pagliara, F., Preston, J. (2004) 'Transport policy impact on residential location', *International Review of Public Administration*, vol 9, no 1, pp. 71–87.

Krzyzanowski, M., Kuna-Dibbert, B. and Schneider, J. (eds) (2005) *Health Effects of Transport-related Air Pollution*, World Health Organization, Regional Office for Europe, Copenhagen.

Lashof, D. A. and Ahuja, D. R. (1990) 'Relative contributions of greenhouse gas emissions to global warming', *Nature*, vol 344, pp. 529–531.

Leung, Y. F. and Marion, J. L. (2000) 'Recreation impacts and management in wilderness: a state-of-knowledge review', USDA Forest Service Proceedings RMRS-P-15, vol 5, pp. 23–48.

Litman, T. (2001) 'Generated traffic: implications for transport planning', *ITE Journal*, vol 71, no 4, pp. 38–47.

Litman, T. (2007) 'Developing indicators for comprehensive and sustainable transport planning', *Transportation Research Record*, vol 2017, pp. 10–15.

Lovelock, B. A. (2010) 'Planes, trains and wheelchairs in the bush: Attitudes of people with mobility-disabilities to enhanced motorized access in remote natural settings', *Tourism Management*, vol 31, no 3, 357–366.

Manning, R. E. (2007). *Parks and Carrying Capacity: Commons Without Tragedy*, Island Press, Washington, DC.

Manning, R., Lawson, S., Newman, P., Hallo, J. and Monz, C. (eds) (2014) *Sustainable Transportation in the National Parks: from Acadia to Zion*, University Press of New England, Lebanon, NH.

Marshall, J. D. and Toffel, M. W. (2005) 'Framing the elusive concept of sustainability: a sustainability hierarchy', *Environmental Science and Technology*, vol 39, no 3, pp. 673–682.

McDonnell, S. and Zellner, M. (2011) 'Exploring the effectiveness of bus rapid transit a prototype agent-based model of commuting behavior', *Transport Policy*, vol 18, no 6, pp. 825–835.

Meehl, G. A., Washington, W. M., Collins, W. D., Arblaster, J. M., Hu, A., Buja, L. E., Strand, W. G. and Teng, H. (2005) 'How much more global warming and sea level rise?', *Science*, vol 307, no 5716, pp. 1769–1772.

Miller, N. P. (2008) 'US national parks and management of park soundscapes: A review', *Applied Acoustics*, vol 69, no 2, pp. 77–92.

Newman, P. W. G. and Kenworthy, J. R. (1989) *Cities and Automobile Dependence: A Sourcebook*, Gower Technical, Brookfield.

Nordhaus, W. D. (2006) 'Geography and macroeconomics: new data and new findings', *Proceedings of the National Academy of Sciences*, vol 103, no 10, pp. 3510–3517.

Orsi, F. and Geneletti, D. (2014) 'Assessing the effects of access policies on travel mode choices in an Alpine tourist destination', *Journal of Transport Geography*, vol 39, pp. 21–35.

Orsi, F., Geneletti, D. and Borsdorf, A. (2013) 'Mapping wildness for protected area management: A methodological approach and application to the Dolomites UNESCO World Heritage Site (Italy)', *Landscape and Urban Planning*, vol 120, pp. 1–15.

Park, L., Lawson, S., Kaliski, K., Newman, P. and Gibson, A. (2010) 'Modeling and mapping hikers' exposure to transportation noise in Rocky Mountain National Park', *Park Science*, vol 26, no 3, pp. 59–64.

Peters, G. P., Andrew, R. M., Boden, T., Canadell, J. G., Ciais, P., Le Quéré, C., Marland, G., Raupach, M. R. and Wilson, C. (2013) 'The challenge to keep global warming below 2 °C', *Nature Climate Change*, vol 3, pp. 4–6.

Pickering, C. M. and Hill, W. (2007) 'Impacts of recreation and tourism on plant biodiversity and vegetation in protected areas in Australia', *Journal of Environmental Management*, vol 85, no 4, pp. 791–800.

Pilcher, E. J., Newman, P. and Manning, R. E. (2009) 'Understanding and managing experiential aspects of soundscapes at Muir Woods National Monument', *Environmental Management*, vol 43, pp. 425–435.

Pope III, C. A., Dockery, D. W. (2006) 'Health effects of fine particulate air pollution: lines that connect', *Journal of the Air & Waste Management*, vol 56, no 6, pp. 709–742.

Prideaux, B. (2000) 'The role of the transport system in destination development', *Tourism Management*, vol 21, no 1, pp. 53–63.

Reigner, N., Kiser, B., Lawson, S. and Manning, R. (2013) 'Using transportation to manage recreation carrying capacity', *The George Wright Forum*, vol 20, no 3, pp. 322–337.

Schäfer, A. (2000) 'Regularities in travel demand: an international perspective', *Journal of Transportation and Statistics*, vol 3, no 3, pp. 1–31.

Schiller, P. L., Bruun, E. C. and Kenworthy, J. R. (2010) *An Introduction to Sustainable Transportation: Policy, Planning and Implementation*, Earthscan, London.

Scuttari, A., Della Lucia, M. and Martini, U. (2013) 'Integrated planning for sustainable tourism and mobility. A tourism traffic analysis in Italy's South Tyrol region', *Journal of Sustainable Tourism*, vol 21, no 4, pp. 614–637.

Solomon, S., Plattner, G.K., Knutti, R. and Friedlingstein, P. (2009) 'Irreversible climate change due to carbon dioxide emissions', *Proceedings of the National Academy of Sciences*, vol 100, no 6, pp. 1704–1709.

Trombulak, S. C. and Frissell, C. A. (2000) 'Review of ecological effects of roads on terrestrial and aquatic communities', *Conservation Biology*, vol 14, no 1, pp. 18–30.

Vaske, J. J. and Donnelly, M. P. (2002) 'Generalizing the encounter-norm-crowding relationship', *Leisure Sciences*, vol 24, no 3–4, pp. 255–269.

Vaske, J. J., Donnelly, M. P. and Heberlein, T. A. (1980) 'Perceptions of crowding and resource quality by early and more recent visitors', *Leisure Sciences*, vol 3, no 4, pp. 367–381.

Wagar, J. A. (1964) 'The carrying capacity of wild lands for recreation', *Forest Science Monograph*, vol 7. Society of American Foresters, Washington, DC.

Weinzimmer, D., Newman, P., Taff, D., Benfield, J., Lynch, E. and Bell, P. (2014) 'Human responses to simulated motorized noise in national parks', *Leisure Sciences*, vol 36, no 3, pp. 251–267.

White, D. D., Aquino, J. F., Budruk, M., and Golub, A. (2011) 'Visitors' experiences of traditional and alternative transportation in Yosemite National Park', *Journal of Park and Recreation Administration*, vol 29, no 1, pp. 38–57.

3 Sustainability potential of various transport modes in natural settings

Francesco Orsi

Introduction

Most of the public debate on sustainable transportation revolves around modes of transport, recognizing the need for a progressive shift from private (e.g. car, motorbicycle) to public modes (e.g. trains, buses) (Newman and Kenworthy, 1999; Banister, 2008). This originates from a widespread awareness of the many negative impacts of private motorized vehicles on both the environment and society, including atmospheric pollution, land use conversions, reduced physical exercise, accidents, etc. In natural and protected areas, the above-mentioned shift, though crucial for the protection of the environment and the recreational experience, is particularly difficult for various reasons. First, private vehicles guarantee an incomparable degree of freedom to visitors by allowing them to go and stop at will, and to reach remote areas in complete autonomy. This is also the reason why automobiles are perceived as a key element of the national park experience (Louter, 2006). Second, private motorized vehicles enable people to carry the heavy and bulky equipment (e.g. canoe, mountain gear) they need to perform their preferred recreational activity. Third, the considerable size of many natural areas along with the scattered distribution of the main attractions (e.g. waterfall, lookout) make most alternative transportation options (e.g. buses) simply inefficient. Finally, the difficulty (or impossibility) to reach many natural areas from urban centres with alternative transportation encourages visitors to use their vehicle to both reach the area and move within it.

Despite all the above-mentioned issues, alternative forms of transport are getting more and more popular in natural areas around the world (Eaton and Holding, 1996; Daigle, 2008). Many national parks and other protected areas in Europe, for example, provide shuttle bus services to move visitors from parking lots to some popular yet sensitive sites. Entire regions, especially in the Alps (e.g. South Tyrol, Switzerland), have built up public transit networks (trains, buses) that extend even within protected areas, allowing both the daily activities of local communities and the excursions of visitors. In the USA, the National Park Service (NPS) directly and indirectly (i.e. concessions, partnerships) provides alternative transportation systems (ATS) in many park units. Various parks in Australia (e.g. Cradle Mountain National Park, Grampians National Park) offer the possibility to ride a shuttle bus to get to some of the most popular attractions. In the Iguazu National Park, between Argentina and Brazil, the rainforest ecological train ('Tren Ecologico de la Selva') constitutes an environmentally friendly way to visit the famous waterfalls. The success of these and many other experiences is mainly due to two aspects. On the one hand, visitors are getting more concerned about the harmful effects of transportation on the environment (e.g. pollution) and the recreational experience (e.g. traffic congestion) (Cullinane and Cullinane, 1999; Guiver *et al.*, 2007; White, 2007). On the other hand, alternative forms of mobility offer an innovative way to experience natural contexts that eliminates some of the nuisances associated with traditional mobility (e.g. need to find a parking space) (Miller and Wright, 1999; White, 2007; Mace *et al.*, 2013).

Nonetheless, the great emphasis that is being placed on alternative transportation may suggest the idea that the sole reliance on modes of transport other than private motorized ones is enough to make a transportation system sustainable. This idea, which is highly rewarding in marketing terms (i.e. a natural area can attract thousands of new visitors by establishing a shuttle bus service and selling itself as a sustainable tourism destination), is clearly misleading. As highlighted in Chapter 2, a sustainable transportation system is in fact one that guarantees the satisfaction of multiple environmental, social and economic requisites across space and over time. Every mode of transport has peculiar characteristics (e.g. speed, range, capacity), which may prove positive for satisfying some requisites, but not as positive for satisfying others. For example, an electric train will certainly minimize carbon emissions and land consumption, but it may bring too many people at one time at a destination, therefore increasing pressure on the local environment (e.g. trampling) and reducing the quality of the recreational experience (e.g. overcrowding). Hence, the sustainability of a transportation system in a given context is dependent on the adoption of the most appropriate mode for that context, along with a careful design and a sound management. This chapter provides an overview of some of the alternative transport modes commonly used in natural settings by describing their strengths and weaknesses as well as the conditions under which their use is truly sustainable.

Modes of transport

Mobility in natural areas worldwide is possible through a range of different modes of transport. Some of these, like buses or bicycles, are commonly used in cities too, whereas others, like cableways, are predominantly adopted in natural settings to allow visitors to reach hardly accessible locations (e.g. upper part of a mountainside) and/or to provide specific recreational experiences (e.g. view of a deep canyon from above). The paragraphs below illustrate, one by one, the alternative modes of transport that are most commonly used in natural settings (i.e. bus, train, boat/ferry, cableway, bicycle), exploring their potential to making transport sustainable, their intrinsic limitations and the design/management conditions under which their use may be actually sustainable. Table 3.1 summarizes the information provided in the text.

Table 3.1 The most common alternative modes of transport adopted in natural areas: buses, trains, boats/ferries, cableways and bicycles. For each one, a list of strengths (i.e. characteristics that contribute to sustainability), weaknesses (i.e. characteristics that thwart sustainability) and sustainability conditions (i.e. design/management conditions under which its use is sustainable) is provided. Strengths and weaknesses are referred to sustainability requisites (in parentheses).

Transport modes	*Strengths*	*Weaknesses*	*Sustainable if…*
Bus	– Considerable capacity (land use conversion) – Uses the same infrastructures as private vehicles (land use conversion, visual impact) – Limited infrastructure required per passenger transported (land use conversion, visual impact) – High flexibility in terms of routes, stops and schedules (visitation impacts on the environment, visitation impacts on the recreational experience)	– Atmospheric emissions (atmospheric pollution) – Acoustic emissions (noise) – Contribution to traffic congestion (atmospheric pollution, visual impact) – Considerable capacity (visitation impacts on the environment, visitation impacts on the recreational experience) – Limited space onboard (freedom of movement)	– Best available technology is used to minimize atmospheric and acoustic emissions – Routes, stops and schedules are designed to deliver visitors at rates and locations that are beneficial to the environment, recreational experience and local economy – Vehicles' interior design allows all visitor groups to use it – Fares guarantee financial sustainability
Train	– Easily powered by electricity (atmospheric pollution) – Limited infrastructure required per passenger transported (land use conversion, visual impact) – Easily accessible by any visitor group (freedom of movement) – Support to businesses in and around stations (local communities)	– Acoustic emissions (noise) – Extended ancillary facilities (land use conversion, visual impact) – Massive construction works (land use conversion) – Considerable capacity (visitation impacts on the environment, visitation impacts on the recreational experience)	– Best available technology is used to minimize atmospheric pollution and acoustic emissions – The infrastructure is designed to be minimally visible – Schedules are conceived to satisfy the demand, maximize 'silent' periods and comply with an area's carrying capacity – Stations are prevented from becoming the seeds of unregulated development – Fares allow the system to be financially sustainable

Transport modes	*Strengths*	*Weaknesses*	*Sustainable if...*
Boat/Ferry	– Moderate above-water acoustic emissions (noise) – Limited infrastructures (land use conversion, visual impact) – Possibility for managers to fully control visitor flows (visitation impacts on the environment, visitation impacts on the recreational experience) – Easily accessible for all visitor groups (freedom of movement)	– Significant carbon emissions (atmospheric pollution) – Significant underwater acoustic emissions (noise) – Considerable capacity (visitation impacts on the environment, visitation impacts on the recreational experience) – Visibility from ashore (visual impact) – High costs involved for operation and maintenance (financial sustainability)	– Best available technology is used to minimize atmospheric pollution and underwater noise – Routes and schedules are carefully designed to avoid impacts on sensitive areas, while assuring quality recreation – Routes serve local settlements – Ticket revenues and other sources of funding cover the significant costs
Cableway	– Powered by electricity (atmospheric pollution, noise) – Minimal infrastructure needed by aerial systems (land use conversion, visual impact) – Very high frequency of service (freedom of movement) – Possibility to connect hardly accessible areas (freedom of movement) – Perceived as an attraction (local communities, financial sustainability)	– Modification of the environment for construction (land use conversion, visual impact) – Possibility to connect hardly accessible areas (visitation impacts on the environment, visitation impacts on the recreational experience)	– Served locations are not highly sensitive to human pressure – The system is run at rates that prevent any issue of overcrowding – Infrastructures are designed to minimize land use conversions and visual impacts

(continued)

Table 3.1 The most common alternative modes of transport adopted in natural areas: buses, trains, boats/ferries, cableways and bicycles. For each one, a list of strengths (i.e. characteristics that contribute to sustainability), weaknesses (i.e. characteristics that thwart sustainability) and sustainability conditions (i.e. design/management conditions under which its use is sustainable) is provided. Strengths and weaknesses are referred to sustainability requisites (in parentheses) (*continued*).

Transport modes	*Strengths*	*Weaknesses*	*Sustainable if...*
Bicycle	– No atmospheric emissions (atmospheric pollution) – No acoustic emissions (noise) – Cycling paths and ancillary facilities cover minimal surfaces (land use conversion) – Visitors are free to establish their travel plans (freedom of movement) – Cycling paths bring economic development over extended areas (local communities)	– Difficulty to control visitor flows (visitation impacts on the environment, visitation impacts on the recreational experience) – Not all visitor groups can use it (freedom of movement) – Large scale impacts depending on supporting transportation systems (atmospheric pollution, noise, visual impact, visitation impacts on the environment, visitation impacts on the recreational experience)	– Cycling paths are designed to minimize crowding at the most sensitive sites and to enable visitors to enjoy the recreational experience – Cycling integrates well with other forms of sustainable mobility

Buses

Buses are probably the most widely employed form of alternative mobility in natural settings. Their extreme popularity is essentially due to the use of the same infrastructure used by private vehicles, good performances on any kind of terrain (e.g. flat, mountainous) and the great adaptability to different routing schemes and schedules. All of this makes buses the most immediate option for providing alternative transportation in a natural area. Two kinds of bus service can be considered: scheduled and shuttle. The former, which is typical of natural areas with a significant resident population, is a complex system including various lines and offering connections between settlements and occasionally between settlements and recreational areas. The latter is an ad hoc system, which is adopted for guaranteeing a connection with places that are served by no other form of public transport. The two services can very well integrate into each other in that the former ensures mobility in the more densely populated portion of a natural area, whereas the latter allows 'last mile' transportation.

Various experiences around the world have already shown the positive contribution of buses to sustainable transportation in parks and other natural areas (Guiver *et al.*, 2007; Lin, 2010; Mace *et al.*, 2013). Technically speaking, the most notable feature of buses is their considerable capacity. A normal bus, which can easily carry up to 50 people, covers an area equivalent to about six cars, which instead can carry only up to 30 people. This means that, given the need to transport a fixed amount of visitors, the use of buses would require slightly over half the road and parking space demanded by the use of cars. In the reality, this figure further diminishes considering that the same bus can be repeatedly used over the day. The limited space required has of course positive repercussions on the conversion of natural land uses into artificial ones (e.g. creation of parking lots) and the visual impact of transport-related facilities. Considering average load factors and fuel efficiency, buses are also the best road transport option in terms of CO_2 emitted per passenger kilometre (Lin, 2010), this very positively contributing to the struggle against climate change. Among other strengths of bus services, is the possibility for managers to accurately design their routes and schedules as well as to easily arrange their stops at locations that help minimize the negative impacts of visitation on natural resources (e.g. visitors are delivered at rates and locations that prevent damages to vegetation), recreational experience (e.g. visitors are delivered at rates and locations that prevent overcrowding) and the local economy (e.g. stops in close proximity of businesses).

While constituting a great asset in terms of service performance, capacity is also a limitation when not properly managed. The possibility to deliver 40 or 50 people at one time at a given location, in fact, may have detrimental effects on the environment and the recreational experience. Carbon and acoustic emissions are other relevant issues: traditional buses may generate significant noise, which affects both animals and visitors. Buses suffer the same problem as other road-based vehicles and, given their large size, they can easily get stuck on a narrow road, quickly generating major issues of traffic congestion. As a consequence of that, buses may contribute to road conditions that are particularly bad in terms of visual perception (i.e. a road plenty of vehicles may reduce the perception of naturalness). Even though buses allow even remote locations to be reached, they may not be the best choice for people with bulky equipment (e.g. baby carriage), as space on board is often limited, therefore strongly constraining the recreational opportunities of some visitor categories (e.g. families).

According to the requisites listed in Chapter 2, the use of buses can be sustainable if various precautions are taken. The best available technology should be adopted to minimize atmospheric and acoustic pollution. A careful design of routes, bus stop locations and schedules, including hours of service and frequency of service, is required to ensure that the transportation system meets the demand and delivers visitors at destinations in a way that is beneficial to the environment, the recreational experience and the local economy. Such design anyway should be linked to adequate fare policies or funding schemes in order to guarantee the financial sustainability of the same bus service. Finally, attention should be paid to the interior design of vehicles and the configuration of bus stops in order to allow all visitor groups (e.g. families, disabled people) to take real advantage of the service.

Trains

Trains have made the history of protected areas in the late nineteenth and early twentieth century, as they were often the first and safest way to reach some of the most famous national parks in North America. While today their use for reaching remote protected areas is marginal, trains provide reliable transportation in many natural regions around the world. To this extent, Switzerland, with its red rail convoys climbing up to 3,000 metres, is a perfect example of how trains can be a great alternative to private vehicles for visiting natural areas.

The potential contribution of trains to the sustainability of a transportation system is significant. More easily than other collective modes of transport (e.g. buses), trains can be powered by electricity, which prevents carbon emissions from being released in the local environment. Compared to their considerable capacity, trains do not need huge linear infrastructures (e.g. a one track-line can easily convey thousands of people per day), thus minimizing land use conversions. Moreover, as long as the service schedule is carefully designed, railways are 'silent' infrastructures for long parts of the day as no traffic occurs between one ride and the next one. This contributes to minimizing disturbance to both animals and visitors and, along with lack of carbon emissions, can make the presence of railways in semi-pristine lands more acceptable. Trains can also do much in terms of equality across visitor groups because their peculiar characteristics (e.g. size, platform shape) allow easy boarding to people with bulky equipment (e.g. bicycles) or disabilities. This would contribute significantly to modal shift from less sustainable forms of transport, as would the fact that, being directly connected to a larger network, the railway encourages people to use the train for both reaching the area and moving within it. Finally, trains offer a chance to sustain the local economy through the activities that can flourish in or around the stations.

Even though the mere rails occupy a limited amount of land, the overall infrastructure (i.e. stations, service roads) may require considerable land conversions and subsequent impacts on both natural resources and a visitor's perception. Such impacts would be particularly significant in areas where the railway is lacking and should be built from scratch. Further, considering the relative difficulty of trains to overcome natural barriers (e.g. steep slopes), massive construction works (e.g. tunnels, excavations) are required to build a railway on a complex terrain. As already mentioned for buses, capacity is both a great asset and a potential threat. The possibility to carry hundreds of people and deliver them all together at a location may cause serious issues of overcrowding with bad effects on soil cover, fauna and the recreational experience. Noise can be a problem, especially if the morphology of the terrain (e.g. canyon) enhances its effects.

Various elements should be considered to guarantee the sustainability of using the trains in natural areas. Regarding specific actions to be taken, much depends on whether the infrastructure is already existing or not. In the former case, proper technologies on both the convoys (e.g. highly efficient locomotives) and the railway (e.g. soundproof barriers) have to be adopted to minimize atmospheric and acoustic emissions. In the latter case, the emphasis is on design, which should aim to minimize the conversion of natural soil, the visibility of the infrastructure (both stations and rails) and the diffusion of noise. Regarding the operation of a line, schedules should be conceived to ensure the satisfaction of the demand, the maximization of 'silent' periods (i.e. periods during which no train is on the line), the delivery of visitors at rates that are compatible with the area's carrying capacity. A sound fare policy is also needed to allow the financial sustainability of the system. Finally, efforts should be made to prevent stations from becoming seeds of unregulated development that could alter the natural balance of places.

Boats and ferries[1]

Whenever a natural area comprises large bodies of water (e.g. sea, lake, river), the transportation of visitors may require the use of boats or ferries. Ferry services allow people to explore areas and reach locations that could hardly be visited on one's own (i.e. few people have a boat and even less people have the necessary skills to undertake extended navigations), therefore playing a major role in the management of visitor flows to such locations. For example, a ferry service on a lake, as the only way for people to move across the lake, ultimately determines which locations will be systematically visited and which locations will not.

The above-mentioned property of boat/ferry services may greatly contribute to the sustainability of a transportation system. In fact, by intervening on such services, managers and administrators have the possibility to thoroughly control where people can go and to ensure that the right amount of people is delivered at each location, therefore guaranteeing the protection of natural resources (e.g. acceptable trampling) and visitors' recreational experience (e.g. no overcrowding). The possibility to control the flows of visitors may be an asset also for economic purposes as ferry services, through adequate routing and schedules, can sustain businesses around harbours. In terms of land use conversions and visual impact, boat-based transportation systems generate limited impacts because harbours tend to occupy moderate land surfaces and to have a minor effect on people's visual perception. Generating reasonable above-water noise, traditional ferries have also limited effects on visitors' acoustic perception, though faster boats (e.g. hydroplanes) may provoke greater disturbance. Finally, large boats represent a fair transport mode in that their size and configuration enable all visitor categories to easily get on board and enjoy the navigation.

While the noise generated by boats may be acceptable above water, it is not so under water. A variety of studies has proved that the impacts of boat-related noise on the aquatic fauna may be significant and possibly leading to permanent damages (Erbe, 2002; Sarà *et al.*, 2007; Slabbekoorn *et al.*, 2010). Beyond noise, the operation and maintenance of boats and ferries have major environmental impacts, including significant greenhouse gas emissions (Byrnes and Warnken, 2006), reduction in vegetation species richness as a consequence of increased water movement and human waste (Eriksson *et al.*, 2004), and large fish killings due to collisions (Vanderlaan and Taggart, 2007). The capacity of a boat, when particularly high (i.e. several hundreds of passengers), can bring considerable impacts on both natural resources and visitors' recreational experience at locations where passengers are disembarked. The visual impact of boats is another major concern for both passengers and people ashore in that excessive boat traffic may reduce an area's scenic beauty and perceived naturalness (Dalton and Thompson, 2013). Finally, maintaining and running an efficient boat service may be a considerably expensive activity, this having serious implications for financial sustainability.

Given the potentially big impact of boat transportation in terms of atmospheric pollution and noise, the use of the best available technology, including highly efficient engines and low carbon fuels, is necessary for sustainability. The same issues can be further tackled through careful design of routes to avoid sensitive areas, and adequate driving behaviour to minimize unnecessary disturbances (e.g. excessive use of engine power in a narrow inlet). Routing schemes and schedules should satisfy the demand while excluding any problem of overcrowding at locations served, assuring a quality experience to people on board and preserving the visual perception of people ashore. All of these precautions are clearly essential when visitor flows are considerable and the locations served are many. Routes and schedules should also be conceived to sustain the local economy, allowing all settlements (e.g. small villages found along a lake's shore) to receive a number of visitors that contributes to their local businesses but does not exceed the social carrying capacity. Achieving the dual goal of preserving the environment and providing a quality recreational experience is obviously subject to the need of covering the costs of the transport system, which in the case of a boat/ferry service may be considerable.

Cableways

Cableway is the general-purpose term indicating any transportation system that relies on cables to move vehicles. Though not as common as buses or trains in normal contexts, cable-based transportation systems are widely used to get through complex terrains such as steep slopes, deep valleys and water bodies. Depending on their characteristics, cableways can be roughly divided in two main categories: aerial and rail-based systems. In the former case cabins are suspended and propelled by cables, whereas in the latter cabins are propelled by cables but supported by rails anchored to the ground. A wide variety of technologies exists for both aerial (e.g. aerial tramways, detachable gondolas, chairlifts) and rail-based systems (e.g. funicular, mini metro) each of which is suitable for a specific context. Cableways are used in many natural areas around the world to allow visitors to reach hardly accessible locations (e.g. top of a mountain) and enjoy unusual perspectives of a site (Sever, 2002). In the Alps, for example, cableways have been part of the transportation networks for years, supporting the tourist activity and guaranteeing the mobility of residents.

Cableways are commonly perceived as sustainable transportation systems. Their peculiar characteristic is to have no engine on-board: this is located in one of the two end stations and is generally powered by electricity. Hence, cableways do not generate in situ atmospheric pollution and minimize acoustic emissions.

Aerial systems also minimize land use conversions as the only fixed structures required are end stations, which may be very limited in size, and pylons, which occupy very small areas. Even though rail-based systems do require a linear infrastructure, this is reasonably small. The minimal visual impact is another asset of cableways, though design requirements and the physical features of a site can eventually increase the visibility of a system. Cableways guarantee a remarkable freedom of movement due to a very high frequency of service (i.e. one ride every few minutes or even seconds in the case of detachable gondolas or chairlifts) and the possibility to connect locations that could not be reached with other means. Such freedom of movement along with the spectacular views offered make cableways very popular. This may contribute to the local economy and financial sustainability, though the latter can only be partial in the low season (Brida *et al.*, 2013).

The possibility of cableways to get through very complex terrains and reach otherwise inaccessible locations is clearly a double-edged sword. While enabling people to visit areas of unique scenic beauty and environmental quality, cableways may in fact increase human pressure on these same areas with significant consequences on natural resources and visitors' recreational experience (Orsi and Geneletti, 2014). This issue is particularly relevant when considering that modern cableways (e.g. detachable gondola lifts) have very high hourly capacity (i.e. up to 5,000 p/h). Even though the land use conversions required by cableways are limited, they are not negligible (e.g. forest cleared to make space for a chairlift, parking lot at an end station) and likely to emphasize the visual impact of a system.

Major areas of intervention for ensuring the sustainability of a cable-based transportation system are: the identification of the location to be served, the operation scheme and the construction design. Cableways should serve sites that, while offering spectacular views or a convenient access to valuable areas, are not highly sensitive to direct human pressure. Given the impact of visitation in and around the locations served, cableways should be run at rates that do not cause any issue of overcrowding. In this respect, studies are needed to find the relationship between a specific rate of delivery and the resulting distribution of visitors around the location served. Finally, a proper design is fundamental to minimize land use conversions and visual impacts associated with the system itself and ancillary facilities (e.g. parking lots).

Bicycles

Motorized transport is not necessarily the best option to visit a natural area. Indeed human-powered vehicles may offer a much better experience as they allow visitors to move slowly enough to appreciate every metre of a natural setting and to perform a physical activity. For this reason, non-motorized vehicles are today the symbol of a new form of tourism commonly referred to as slow travel, which denotes an alternative to the traditional air or car travel, where people travel less and more slowly (Dickinson *et al.*, 2010; Dickinson and Lumsdon, 2010). Among other human-powered vehicles (e.g. canoes) used in parks and natural areas, bicycles are the most common one because they are easy to use and allow the average person to cover considerable distances over the day. Following the growing interest in slow active travel and in an attempt to stimulate alternative mobility, many natural areas worldwide are taking action to make the use of bicycles easier and safer through the construction of cycling paths, the introduction of bicycle rentals and the definition of ad hoc itineraries (Weston and Mota, 2012).

The diffusion of bicycles is very positive from the sustainability point of view. Bicycles have a minimal environmental impact as their use does not cause atmospheric or acoustic emissions and their infrastructures (e.g. cycling paths) occupy only minimal surfaces. The limited size of infrastructures also minimizes visual impacts and subsequently safeguards perceived naturalness. As already suggested, cycling favours an exceptional freedom of movement, giving visitors the ability to plan individual itineraries of visitation and to stop at their most preferred locations. Finally, cycling tourism in natural areas is a conveyor of local economic development as cyclists need a wide variety of services (e.g. bars, hotels, repair shops) on an extended land stripe (e.g. 50 km cycling path).

As opposed to the other modes of transport described in this chapter, the bicycle is a private kind of transportation. Hence, while guaranteeing an exceptional freedom of movement, it also reduces the actual possibility to control the distribution of visitors across space and over time. Uncontrolled visitation patterns could result, for example, in too many people being simultaneously at a sensitive site with negative consequences on the protection of the site and the recreational experience (Arnberger *et al.*, 2010). In terms of freedom of movement, bicycles can be criticized for being a form of mobility that not all visitor groups can actually use (e.g. disabled people, elders). The actual sustainability of cycling tourism, however, can only be evaluated considering which support transportation (i.e. transport mode used to reach the starting location of an itinerary or to go back to the starting point) visitors rely on to perform their cycling activity. The use of a largely unsustainable support transportation (e.g. cars), in fact, may detract from the sustainability of the cycling activity.

The sustainable use of bicycles in natural areas implies actions on the two above-mentioned elements: management of visitor flows and support transportation. Managers need to conceive a cycling network and associated access points that prevent overcrowding issues at the most sensitive sites, while allowing different visitor groups to maximally enjoy their recreational experience (e.g. expert riders should be offered relatively empty paths). This requires a careful study of visitor preferences and usual routes to better understand the relationship between the inflow of people and their resulting distribution across space and over time, and eventually define the management of the network. Comprehensive mobility plans must also be designed that clarify how cycling integrates with other transport modes. Today one of the most interesting forms of integrated mobility is given by the combination of trains and bicycles on itineraries where the cycling path runs beside the railway. This allows cyclists to easily shape their own itinerary by jumping on and off the train at the most preferred locations.

Conclusion

Managers and administrators have a variety of modes of transportation to choose from to ensure sustainable mobility in natural settings. This chapter has provided an overview of the five most commonly used, describing their potential contribution to sustainable mobility as well as their limitations. Even though transportation not based on private motorized vehicles is often perceived as inherently sustainable, this chapter has conveyed the message that there is no sustainable mode, only a sustainable use of a given mode. Such use essentially relies on a sound design and management.

A sound design starts from the adoption of the best available technology (e.g. efficient bus engines) to minimize negative environmental impacts such as atmospheric pollution and acoustic emissions. The design of infrastructures (e.g. railways, roads) should aim to limit land use conversions and camouflage transport facilities within nature so that visitors' perception is minimally affected. Routes and stops of public transport, and the network of cycling paths should be conceived to guarantee freedom of movement to visitors as well as accessibility to important natural sites and villages. Freedom of movement of all visitor categories is enhanced if the interior design of vehicles takes into account the special needs of some categories (e.g. disabled people need to carry wheelchairs). Finally, sound design is about adequately integrating a given transportation mode within a larger transportation system to ensure the actual usability of the mode on the one hand and to prevent unintended consequences on the other (e.g. a cableway whose end station is not served by any public transit will either be scarcely used or force visitors to rely on their cars to reach it). This consideration brings us back to what already discussed in Chapter 2: the effects of transportation spreads across space and over time. Hence, a truly sustainable transport mode should be designed in a way that does not generate negative outcomes on other parts of a complex transport network (e.g. a very attractive cycling loop generating unacceptable car traffic on the roads that provide access to it).

Management is fundamental to prevent or correct the unintended consequences of adopting a specific mode of transport. Intervening on the schedules of a public transit, for example, is a way to keep visitor flows at a level that guarantees the preservation of natural resources and the quality of the recreational experience. Public transit fares instead may be used to control the demand for a given transport option over time, while ensuring the financial sustainability of the same option. Similarly, road tolls, access restrictions, quotas, parking fees can all be used, individually or in combination, to minimize the negative impacts of a given transportation system on natural resources, society and economy.

Designing and managing a mode of transport for sustainability is a complex task, which may require multiple attempts and modifications prior to meeting the relevant standards. It is a complex task primarily because it is affected by human behaviour: the impact of a given transport mode does not only depend on its atmospheric emissions or economic cost, but also on how many people are willing to shift to it from other transport modes and how frequently they will use it.

Hence, it is very important that this process be informed by an in-depth knowledge about visitors' preferences towards different transport characteristics, environmental conditions and management policies (see Chapters 4 through 7). Today simulation models can very well support such process as they enable managers to predict in great detail the likely effects of a transport mode before it is actually introduced (Lawson *et al.*, 2011). Nevertheless, these tools should be applied considering sufficiently large spatial and temporal scales in order to detect all possible consequences of a transport mode and therefore plan the necessary mitigation measures.

Note

1 Only public boat services are considered: the sustainability of private boats is not analyzed.

References

Arnberger, A., Aikoh, T., Eder, R., Shoji, Y. and Mieno, T. (2010) 'How many people should be in the urban forest? A comparison of trail preferences of Vienna and Sapporo forest visitor segments', *Urban Forestry & Urban Greening*, vol 9, no 3, pp. 215–225.

Banister, D. (2008) 'The sustainable mobility paradigm', *Transport Policy*, vol 15, no 2, pp. 73–80.

Brida, J. G., Deidda, M. and Pulina, M. (2013) 'Tourism and transport systems in mountain environments: analysis of the economic efficiency of cableways in South Tyrol', *Journal of Transport Geography*, vol 36, pp. 1–11.

Byrnes, T. A. and Warnken, J. (2006) 'Greenhouse gas emissions form marine tours: a case study of Australian tour boat operators', *Journal of Sustainable Tourism*, vol 14, no 3, pp. 255–270.

Cullinane, S. and Cullinane, K. (1999) 'Attitudes towards traffic problems and public transport in the Dartmoor and Lake District National Parks', *Journal of Transport Geography*, vol 7, no 1, pp. 79–87.

Daigle, J. J. (2008) 'Transportation research needs in national parks: a summary and exploration of future trends', *The George Wright Forum*, vol 25, no 1, pp. 57–64.

Dalton, T. and Thomspon, R. (2013) 'Recreational boaters' perceptions of scenic value in Rhode Island coastal waters', *Ocean & Coastal Management*, vol 71, pp. 99–107.

Dickinson, J. and Lumsdon, L. (2010) *Slow Travel and Tourism*, Earthscan, London, UK.

Dickinson, J. E., Robbins, D. and Lumsdon, L. (2010) 'Holiday travel discourses and climate change', *Journal of Transport Geography*, vol 18, no 3, pp. 482–489.

Eaton, B. and Holding, D. (1996) 'The evaluation of public transport alternatives to the car in British National Parks', *Journal of Transport Geography*, vol 4, no 1, pp. 55–65.

Erbe, C. (2002) 'Underwater noise of whale-watching boats and potential effects on killer whales (Orcinus Orca), based on an acoustic impact model', *Marine Mammal Science*, vol 18, no 2, pp. 394–418.

Eriksson, B. K., Sandström, A., Isæus, M., Schreiber, H. and Karås, P. (2004) 'Effects of boating activities on aquatic vegetation in the Stockholm archipelago, Baltic Sea', *Estuarine, Coastal and Shelf Science*, vol 61, no 2, pp. 339–349.

Guiver, J., Lumsdon, L., Weston, R. and Ferguson, M. (2007) 'Do buses help meet tourism objectives? The contribution and potential of scheduled buses in rural destination areas', *Transport Policy*, vol 14, no 4, pp. 275–282.

Lawson, S., Chamberlin, R., Choi, J., Swanson, B., Kiser, B., Newman, P., Monz, C., Pettebone, D. and Gamble, L. (2011) 'Modeling the effects of shuttle service on transportation system performance and quality of visitor experience in Rocky Mountain National Park', *Transportation Research Record*, vol 2244, no 1, pp. 97–106.

Lin, T. P. (2010) 'Carbon dioxide emissions from transport in Taiwan's national parks', *Tourism Management*, vol 31, no 2, pp. 285–290.

Louter, D. (2006) *Windshield Wilderness: Cars, Roads, and Nature in Washington's National Parks*, University of Washington Press, Seattle, WA.

Mace, B. L., Marquit, J. D. and Bates, S. C. (2013) 'Visitor assessment of the mandatory alternative transportation system at Zion National Park', *Environmental Management*, vol 52, no 5, pp. 1271–1285.

Miller, C. A. and Wright, G. R. (1999) 'An assessment of visitor satisfaction with public transportation services at Denali National Park and Preserve', *Park Science*, vol 19, pp. 18–21.

Newman, P. and Kenworthy, J. (1999) *Sustainability and Cities: Overcoming the Automobile Dependence*, Island Press, Washington, DC.

Orsi, F., Geneletti, D. (2014) 'Assessing the effects of access policies on travel mode choices in an Alpine tourist destination', *Journal of Transport Geography*, vol 39, pp. 21–35.

Sarà, G., Dean, J. M., D'Amato, D., Buscaino, G., Oliveri, A., Genovese, S., Ferro, S., Buffa, G., Lo Martire, M. and Mazzola, S. (2007) 'Effect of boat noise on the behaviour of blufin tuna Thunnus thynnus in the Mediterranean Sea', *Marine Ecology Progress Series*, vol 331, pp. 243–253.

Sever, D. (2002) 'Some new methods to assure harmonization of sustainable development of mountain resorts', *Promet*, vol 14, no 5, pp. 213–220.

Slabbekoorn, H., Bouton, N., van Opzeeland, I., Coers, A., ten Cate, C. and Popper, A. N. (2010) 'A noisy spring: the impact of globally rising underwater sound levels on fish', *Trends in Ecology & Evolution*, vol 25, no 7, pp. 419–427.

Vanderlaan, A. S. M. and Taggart, C. T. (2007) 'Vessel collisions with whales: the probability of lethal injury based on vessel speed', *Marine Mammal Science*, vol 23, no 1, pp. 144–156.

Weston, R. and Mota, J. C. (2012) 'Low carbon tourism travel: cycling, walking and trails', *Tourism Planning & Development*, vol 9, no 1, pp. 1–3.

White, D. D. (2007) 'An interpretive study of Yosemite National Park visitors' perspectives toward alternative transportation in Yosemite Valley', *Environmental Management*, vol 39, pp. 50–62.

Part II

Preferences – Attitudes towards sustainable transportation

4 Visitor preferences toward scheduled bus use in natural and protected areas

Jo Guiver, Nick Davies and Richard Weston

Introduction

A protected area's peace and tranquillity and the value of its natural or historic landscape can easily be diminished by the traffic created by its visitors, threatening the very qualities they seek in the area. Yet car travel is the mode of choice for the majority of visitors to national parks and rural areas with 70 to 90 per cent of visitors arriving by car. Car offers the means to get to the area as well as travel around it, flexibility of timing, route and destination, protection if the weather turns bad and capacity to carry provisions and equipment. However, while car travel may offer its users a number of advantages, it imposes tremendous costs on the area, its residents and other visitors by occupying land for roads and parking, creating noise, pollution, congestion and danger, discouraging more sustainable forms of travel such as walking and cycling and vastly inflating the carbon count of tourism. In this age of individualised private mobility, the potential of public transport to reduce traffic and enhance a visit can be overlooked by both visitors and planners.

Using the same network as cars, buses offer a way of minimising the land use, pollution, noise, danger and congestion of car travel while taking away the need to navigate unfamiliar roads and affording the opportunity to gaze on the scenery. Additionally, they present opportunities to enhance the travel experience through interpretation, contact with local people and novelty and can encourage the use of more sustainable modes to reach the destination area, so reducing the global impact of tourism in the area. For walkers, they offer the chance to do linear walks without having to return the same way, particularly important along linear features such as coasts, rivers and hill ridges. They also open rural tourism to people without their own cars. In the UK, for example, 25 per cent of households are without cars (Department for Transport, 2013, p4), a growing number of young people have not learnt to drive or bought a car (Department for Transport, 2013, p4), many people do not drive for economic or ecological reasons, and 70 per cent of overseas visitors (VisitEngland, 2008) arrive without a car, many of whom do not want to attempt to drive on the 'other side of the road'.

While there has been considerable research into encouraging modal shift from cars to more sustainable forms of mobility for utility travel (see for example Jones and Sloman, 2003; Haq *et al.*, 2008), there has been relatively little exploration of the necessary inducements for people to get out of their cars for leisure travel (Dickinson and Dickinson, 2006). Leisure travel differs from utility travel in a number of ways. By definition, it is discretionary and this refers not only to the decision about whether or not to travel, but also where, when, how and with whom to travel. Because destinations, as well as travel modes, can change, tourism providers are wary of discouraging car travel for fear of displacing trade to other areas. However, because, unlike utility travel, the journey is often an integral part of the tourism experience, there are opportunities to enhance the experience through the provision of novel, innovative or customised public transport services, which offer something better or different from private motoring.

This chapter first describes the benefits of scheduled bus services to visitors, tourism providers and the area's residents and economy, before discussing the key findings from our research into the attitudes and preferences of passengers using bus services in rural tourist areas in the UK. The next section discusses issues beyond the survey findings that affect the success of such bus services. The conclusions stress the importance of local knowledge in applying principles learnt from other areas. The chapter is illustrated with comments from respondents of surveys conducted in various protected areas around the UK.

Benefits of buses in protected areas

Apart from the benefits of being able to do a linear walk and being able to travel within a destination area without needing a car or driving licence, well-planned visitor bus services can offer a number of advantages to their passengers. Car drivers particularly can enjoy the luxury of being able to take in the view while a professional steers and navigates, with no need to seek or pay for parking. The height of a bus, especially a double decker bus, also affords much better views than cars in areas of hedges and stone walls. Even in a wind-swept, rainy country such as the UK, for example, there is a joy of riding an open-top double decker bus, with the wind in one's hair and the smells of the countryside (see Cornish Wools, 2011).

> *'Being from Germany I enjoy the English bus service a lot, especially as you have double decker buses also with open tops; super for sightseeing. The friendliness and helpfulness of the drivers is also fantastic.'*

Bus services provide potential to offer interpretation, through leaflets, guides or audio. Increasingly, packages are being promoted that offer bus travel and the use of other modes such as boat or ski lift within the area for a period of time (see golakes, the Lake District, 2014), as well as included or discounted entry into attractions (see Eden Project, 2014). Several tourist services provide or sell leaflets about the walks that are possible from the bus route (see Brighton and Hove City Council, 2014).

Residents benefit from reduction in demand for parking, road space and all the other externalities generated by road traffic. In addition, they can usually take advantage of improvements in bus services, be they cheaper fares, improved frequencies or better quality vehicles.

'I use the bus frequently and miss it during the winter months.'

Any improvement to the tourism offer in an area has the potential to attract more visitors. Providing opportunities for a visit without needing a car opens the area to market segments who might otherwise not consider it, including young people and an increasing number of retired people, many of whom have considerable spending power. Attractions served by local buses can increase their revenue without provision of extra parking and frequently there are opportunities for joint marketing of both the bus service and the attractions along its route.

Apart from the obvious benefits to the local environment of reductions in pollution, parking space and congestion, moving people from private cars to public transport changes their spending patterns. Paying for bus fares rather than fuel helps keep money in the local economy and can increase local employment. Not needing a car for day trips makes it possible to arrive in the area by public transport, which reduces the carbon footprint of the whole holiday (Le Klähn *et al.*, 2014) (see also Cumbria County Council and Lake District National Park Authority, 2011).

There are a number of economies of scale with bus services. Once a bus is running, it will be more efficient and sustainable to run it full rather than empty ('bums on seats'). Encouraging more passengers onto existing services can make them more economically viable and reduce the fuel and emissions per passenger kilometre. In Bavaria and the Black Forest (Germany), the introduction of a free travel for tourists on local public transport, financed by a modest bed-tax, is popular with tourists, transport providers and destination managers (Hilland, 2011; Wibner, 2012; Guiver and Stanford, 2014). Additional economies of scale result from denser networks, offering more destinations through interchange and integrated timetabling, ticketing and promotion between operators and modes.

Evidence from research in the UK

Between 2005 and 2011 the Institute of Transport and Tourism at the University of Central Lancashire conducted 37 surveys of bus passengers in 24 rural tourist areas. These included national parks (such as the Lake District, New Forest, Snowdonia and Loch Lomond and the Trossachs), Areas of Outstanding Natural Beauty such as the Forest of Bowland, Northumberland and the North Norfolk coast and other areas attracting day visitors and tourists (see Guiver and Lumsdon, 2006; Guiver and Davies, 2007; Guiver, 2012). Although there were differences in questionnaires, several basic questions remained the same, allowing the data to be compared. The findings quoted in this chapter refer to the most recent survey of nine tourist areas in 2010 and 2011. The respondents' quotations also originate from this survey.

Visitors and their preferences

Passengers tend to be older than the national population, with 52 per cent older than 60 and 19 per cent over 70 years old. Pairs and solo travellers predominate. The main purposes for their trips are walking, visiting a place or attraction and

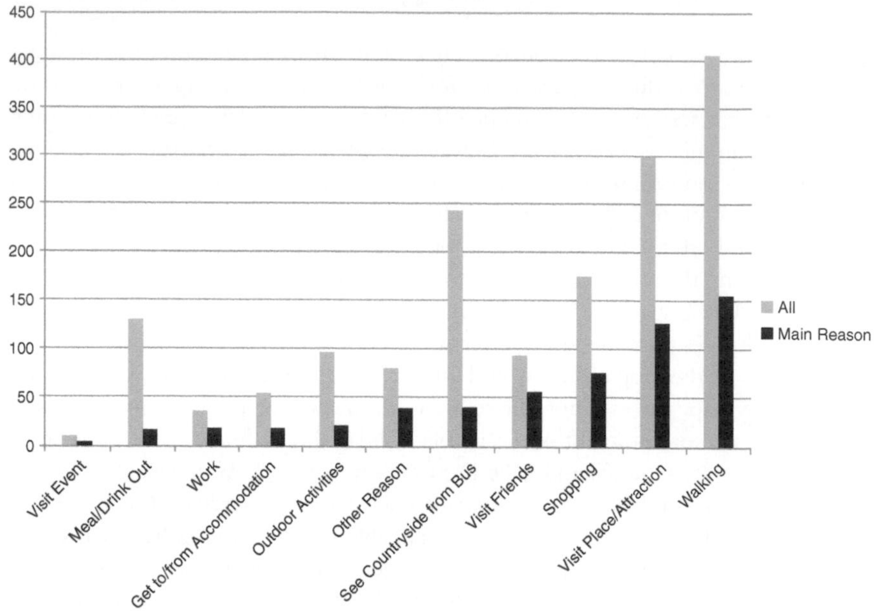

Figure 4.1 Main and secondary reasons for using the bus.

seeing the countryside (although this is mainly a secondary reason for using the bus) (Figure 4.1).

The main way of knowing about the bus is having used it before (33 per cent), yet 30 per cent of respondents say they have never caught that bus before and 19 per cent have never visited the area before. This demonstrates the need to cater for both the 'bus-virgin' and the 'bus-veteran'. There are a variety of other ways of knowing about it: bus stop information, word of mouth and leaflets (each 11 per cent), seeing bus (10 per cent) and Internet search (9 per cent) and many people know about it through more than one channel, so it is important for transport providers to use several means of communication. A distinctive vehicle can be the best advertisement for a visitor bus service. This has been exploited for example by the Hadrian's Wall Bus in the North of England with its customised livery.

One of the greatest issues in bus planning is understanding which alternatives potential passengers would choose if the service were not running. Passengers were asked what they would have done had the bus service not been available. The results show that roughly equal proportions of the passengers would have stayed at home or in their holiday accommodation, gone to the same destination or another destination (Figure 4.2). This indicates that the buses have a social inclusion role, by providing travel opportunities for people who otherwise would not have travelled and helping to reduce car travel as one quarter of respondents would have used the car and contributed to local traffic in the absence of the bus service.

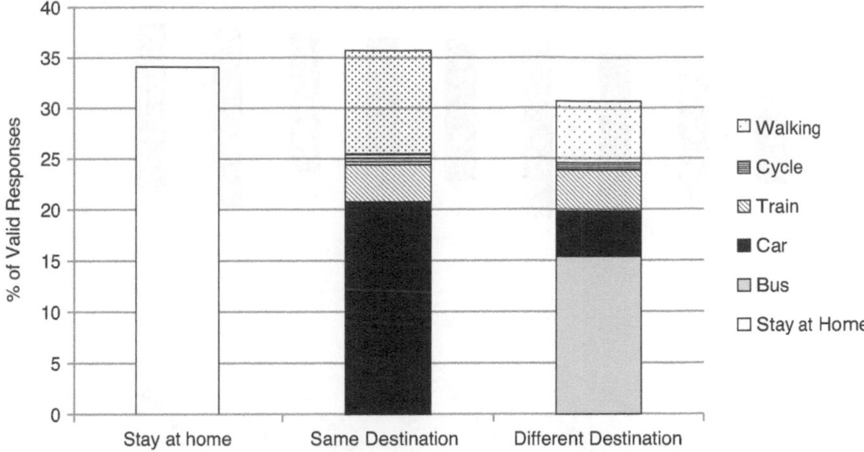

Figure 4.2 Choice of visitors if the bus had not been running. In case the same or a different destination is chosen, the preferred mode of transport is specified.

Since the introduction of nationwide, free concessionary bus travel for people over 60 years old in the UK in 2008, the proportion of people who would go to other destinations by bus has increased. This creates a growing market of mode-dependent over destination-dependent visitors: that is, people who go where a bus allows them to go, rather than choosing a destination and then finding the best mode to reach it. Other related changes have been an increasing proportion of retired passengers and a greater number of men using the buses.

The survey results show a high level of overall satisfaction, which tends to increase with the age of the respondent. This is endorsed by the 89 per cent of valid responses indicating that people would recommend the service to their friends. Figure 4.3 shows the relative ratings of all the attributes asked about. It is evident that satisfaction is high for all the attributes, although 'Frequency of Service' generally falls below the average. Perhaps this is not surprising as many of these buses run at two-hourly or longer intervals.

'Excellent service. Frequency has improved from two hours to one hour.'

Low frequency not only reduces choice about departure/arrival times, it can severely restrict flexibility about duration of stay at any destination. For example, in one of the areas surveyed, a leisure route operating every two hours gave passengers the potential of stays of one hour and ten minutes, four and a half hours or seven and a half hours at one of the destinations. An hourly service would have offered a much greater range of potential stays. When current bus users complain about the poor frequency, it is highly probable that this has already deterred other potential passengers. Unfortunately, recent rounds of local authority financial cuts in the UK

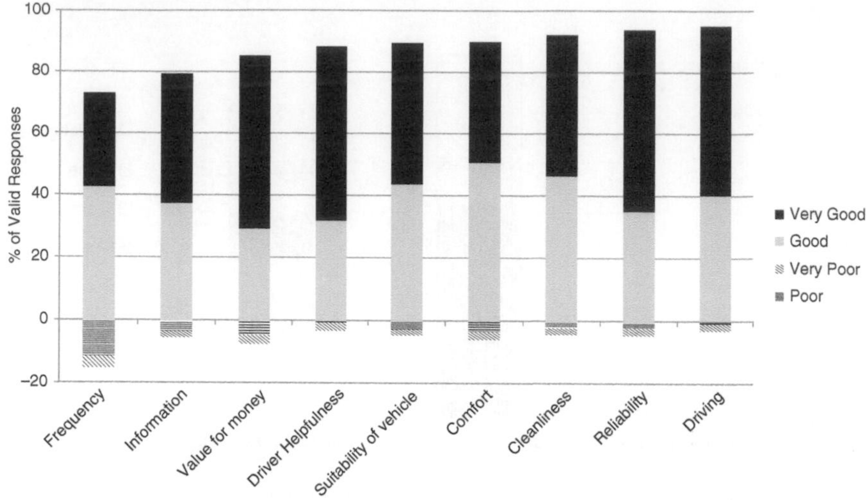

Figure 4.3 Evaluation of the bus service's attributes based on four classes (i.e. very poor, poor, good, very good). The percentages associated with the classes 'very poor' and 'poor' are shown below the zero.

have further reduced many bus frequencies, which is likely to make them even less acceptable to passengers and result in a disproportionately high reduction in patronage. Not surprisingly, 'more frequent buses' is the factor indicated as most likely to encourage bus use (73 per cent) closely followed by 'going to more places' (72 per cent). Some areas, (Brecon Beacons and North York Moors) have improved the efficiency of their services and expanded the number of destinations for their passengers by operating a hub and spoke service, whereby buses congregate at a central point at certain times in the day enabling passengers to transfer. In the North York Moors National Park, prior to public service financial cuts in 2014, the service had been considerably enhanced by the presence of bus co-ordinators, who could answer passengers' questions, give them times and direct them to the right bus stops, etc. Several respondents requested later buses and it is clear that having last buses leaving in the afternoon truncates the day's activities, especially if passengers use the last bus as their 'insurance' and try to catch the last, but one, as often happens with public transport services. Confidence in catching the last bus can be increased by offering a free taxi service (with the number provided) if the bus fails to turn up (Lumsdon and Caffyn, 2013). Interestingly, people with a car available tend to be more satisfied than people without the option of using a car, possibly because bus is their chosen mode and they do not feel captive to it. It seems that the provision of buses in rural protected areas is valued by car drivers and may even serve as an attractor to an area.

'More regular and frequent timings would be helpful.'

'The bus service in this area after 6 pm is very poor on any route. Getting back from your destination is difficult.'

Regarding fares and ticketing, understandably, people qualifying for free concessionary bus travel are satisfied with the 'value for money', while the least satisfied group tend to be young people. Larger groups also find bus travel costs compare unfavourably with travelling together in a car. Rover tickets, offering unlimited travel for a day or longer, are generally popular.

> *'A teen ticket to subsidise 16-19 year olds.'*

> *'One day ticket would be better.'*

Most passengers are extremely happy with the quality of bus service. Many comment on the excellence of the drivers, how they help passengers and how they cope with difficult circumstances. However, it is clear that poor driver service can spoil the whole trip, while an unexpected consideration can increase enjoyment of the trip and make it a social occasion.

> *'Not all drivers helpful. This driver very good.'*

> *'Great service, lovely drivers (mostly), friendly atmosphere.'*

> *'The driver on our bus was sociable, friendly and very helpful! It was a great journey and a comfortable ride with all road rules being observed.'*

> *'Bus driver very helpful, also on previous trip he paused for a couple of minutes so that we could watch a deer in adjacent field.'*

Service design issues

The research also revealed numerous issues relating to the design and practices of providing bus services in tourist areas in addition to the survey results. Among other issues, marketing and information are crucial to encouraging the use of bus services. In fact, people cannot use a bus they do not know exists and even if they know it is available they need to know where and when to catch it and something about the destination and journey. Decisions about the destination for some day trips during a holiday may have been taken before leaving home, though very few people will want precise times and stopping places much before they set out on their trip. Marketing materials mentioning or picturing the potential of travelling by bus, and any advantages or special attraction of using buses (e.g. open-top, possibilities of linear walks) will help sow the seed about using public transport. Travel articles by journalists, bloggers, etc. who have sampled the local public transport will also raise its profile. People arriving by public transport also need advice about suitable accommodation close to useful bus routes. Journey planners offer information about possible modes, when you know where you want to go: public transport users often want to specify the mode and be told the range of potential destinations. They also need information about how to get from the railway or coach station to their accommodation. This may need to be as precise as which bus stand to wait at, times of services and the ultimate destination of the bus or the telephone number for the local taxi company.

Once within the area, tourists may be searching for ideas about day trips. The destination is probably decided before or with the mode, hence it needs to be the focus of marketing. However, information about the cost, ease or advantages of using the bus can accompany the destination description. Once a decision has been made to use the local buses, much more precise information needs to be delivered in as concise and customised way as possible. Large timetables are daunting and most passengers only want to know the time the bus leaves the stop nearest to their accommodation, the journey time and/or the time of arrival at the destination with similar information for the return journey. Although large timetable booklets can be useful for planning multiple trips, they become an encumbrance if one needs to carry them around, especially if they do not fit into pockets or handbags. People with smartphones may be able to access information through the Internet, but in many protected areas, there is poor or no signal or Wi-Fi. Many leisure trips take place at the weekend or on public holidays, when bus service is limited or absent. There can also be confusion about whether weekday or weekend services run on public holidays. This needs to be clearly marked on every timetable and, whenever possible, the service during holidays should be as similar as possible to that of weekdays. The difficult logistics of providing comprehensive and current information about bus services at bus stops (e.g. need to replace timetables periodically) was addressed in one area by solar-powered information devices that both display information and allow passengers to speak to an information officer. Such tools are particularly helpful for people who, being new to the area or to using buses, may not know the geography of the area, the name of the terminal destination or even on which side of the road to wait.

> *'All stops should be mentioned in the timetable. All stops need a sign. All signs need to indicate direction of travel.'*

According to passengers, the most important improvement to a bus service would be an increased frequency of service. However, that depends on the ability to generate additional patronage; otherwise the result would be the same number of people using more buses. Yet there are a number of ways in which bus services could serve the needs of tourists better, and therefore attract a greater number of passengers. Time anchors, whereby people have to be at their destinations by specific times, are rarer for discretionary travel than utility travel, although there are some time constraints. People staying in serviced accommodation should not have to choose between a trip out and their breakfast. This means that, if there is only one outward journey in the morning, it should leave after people in hotels and guesthouses have breakfasted. Ideally the timetable should offer the option of a half or full day at the destination, although this obviously depends on the length of the journey, the type of activity undertaken at the destination and the resources available. Many complain, for example, that bus services finish very early in the day even when daylight lasts late into the evening, as it happens in northern summers. Downward and Lumsdon (2004) argue this limits the daily expenditure of bus users who, unlike motorists, cannot stop in the area for a drink and meal, if the last bus leaves at 4 pm.

Connections with incoming public transport links open up the area to day trippers within the region, but need to allow passenger time to transfer on both the inbound and outbound stages of the journey.

> *'Not enough time to get from train station to catch bus – about seven minutes. Poor connection times.'*

> *'First bus of the day connects with train at Llandovery. Pity the other buses do not connect. For example, the connection for Shrewsbury: you only have one minute if the bus is on time.'*

With an area network, a variety of destinations can be made accessible from different origins through co-ordinated timetables, including the hub and spoke design. Failure to ensure scheduled connections, however, can cause anxiety and frustration, rather than the positive emotions which generate repeated visits. Clockface timetables, with buses departing each stop at the same number of minutes past the hour throughout the day, are much easier to use for passengers and require less reliance on printed timetables, but are difficult to design. Many areas with limited funds have to trade off more frequent buses with the length of the season or the number of days the bus operates. Improvements in this respect can be made step by step. For example, the CoastHopper in Norfolk first proved its popularity, then increased its frequency and eventually extended its season.

> *'I would like this bus to run throughout the year.'*

Tourists, particularly international visitors, to rural areas may book their holidays several months in advance and want some idea about how they can travel within the areas. With short-term funding, it has often proved impossible to issue definitive timetables in advance, which can lead some tourists to choose other areas, arrange a hire car or resolve to use their own car.

Like in many other areas around the world, the stereotypical view of buses in Britain is that they are unreliable and this is often given as the reason for not using buses (Guiver, 2007). Catching a bus in an area you do not know can be nerve-racking, especially if it does not turn up when you expect it or you are waiting in uncomfortable surroundings (e.g. bus shelters are sometimes not permitted in protected areas). Lack of familiarity can also create perceptions of unreliability, which can be allayed with more detailed or readily available information.

> *'All the bus stops need timetables.'*

> *'At the bus stop in Crickhowell I couldn't see the x43 bus timetable so was glad I had printed it off the Internet.'*

Unreliability often occurs for reasons outside the bus operator's control. Narrow country roads, accidents, local festivals, haymaking, animal movements, motorists' poor driving and parking can impact on reliability, which in turn diminishes confidence in using the service.

> *'If the 27 would run more to the timetable I would travel more often. Always miss the connection for Breeze up to the Downs or the 13x.'*

> *'Sunday 12/9 bus 10.03 from Hexham Station broke down. Driver apparently unable to communicate with passengers.'*

Fares are also a crucial issue in the design of a bus service. People on low incomes (often without a car) tend to be the most price sensitive, so high fares may prevent them from travelling. However, car users also do not usually make a full estimate of the costs of fuel, parking, etc. (Gardner and Abraham, 2007), so may see car travel as cheaper than highly priced bus services. This is particularly true for families and groups.

> *'Bus fares for one person are ok. Two adults or family travelling is probably more expensive than a car.'*

Among the ways to reduce the fare burden are group tickets, combined parking and bus tickets, packages of fare and entrance fees for specific attractions and rover tickets for multiday unlimited travel within an area. The latter have proved extremely popular and can lift an uncommercial service into breaking even.

While bus users' opinions are useful, those of the potential market of car users are possibly more important. The results of a parallel survey on motorists (Lake District, 2006) showed that most of these believe they chose their destination and then selected the best way to get there. However, although they knew where to find bus information, very few consulted it. This suggests they actually will go to the destination by car, unless information making public transport look more attractive intervenes in their decision-making process. It also confirms the views of tourist attraction managers who see the mode markets as completely separate, with almost no transfer between modes. That is why managers are happy to improve facilities for walkers, cyclists and bus users who represent new markets, but are unwilling to instigate any measures that might deter car-users, as these would probably redirect them to another more welcoming area and might even lengthen their journeys.

Conclusions

Bus services running on the same road network as private cars provide the most readily available alternative to private cars in protected areas. Encouraging the use of public transport instead of cars helps reduce pollution, noise, danger and emissions within protected areas, thus enhancing the experience of being in the natural landscape for everyone. Using buses can also benefit travellers by reducing the stress of driving, offering novelty, interpretation, greater connection with the area and the opportunity to do linear walks. Buses also open up protected areas to people without cars, whose taxes also contribute to the preservation of such areas. However, providing attractive services for people who may be unfamiliar with the area or with using buses needs attention to details. Visitors' journeys often begin before they arrive in the area, finding information in brochures, magazines and on the Internet. Many choices are interlinked: travel to and within the area, travel mode and location of the accommodation, travel and potential activities and plans which allow or rule out the use of public transport can be made early in the process.

The most desired quality wanted by passengers is a high frequency of service. This has to be provided intelligently, with a view to the destinations and activities that potential passengers can access and the length of time they might want to spend there. Once the service is in place, it needs to be marketed and made visible to visitors to the area. The bus itself is often its own advertisement, especially when it is full of happy passengers. Monitoring the performance of the bus, how it is being used, whether passengers are being attracted to the area or out of their cars and how much they are spending is important for the long-term survival of the service. However, each area is unique, part of the 'placeness' visitors hope to sample on their visit. Visitors are equally very different, with different expectations, needs and desires. Thus, while the generic qualities of a good bus service for visitors can be loosely described, there is no substitute for applied and detailed local knowledge, the imagination to see what could be achieved and the persistence to make others believe it.

References

Brighton and Hove City Council (2014) 'Breeze up to the Downs!' [Online]. Brighton. www.brighton-hove.gov.uk/content/parking-and-travel/travel-transport-and-road safety/breeze-downs-0, accessed 5 March 2014.

Coulson, B.1 (2010) Personal communication to Guiver, J.

Coulson, B. (2011) 'Moving around the North Norfolk coast', *Funding Buses in Tourist Areas*. University of Central Lancashire, Preston.

Cornish Wools (2011) 'Whee!!! and off we go – A day out on the "Open Top Bus"', news. cornishwools.co.uk/?p=952, accessed 3 March 2014.

Cumbria County Council and Lake District National Park Authority (2011) Lake District Sustainable Visitor Transport Beacon Area. Bid to Department for Transport.

Department for Transport (2013) *Statistical Release: National Travel Survey: 2012*. London: Department for Transport.

Dickinson, J. E. and Dickinson, J. A. (2006) 'Local transport and social representations: challenging the assumptions for sustainable tourism', *Journal of Sustainable Tourism*, vol 14. no 2, pp.192–208.

Downward, P. and Lumsdon, L. (2004) 'Tourism transport and visitor spending: a study in the North York Moors National Park', UK. *Journal of Travel Research*, vol 42. no 4, pp. 415–420.

Eden Project (2014) 'Tickets and Tours', www.edenproject.com/visit-us/tickets-and-tours, accessed 5 May 2014.

Gardner, B. and Abraham, C. (2007) 'What drives car use? A grounded theory analysis of commuters' reasons for driving.' *Transportation Research Part F: Traffic Psychology and Behaviour*, vol 10, no 3, pp. 187–200.

golakes: the lake district. 2014. Combined Travel Tickets, www.golakes.co.uk/travel/tickets -multi-model.aspx accessed 5 May 2014.

Gregory, C. (2011) *The New Forest Tour: Funding Models and Lots More. Funding for Buses in Tourist Areas*. University of Central Lancashire, Preston.

Guiver, J. (2012 a). *Measuring the Costs and Benefits of Buses used for Leisure Trips*. Transport Practitioners Conference. Liverpool.

Guiver, J. (2012 b). 'How can you estimate the value of a bus service? Evaluating buses in tourist areas,' Association for European Transport, Glasgow.

Guiver, J. and Davies, N. (2007) *Tourism on Board 2006*.

Guiver, J. and Lumsdon, L. (2006) *Tourism on Board*. Institute of Transport and Tourism, Preston.

Guiver, J. and Stanford, D. (2014) 'Why destination visitor travel planning falls between the cracks', *Journal of Destination Marketing & Management*, vol 3, no 3, pp. 140–151.

Guiver, J. (2007) 'Modal talk: discourse analysis of how people talk about bus and car travel', *Transportation Research Part A: Policy and Practice*, vol 41, no 3, pp. 233–248.

Guiver, J., Lumsdon, L., Weston, R. and Ferguson, M. (2007) 'Do buses help meet tourism objectives? The contribution and potential of scheduled buses in rural destination areas', *Transport Policy*, vol 14, no 4, pp. 275–282.

Haq, G., Whitelegg, J., Cinderby, S. and Owen, A. (2008) 'The use of personalised social marketing to foster voluntary behavioural change for sustainable travel and lifestyles', *Local Environment*, vol 13, no 7, pp. 549–569.

Hilland, S. (2010) *The Black Forest and its KONUS Guest Card. Funding Buses in Tourist Areas*. University of Central Lancashire, Preston.

Jones, P. and Sloman, L. (2003) 'Encouraging behavioural change through marketing and management: what can be achieved', 10th International Conference on Travel Behaviour Research, Lucerne, Switzerland, pp. 10–15.

Le-Klähn, D. T., Gerike, R. and Hall, C. M. (2014) 'Visitor users vs. non-users of public transport: The case of Munich, Germany', *Journal of Destination Marketing & Management*, vol 3, no 3, pp. 152–161.

Lumsdon, L. and Caffyn, A. (2012) *Brecon Beacons and Powys Visitor Transport Plan*. Brecon Beacons National Park Authority.

Peak District National Park Authority (2012) *Sustainable Transport Action Plan*. Bakewell, Derbyshire.

Research Team (2013) *Visitor Survey 2013*. New Forest National Park.

Speakman, C. (2011) *Interview for seasonal buses project*. Personal communication to Guiver, J.

Transport For Leisure Ltd. (2000) *Transport Tourism and the Environment in Scotland*, Scottish Natural Heritage, Inverness.

VisitEngland (2008) *Destination Manager's Toolkit: Sustainable Visitor Transport*. VisitEngland.

Wibmer, C. (2012) 'Where public transport runs on GUTi : GUTi Gasteservice Umwelt-Ticket Guest-Service-Ticket for sustainable environment', *Association of European Transport Conference*. Glasgow.

5 The shift from automobiles to alternatives and the role of intelligent transportation systems

Kourtney K. Collum and John J. Daigle

From automobiles to alternatives: historical context

The modern day national park, and particularly the 'national park experience', is inextricably linked to the automobile (Dilsaver and Wyckoff, 1999; Sutter, 2002; Sims *et al.*, 2005; Louter, 2006). In 1908, Mount Rainier National Park became the first park in the United States to officially admit automobiles (Louter, 2006). This event occurred eight years before the National Park Service was officially established by Congress (Dilsaver and Wyckoff, 1999). As nature tourism began to flourish in the United States in the early part of the twentieth century, visitors flooded to parks such as Yellowstone, Yosemite and Mount Rainier by way of railroad, wagons and travel by horse and foot (Louter, 2006; Youngs *et al.*, 2008). But these modes of transportation were relatively short lived. The construction of the interstate highway system and the growing affordability of the automobile moved auto tourism from the realm of the wealthy to the realm of the middle class (Shaffer, 2001). The freedom and control afforded by the automobile had powerful implications for nature tourism. Specifically, it: (1) gave strength to grassroots movements dedicated to establishing more national parks and protected lands; (2) initiated the rise of auto tourism and the joining of government and private industry to meet public demand for recreation opportunities; and (3) launched the rapid movement of highways into the heart of America's most sublime landscapes (Shaffer, 2001; Sutter, 2002; Louter, 2006).

Many researchers have examined the impact of the automobile on the national park experience, emphasizing the influence on park design and infrastructure (Colten and Dilsaver, 2005; Dilsaver and Wyckoff, 1999; Hallo and Manning, 2009; Louter, 2006; White, 2007; Youngs *et al.*, 2008). For example, Dilsaver and Wyckoff (1999) exposed the deleterious ramifications of automobile infrastructure, describing the National Park Service (NPS) approach to transportation management as a 'process of cumulative causation', a type of positive feedback loop where each infrastructural addition encourages additional use, which in turn requires additional infrastructure. David Louter (2006) documented the changing aesthetics of national parks since the beginning of the twentieth century as a result of shifting attitudes towards automobiles. Using case studies of Mt. Rainier, Olympic and North Cascades National Parks, he highlighted three distinct phases of landscape design employed by the National Park Service: roads running through, roads designed to travel around and roads built completely outside of designated wilderness areas.

Currently, we are seeing a further shift in transportation management from infrastructure growth to infrastructure management. This current approach to management places emphasis on providing alternatives to travel by private automobile. The goal of this new paradigm is to reduce the harmful effects of automobiles on park resources while maintaining or improving the level of visitor satisfaction.

As alternative transportation systems (ATS) have gained popularity over the last 40 years, researchers have begun to explore the various components of the recreation experience that are affected by these systems (Harrison, 1975; Sims *et al.*, 2005; White, 2007; Holly *et al.*, 2010). The majority of studies have focused on visitor attitudes towards changes in existing transportation systems. Harrison (1975) was one of the first to do this by surveying visitors at Denali National Park about newly implemented restrictions on private automobiles. Cars were banned in certain areas within the park and a fare free shuttle bus was introduced. Contrary to expectations, 84 per cent of those surveyed approved of the new policy. Although support for the new policy was relatively high, respondents who utilized the bus service indicated stronger support of the policy than those who used a private automobile. This suggests that if visitors can be influenced to try park shuttles, they may find them less of an inconvenience than previously anticipated, thereby increasing support for such systems. Harrison (1975) stressed, however, that shuttles must offer amenities equal to those available from a private automobile, or offer a unique service, if they are to succeed as a competitive alternative.

By modeling visitor acceptance of a proposed shuttle system at Cades Cove in Great Smoky Mountains National Park, Sims *et al.* (2005, p25) were able to explore the assumption that based on a historic perception of automobiles as the primary and best way to experience national parks, 'the establishment of shuttle systems could potentially result in greater impact on visitor experience than that resulting from the increase in traffic congestion'. However, the results of the study revealed higher support for a mandatory shuttle system than managers had anticipated, with 75 per cent of respondents indicating that they would support a mandatory shuttle system if there were no fee for parking or riding. This study showed that the value of reduced traffic congestion to visitors was significant.

White (2007) conducted an interpretive study of visitor attitudes towards the shuttle system at Yosemite National Park and concluded that visitors primarily value convenience and freedom when considering travel modes. Visitors using private automobiles praised the convenience of their automobiles and used rationalization as a cognitive coping mechanism when confronted by congestion and crowding. Interestingly, visitors using alternative transportation also referred to their travel mode as 'convenient', precisely because it allowed them to avoid traffic congestion. This finding suggests that visitors share similar values, such as freedom and convenience, yet often disagree on which travel modes best suit their values.

In recent years, an increasing number of parks, wildlife refuges, and national forests worldwide have attempted to address crowding, congestion and resource degradation by implementing ATS, which combine various travel modes such as bicycles, buses and hiking trails in order to reduce visitor reliance on private automobiles. ATS have been successfully implemented at parks across the United States of America (USA), but there is further need to promote these systems and convince visitors to switch from the car to other available modes. An emerging strategy is to apply intelligent transportation systems (ITS), an approach to transportation management that uses information technologies to provide visitors with relevant and real-time traffic information. Focusing on ITS application in Rocky Mountain National Park (USA), this chapter explores how specific forms of ITS can encourage shuttle use in protected areas. The next section provides a brief overview of ITS and their pioneering applications in US national parks. The central part of the chapter presents a case study of a pilot ITS project at Rocky Mountain National Park, with special attention to visitor valuations of ITS.

Emergence of intelligent transportation systems

Transportation experts and national park managers have identified four ITS applications that offer the most valuable solutions to auto-related problems: (1) provide traveler information about road conditions so as to reduce congestion; (2) provide relevant information about transit options so that visitors can make informed decisions; (3) provide real-time information on weather, traffic, and parking lot conditions; and (4) direct visitors to areas with reduced congestion (Dilworth, 2003).

Technologies designed to address these key areas have been tested in various national parks in the USA including Acadia, Kings Canyon, Sequoia, Grand Canyon and Arches, among others (Dilworth, 2003; Daigle and Zimmerman, 2004a, 2004b; Ye *et al.*, 2010). A study conducted at two park units in California showed that visitors reported willingness to use two ITS technologies - electronic message signs (EMS) and highway advisory radio (HAR) - to access information about road closures and parking and weather conditions (Dilworth, 2003). Willingness to use these technologies was highest among respondents with higher past experience using them and with positive attitudes regarding the appropriateness of ITS tools in parks. At Grand Canyon National Park, EMS and HAR were also evaluated and support vector regression analysis suggested that the two ITS technologies were responsible for a 30 per cent increase in shuttle ridership (Ye *et al.*, 2010).

ITS have also been tested in urban park settings. In an effort to improve visitor safety and inform motorists' decision-making, portable changeable message signs (PCMS) were installed at Golden Gate National Recreation Area in California. Using a combination of qualitative and quantitative methods, researchers evaluated the influence of the signs. While visitor surveys indicated only small influences, traffic counts indicated a 12-14 per cent reduction in traffic volumes on weekdays, and up to 19 per cent on weekends. Overall, the PCMS appeared to positively influence shuttle ridership, notwithstanding operations and maintenance challenges (Strong *et al.*, 2007).

Despite these studies, park and recreation researchers have stated the need for additional transportation focused research (Daigle and Zimmerman, 2004a; Sims *et al.*, 2005; White, 2007; Daigle, 2008). While preliminary research reveals that ITS technologies are viable, further research is needed to determine what specific technologies are most effective for shifting visitors from private automobiles to alternatives. Research is also needed to identify the effectiveness of ITS given specific geographic areas, user groups, levels-of-use and capital and resource constraints. The best technologies must be identified so that a switch in travel mode does not necessitate a decrease in visitor experience. Recent research conducted at Rocky Mountain National Park explored the utility of an ITS to increase shuttle ridership to and within the park, and sought to identify potential strategies for increasing awareness and use of shuttles.

Research on ITS at Rocky Mountain National Park

The context

Rocky Mountain National Park (ROMO) is the most visited park in Colorado and is challenged by consistently high visitation concentrated within the peak summer season. The months of June, July, and August alone see more than half of ROMO's three million annual visitors (NPS, 2014). In the 1970s a fare-free visitor transportation system was introduced to help manage the influx of visitors. The shuttle service has since grown, with nearly half a million rides provided in 2010 (Villwock-Witte and Collum, 2012). Despite this, private automobiles remain the preferred travel mode by the majority of visitors. Symptoms of this high visitation rate include parking lots filled to capacity early in the day, traffic congestion within the park, and pressure on natural and managerial resources. The issue is further exacerbated by bottlenecking at the park's primary access point, where two US highways and one state highway converge at a major intersection in downtown Estes Park. As an interim solution, a pilot ITS was implemented to direct day visitors to a new park-and-ride lot located east of downtown Estes Park, where visitors could board a shuttle that provided a five minute ride to the Estes Park Convention andVisitors Bureau (CVB). Once at the CVB, visitors could gather information for their trip and transfer to any of four shuttle routes servicing Estes Park and ROMO. The pilot ITS was comprised of dynamic message signs (DMS) and highway advisory radio (HAR). The DMS displayed short, concise messages to passing motorists (Figure 5.1), while the HAR broadcast longer messages on a continuous loop to warn approaching visitors of the status of parking within ROMO, particularly at Bear Lake, one of the most heavily visited areas of the park (Collum, 2012). Once parking lots reached capacity, messages were displayed to inform visitors that the Bear Lake lot was full and to recommend use of the park-and-ride lot and shuttle. The DMS also displayed the station number for the HAR, which played a recorded message informing callers of travel conditions, parking options, and shuttle services.

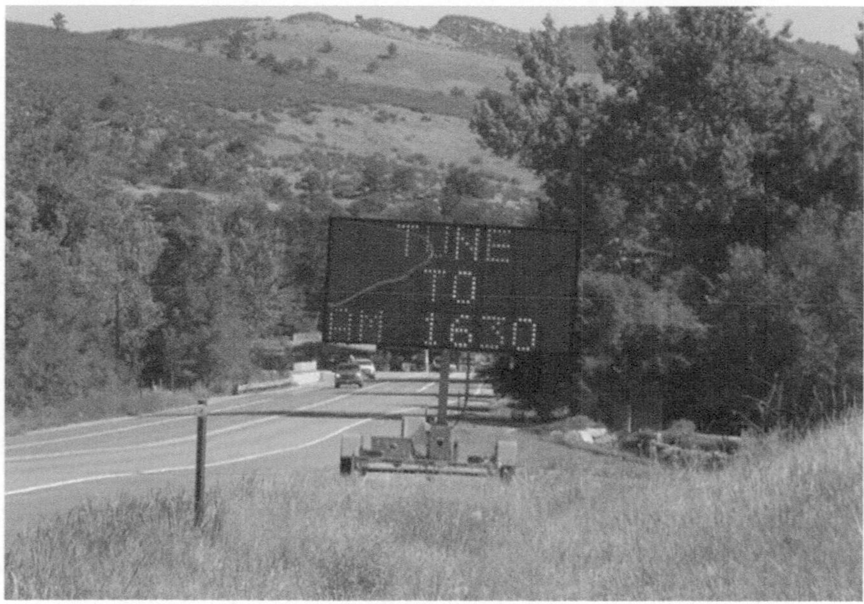

Figure 5.1 Dynamic message signs (DMS) displaying short, concise messages can be used to inform motorists about road and weather conditions, and/or let them know the station number of a highway advisory radio (HAR) that provides further information on travel conditions (source: photo by Natalie Villwock-Witte).

The research conducted at ROMO sought to answer two questions:

1 How useful are the pilot ITS technologies for visitors and residents?
2 To what extent do ITS influence visitors' travel mode choices?

Two questionnaires were developed for this study: an on-board shuttle survey for the new park-and-ride shuttle, and an intercept visitor survey in town to capture visitors not using the park-and-ride or other alternate mode of travel. More details about the study approach, questionnaire, non-response bias and data analyses can be found in a thesis by Collum (2012). Of the shuttle survey respondents, 80 per cent indicated that they saw a DMS. Of those who saw a DMS, the majority (86 per cent) reported that the DMS displayed a message prompting them to tune to the HAR, and 67 per cent of those prompted indicated that they did tune to the HAR. In contrast, among the visitor survey respondents, 65 per cent stated that they saw a DMS, and of those, 68 per cent reported that the DMS displayed a message prompting them to tune to the HAR. Of those who reported seeing a prompt for the HAR, only 28 per cent actually tuned to the HAR.

Evaluation of the highway advisory radio

Visitors rated the HAR on a number of criteria, such as how accurate and useful the information was, whether the information saved them time and helped them get around or avoid traffic congestion, and whether they would use the information again. Overall, shuttle survey respondents indicated high levels of satisfaction with the HAR. All respondents who used and evaluated the highway advisory radio strongly agreed or somewhat agreed that the information was accurate. When asked if the information saved them time, 65 per cent agreed, and 72 per cent indicated that they were able to get around easier with the information. Similarly, 75 per cent agreed that the information helped them avoid traffic congestion. A high proportion of users (89 per cent) agreed that the information was useful to them, and 85 per cent agreed that they planned to use the information if visiting again. Interestingly, 79 per cent agreed that they needed more information, despite the high levels of satisfaction with the available information (Table 5.1). Unfortunately, respondents provided little to no information regarding the type of information they needed.

Satisfaction with the HAR among visitor survey respondents was much lower. Though a high proportion of respondents (86 per cent) agreed that the information was accurate, less than half (44 per cent) agreed that the information saved them time and helped them avoid traffic congestion. A slightly higher percentage (64 per cent) agreed that they were able to get around easier with the information, and 75 per cent agreed that the information was useful to them. Despite the lower satisfaction among respondents in some areas, 82 per cent agreed that they would plan to use the information again. Just over half (52 per cent) agreed that they needed more information (Table 5.1).

Table 5.1 Visitors' evaluation of the highway advisory radio (HAR) based on responses to the shuttle survey and the visitor survey.

	Shuttle survey respondents				*Visitor survey respondents*			
	Mean[a]	*SD*	*Agree*[b] *(N)*	*Agree*[b] *(%)*	*Mean*[a]	*SD*	*Agree*[b] *(N)*	*Agree*[b] *(%)*
The information was accurate	4.74	0.45	19	100	4.34	0.72	25	86.2
The information saved me time	3.90	1.33	13	65.0	3.56	0.80	12	44.4
I was able to get around easier with the information	4.11	1.08	13	72.2	3.86	0.85	18	64.3
I would plan to use the information if visiting again	4.20	1.11	17	85.0	4.25	0.84	23	82.1

	Shuttle survey respondents				*Visitor survey respondents*			
	Mean[a]	*SD*	*Agree*[b] *(N)*	*Agree*[b] *(%)*	*Mean*[a]	*SD*	*Agree*[b] *(N)*	*Agree*[b] *(%)*
The information was useful to me	4.32	1.00	17	89.4	3.96	1.17	21	75.0
The information helped me avoid traffic congestion	4.00	1.21	15	75.0	3.52	0.85	12	44.4
I needed more information	3.93	1.21	11	78.6	3.76	1.26	11	52.4

a Mean based on 5-point Likert scale ranging from 1 (*strongly disagree*) to 5 (*strongly agree*).
b Responses of 4 and 5 were collapsed into the category 'agree'.

Evaluation of the park-and-ride shuttle

Overall, the respondents indicated high levels of satisfaction with the shuttle. When asked if they enjoyed their experience using the shuttle, 92 per cent strongly agreed or somewhat agreed. In addition, 92 per cent agreed that they would use the shuttle again. More than 90 per cent of respondents agreed that the shuttle was convenient (95 per cent) and easy to use (97 per cent). Additionally, 79 per cent agreed that the shuttle saved them time. Only 5 per cent felt that the shuttle was confusing, and less than 5 per cent felt that it was physically challenging for them or someone in their group to get on/off the shuttle (3 per cent), that the shuttle did not have sufficient room for their gear (3 per cent), and that it seemed difficult to travel with children on the shuttle (2 per cent). However, 20 per cent of respondents felt that they had to switch shuttles too many times to get to their desired destination, 14 per cent said the shuttle did not run frequently enough for their needs, and 10 per cent had trouble finding the shuttle schedule (Table 5.2).

There are many potential reasons why visitors may have chosen not to use the park-and-ride shuttle. The visitor survey allowed an assessment of what reduces visitors' willingness to choose a shuttle service. Seventy one per cent (n = 215) of respondents who provided a reason (or 44 per cent of total visitor survey respondents) indicated that they were not aware of the park-and-ride shuttle. Furthermore, because the park-and-ride was designed to provide an alternative for day visitors, 62 per cent of visitor survey respondents who were overnight visitors had no reason to use the shuttle, as they could leave their vehicles at their lodging and board a different shuttle from there. For those who were aware of the shuttle but did not use it, written comments indicate that the majority were staying overnight and did not need the shuttle or were simply 'not interested'.

Sources of information about shuttles

Although the primary goal of the pilot ITS was to encourage day visitors to use the new park-and-ride lot and associated shuttle, it was also expected that the ITS

Table 5.2 Visitors' evaluation of the park-and-ride shuttle based on 12 criteria (only the responses to the shuttle survey are reported).

	Mean[a]	*SD*	*Agree*[b] *(N)*	*Agree*[b] *(%)*
The shuttle is convenient	4.71	0.62	55	94.8
I would use the shuttle again	4.68	0.91	58	91.9
The shuttle is easy to use	4.66	0.80	57	96.6
I enjoyed my experience using the shuttle	4.63	0.86	55	91.6
The shuttle saved me time	4.33	1.03	50	79.4
I had to switch shuttles too many times to get to my desired destination	2.14	1.21	12	20.3
The shuttle does not run frequently enough for my needs	2.05	1.18	8	13.7
It seems difficult to travel with children on the shuttle	1.76	0.95	1	2.0
I had trouble finding the shuttle schedule	1.74	1.16	6	10.4
The shuttle does not have sufficient room for my gear	1.66	0.93	2	3.4
The shuttle schedule is confusing	1.56	0.93	3	5.3
Getting on/off the shuttle is physically challenging for me or someone in my group	1.29	0.73	2	3.4

a Mean based on 5-point Likert scale ranging from 1 (*strongly disagree*) to 5 (*strongly agree*).
b Responses of 4 and 5 were collapsed into the category 'agree'.

would increase awareness of all shuttle options among day visitors and overnight visitors alike. To evaluate this, respondents were asked to indicate how they learned about the shuttles. The most frequent source of information cited by shuttle survey respondents was the DMS: 41 per cent indicated that they learned about the shuttles from this source (Table 5.3). The HAR was cited the second most frequently (22 per cent). Four information sources were used by less than 5 per cent of shuttle survey respondents: hotel/lodge/campsite staff, the town website, employment with ROMO, and employment with a business in Estes Park. No shuttle survey respondents reported that they learned about the shuttles from the ROMO website. Among visitor survey respondents, the most frequently cited source of information was visitor center staff, with 24 per cent indicating that they learned about the shuttles from this source (Table 5.3). The second most cited information source was the DMS (20 per cent), followed closely by 'previous visits' (17 per cent). Less than 5 per cent of visitor survey respondents indicated that they learned about the shuttles from the

Table 5.3 Sources of information about shuttles for the shuttle survey respondents and the visitor survey respondents.

	Shuttle survey respondents		*Visitor survey respondents*	
Information source	*N*	*%*	*N*	*%*
Dynamic message signs	27	41.5	85	20.1
Highway advisory radio	14	21.5	14	4.0
Family or friends	9	13.8	54	12.8
Visitor center staff	8	12.3	103	24.4
A newspaper article	7	10.8	15	3.5
Previous visits	4	6.2	72	17.0
Hotel/lodge/campsite staff	3	4.6	35	8.3
The Town of Estes Park website	2	3.1	16	3.8
Through employment with a business in Estes Park	2	3.1	6	1.4
Through employment with ROMO	1	1.5	4	0.9
The ROMO website	0	0.0	43	10.2

Note: Totals amount to more than 100% as respondents were instructed to indicate all sources of information used.

HAR, a newspaper article, the town website, through employment with ROMO, and through employment with a business in Estes Park.

Effect of shuttle use on visitor experience

Respondents of both the shuttle and visitor survey were asked to rate their overall experience visiting the town and national park, as well as their overall travel experience (i.e. driving, navigating and parking). Overall experience with the town and national park was rated very highly, whereas the travel experience was rated slightly lower. Nearly all respondents rated their overall experience 'good' or 'very good' (97 and 96 per cent among the shuttle survey and the visitor survey respondents, respectively), and most respondents (85 and 82 per cent, respectively) gave a 'good' or 'very good' rating for their travel experience. Interestingly, shuttle-users did not rate their overall experience or travel experience significantly higher than respondents who did not use shuttles. Furthermore, respondents who used ITS did not rate their overall experience or travel experiences significantly higher than respondents who did not use ITS (Collum, 2012).

Visitor response to ITS

The results indicate that 80 per cent of shuttle survey respondents who approached the area where the DMS were located saw one or more DMS. Additionally, 42 per cent of shuttle survey respondents reported that they learned about the park-and-ride shuttle from the DMS. It was originally anticipated that the DMS would simply encourage visitors to tune to the HAR, and the HAR would then influence visitors to use the park-and-ride and shuttle. However, 43 per cent of the visitors who used the park-and-ride shuttle saw a DMS but did not tune to the HAR. This suggests that the information on the DMS was enough for many visitors to decide to use the park-and-ride, without needing to tune to the HAR for further information. Additionally, 65 per cent of visitor survey respondents who approached the area where the DMS were located reported seeing a DMS. Importantly, 20 per cent indicated that they learned about the town and park shuttles from this source. Thus, the DMS appear to have successfully increased awareness of the various town and park shuttles.

The utility of the HAR is less evident. Only 12 per cent of visitor survey respondents who approached the area where the HAR was located reported using the HAR, and less than half of those visitors felt that the information saved them time or helped them avoid traffic congestion. However, use of the HAR was much higher (44 per cent) among shuttle survey respondents, and the effect of the information on visitor mobility much greater. Sixty-five per cent or more of those who used the HAR felt it saved them time and helped them avoid traffic congestion and nearly 90 per cent found the information useful. With more than 80 per cent of shuttle and visitor survey respondents indicating that they would use the information again, it is clear that for this segment of visitors, the HAR provided a valuable service. Though satisfaction was fairly high among users of the HAR, it is a fact that few people chose to use it. When choosing appropriate technologies for an ITS, managers must not only consider visitor satisfaction with the technologies but also the appeal of these technologies to a broad user base, as well as the cost of operating and maintaining the devices. While the HAR did contribute to awareness and use of the park-and-ride shuttle, it was not widely used by other visitors. Furthermore, the HAR failed to transmit as far as anticipated and several visitors indicated that the channel was overpowered by static and difficult to hear at times. The geography of the area presented considerable challenges for transmission of the radio signal. Further research is necessary to determine whether the DMS alone can increase awareness and use of shuttles or if the HAR is a vital component for a significant number of visitors.

Examining the two survey groups individually, we found that 95 per cent of shuttle survey respondents who tuned to the HAR indicated that the information on the HAR influenced them to use the park-and-ride. This is substantial, especially considering that such a large proportion (35 per cent) of shuttle survey respondents were seasonal and fulltime residents who had no occasion to use the ITS. Although the HAR appears to have been of great value to this segment of visitors, in terms of sheer numbers, the effect is negligible. Furthermore, the HAR had a small impact on visitor survey respondents. Only 57 per cent of those who tuned to the HAR, or less than 6 per cent of total visitor survey respondents who approached from where this technology was located, agreed that the information influenced their mode choice.

Although the HAR failed to have broad impact, the DMS cannot be expected to serve as the sole technology of an ITS. These technologies are meant to work in tandem, especially considering the limited text which can be communicated via the DMS. If DMS are not paired with HAR, other technologies must be implemented or expanded in order to provide visitors with adequate information to inform their travel decisions. Electronic signs that display real-time departure time for shuttles have been successfully demonstrated at other parks (Daigle and Zimmerman, 2004b). A recent study at Acadia National Park suggested that real-time technology be linked with transportation-related indicators and standards of quality (Pettengill *et al.*, 2012). For example, automated traffic counters, along with monitoring of ridership by drivers on buses, could provide real-time information to visitors about when and where transportation-related conditions are not meeting visitor-based standards of quality (e.g. seeing more than eight cars per 125 meters of park road or ten available seats on buses), offering visitors a chance to adjust their use of the park accordingly.

Conclusions

Alternative travel modes must account for a larger percentage of visitor transportation at parks and other public lands that are experiencing significant crowding and congestion. ITS offer a valuable tool for travel management as they can provide relevant information to help visitors make informed travel decisions. The example of the pilot ITS study at Rocky Mountain National Park (USA) demonstrates the utility of ITS to increase visitor awareness and use of shuttles in protected areas. Although fare-free shuttle services are popular in many parks around the world, it remains uncertain if voluntary systems will be enough to result in a significant decrease in crowding, congestion and resource degradation. That is also the reason why in a growing number of parks, shuttles have been made compulsory at the most popular and congested areas (Harrison, 1975; Sims *et al.*, 2005; Manning *et al.*, 2014). Without a combination of ITS and other powerful incentives, mandatory shuttle use and/or restrictions will be necessary in the future to better preserve natural resources and the recreational experience.

References

Bishop, P. (1996) 'Off road: Four-wheel drive and the sense of place', *Environment and Planning D: Society & Space*, vol 14, no 3, pp. 257–271.

Collum, K. (2012) From automobiles to alternatives: Applying attitude theory and information technologies to increase shuttle use at Rocky Mountain National Park, MS thesis, University of Maine, USA.

Daigle, J. J. (2008) 'Transportation research needs in national parks: A summary and exploration of future trends', *The George Wright Forum*, vol 25, no 1, pp. 57–64.

Daigle, J. J. and Zimmerman, C. A. (2004a) 'Alternative transportation and travel information technologies: Monitoring parking lot conditions over three summer seasons at Acadia National Park', *Journal of Park and Recreation Administration*, vol 22, no 4, pp. 81–102.

Daigle, J. J. and Zimmerman, C. A. (2004b) 'The convergence of transportation, information technology, and visitor experience at Acadia National Park', *Journal of Travel Research*, vol 43, no 2, pp. 151–160.

Dilsaver, L. M. and Wyckoff, W. (1999) 'Agency culture, cumulative causation and development in Glacier National Park, Montana', *Journal of Historical Geography*, vol 25, no 1, pp. 75–92.

Dilworth, V.A. (2003) Visitor perceptions of alternative transportation systems and intelligent transportation systems in national parks. PhD thesis, Texas A&M University, USA.

Harrison, G. S. (1975) 'The people and the park: Reactions to a system of public transportation Mt. McKinley National Park, Alaska', *Journal of Leisure Research*, vol 7, no 1, pp. 6–15.

Holly, M. F., Hallo, J. C., Baldwin, E. D. and Mainella, F. P. (2010) 'Incentives and disincentives for day visitors to park and ride public transportation at Acadia National Park', *Journal of Park and Recreation Administration*, vol 28, no 2, pp. 74–93.

Louter, D. (2006) *Windshield Wilderness: Cars, Roads, and Nature in Washington's National Parks*, University of Washington Press, Seattle, WA.

Manning, R. E. (2011) *Studies in Outdoor Recreation: Search and Research for Satisfaction* (3rd ed.). Oregon State University Press, Corvallis, OR.

Manning, R., Lawson, S., Newman, P., Hallo, J., and Monz, C. (eds) (2014) *Sustainable Transportation in the National Parks*. University Press of New England, Hanover and London.

NPS (2014) National Park Service public use statistics office, National Park Service, Washington, DC, www.nps.gov/romo/parkmgmt/statistics.htm, accessed 16 August 2012.

Pettengill, P.R., Manning, R.E., Anderson, L.E., Valliere, W., and Reigner, N. (2012) 'Measuring and managing the quality of transportation at Acadia National Park', *Journal of Park and Recreation Administration*, vol 30, no 1, pp. 68–84.

Sheldon, P. J. (1997) *Tourism Information Technology*. Cab International, New York, NY

Sims, C., Hodges, D., Fly, J. and Stephens, B. (2005) 'Modeling visitor acceptance of a shuttle system in the Great Smoky Mountains National Park', *Journal of Park and Recreation Administration*, vol 23, no 3, pp. 25–44.

Strong, C. (1999) 'Overview of National Park Service Challenges', Paper presented at National Parks: Transportation Alternatives and Advanced Technology for the 21st Century, Big Sky, Montana, USA.

Strong, C., Eidswick, J., Gartland, C. and Lindsay, A. (2007) *Evaluation of Portable Changeable Message Signs at Golden Gate National Recreation Area*, Western Transportation Institute, Bozeman, MT.

Turnbull, K. (2010) 'Transportation to enhance America's best ideas: Addressing mobility needs in America's national parks', *TR News*, vol 267, pp. 20–26.

Vaske, J. J. (2008) *Survey Research and Analysis: Applications in Parks, Recreation and Human Dimensions*, Venture Publishing, State College, PA.

Villwock-Witte, N. and Collum, K.K. (2012) Evaluation of an Intelligent Transportation System for Rocky Mountain National Park and Estes Park: Final Report. Paul S. Sarbanes Transit in Parks Technical Assistance Center, Bozeman, MT.

Waitt, G. and Lane, R. (2007) 'Four-wheel drivescapes: Embodied understandings of the Kimberley', *Journal of Rural Studies*, vol 23, no 2, pp. 156–169.

White, D. D. (2007) 'An interpretive study of Yosemite National Park visitors' perspectives toward alternative transportation in Yosemite Valley', *Environmental Management*, vol 39, no 1, pp. 50–62.

Ye, Z., Albert, S., Eidswick, J. and Law, S. (2010) 'Improving shuttle ridership using intelligent transportation system technologies: National park case study', *Transportation Research Record*, vol 2174, pp. 44–50.

Youngs, Y. L., White, D. D. and Wodrich, J. A. (2008) 'Transportation systems as cultural landscapes in national parks: The case of Yosemite', *Society & Natural Resources*, vol 21, no 9, 797–811.

6 Park visitor and gateway community perceptions of mandatory shuttle buses

Joshua D. Marquit and Britton L. Mace

Introduction

Over the past 50 years, national parks worldwide have experienced an exponential increase in the number of visitors (e.g. Ansson, 1996). For example, according to the US National Park Service Annual Visitation Summary Report for 2013, the most significant increase has occurred over the past decade, with an estimated 270 million annual visitors to the 394 park units in the USA (NPS, 2014a). Consequently, the efforts of many parks around the world to protect and preserve natural resources and the visitor experience have become increasingly difficult and complex. For example, visitors and their vehicles bring social and environmental consequences to the parks and surrounding gateway communities. Some of these impacts include crowding (Manning *et al.*, 2000); parking issues (Clarke, 2001); traffic problems including long lines of idling cars, congestion and accidents (Runte, 1979; Reeves, 2006); air pollution and associated problems including acid rain and soil erosion (Turk and Campbell, 1997; Mace *et al.*, 2004;); and noise pollution (NPS, 1994; Mace, *et al.*, 2004). Further complicating these problems are the ever-decreasing financial resources necessary to maintain existing infrastructure and build new roadways and parking lots (Ansson, 1996). Loomis (2002) warns that if these visitation trends continue, they may lead to irreversible damage to the health of the parks' ecosystems and quality of life in the surrounding gateway communities. For these reasons, park managers have explored a variety of programs and initiatives to protect natural resources, enhance the visitor experience, and establish symbiotic relationships between parks and gateway communities.

In many parks around the world, park and community managers have implemented alternative transportation systems that can potentially minimize the adverse effects of vehicles while protecting wildlife, plants, air quality and the natural soundscape, while also enhancing the visitor experience (e.g. NPS, 2014b). Additionally, many of these alternative transportation systems, such as shuttle buses, do not require drastic alterations of existing park infrastructure. For example, a number of alternative transportation systems are functional in some of the busiest parks in the US, including Denali, Acadia, Yosemite, Grand Canyon and Zion. In fact, over 50 parks in America have instituted and experimented with alternative transportation systems to counter the effects of overuse (Gallegos, 2005). Unfortunately, there is a limited amount of empirical data on alternative transportation systems in parks, which complicates the planning and decision-making process (Daigle, 2008; Pettengill *et al.*, 2012). Recent research has found that those parks that have used voluntary alternative transportation options have exhibited many signs of improvement, including the return of indigenous wildlife, improved visibility, fewer parking and traffic-related issues and improved visitor experiences (White, 2007; White *et al.*, 2011). Other parks with voluntary transportation systems, however, have encountered unforeseen problems, such as increased erosion on remote trails, increased wait times, difference in preference based on the form of alternative transportation offered, prohibitive ridership fees and dramatic changes of visitor flow on trailheads near transportation stops (e.g. Morgan, 1985; Shiftan *et al.*, 2006). Furthermore, there are additional transportation-related factors, including convenience, comfort, accessibility, number of stops made, and the perceived freedom to move through the park that influence voluntary ridership (e.g. Laube and Stout, 2000; Dilworth and Shafer, 2004; Hallo and Manning, 2009; White, 2007; Youngs *et al.*, 2008; Pettengill *et al.*, 2012).

To date, there are a limited number of studies on mandatory shuttle systems like those found in Zion National Park in Utah, Denali National Park in Alaska, Bayerischer Wald National Park in Germany, and a few of the parks in the Great Britain. Existing studies have found favourable visitor response to shuttle use, with improved freedom, convenience, and quality of the park experience (e.g. Harrison, 1975; Eaton and Holding, 1996; Holding and Kreutner, 1998; Miller and Wright, 1999; Mace *et al.*, 2013). Still fewer studies have examined the impact a mandatory shuttle has on gateway communities that are located adjacent to national parks. Gateway communities play an integral role in supporting park goals to protect natural resources and enhance the overall visitor experience by providing a variety of amenities and visitor services. Initial research by Howe *et al.* (1997) found that park-related decisions including transportation systems can have a significant impact on environmental, cultural and social issues in gateway communities. Additional research is needed to investigate the impacts of mandatory shuttle systems on local residents and the economic vitality of gateway communities. Starting from research conducted about the shuttle system in Zion National Park (USA), this chapter explores the perceptions of visitors and gateway communities towards mandatory shuttle services. Particular emphasis is placed on how business owners and employees in gateway communities perceive they are affected by the shuttle service. Key findings and areas for future research are described at the end of the chapter.

Gateway communities: balancing natural resource protection and economic viability

Gateway communities are towns and cities that border national and state parks, wildlife refuges, forest and other public lands (Howe *et al.*, 1997). In the USA, the National Park Service (NPS) expands this definition to include any community that lies within 60 miles (96.5km) of a park (NPS, 2014c). Gateway communities are important not only for their role in providing tourists with much-needed amenities, accommodations and public services, but also because they act as 'portals' to some of the most unique and stunning landscapes on the planet (Howe *et al.*, 1997). As such, these communities are vital to the preservation of the natural environment and the cultural history in their regions.

Gateway communities also play an instrumental role in the establishment and maintenance of an enjoyable experience for park visitors. The natural resources in these regions have become a significant part of the economic livelihood and cultural identity of communities. For example, in the 2012 National Park Visitor Spending Effects: Economic Contributions to Local Communities, States, and the Nation Report, the NPS estimated that visitors spent US $14.7 billion in local gateway communities, which contributed to over 243,000 jobs (NPS, 2014c). A closer examination of those numbers shows US $9.3 billion in labour income including employee wages, salaries and benefits, US $15.9 billion in value added or Gross Domestic Product (GDP), and US $26.8 billion in output such as sales to consumers and exports or business-to-business exchanges (NPS, 2014c). The lodging and restaurant/bar sector saw the highest direct contributions with 40,000 and 30,000 jobs and US $4.5 and $3 billion to local gateway region economies, respectively (NPS, 2014c). It is clear, therefore, that gateway communities support and are supported by their adjacent parks and should be partners in the planning process, as both are affected by major decisions, including those on transportation systems.

Further, the relationship between local residents in gateway communities and public land management is vital to the success of these initiatives and programs aimed at the protection of natural resources. In recent years, park managers have asked gateway communities to assist in their efforts to protect natural resources and preserve the cultural history of the surrounding land and people (Wilson, 2000). The relationship between gateway communities and public land management agencies is symbiotic, multi-faceted and complex. Public land management depends heavily upon gateway communities to provide accommodations for tourists, assist in protecting the scenic beauty and resources of the natural landscape, and improving the visitor experience. At the same time, gateway communities are dependent on the public lands for economic, social and cultural reasons. Consequently, when major decisions are made about the park, such as the implementation of an alternative transportation system, residents of gateway communities should be included and encouraged to participate in the decision-making process.

A pioneer mandatory shuttle service: the Zion National Park Alternative Transportation System

Zion National Park is located in southwestern Utah and encompasses approximately 229 square miles of desert landscape, world-renowned slot canyons, and forested ecosystems that are part of the Colorado Plateau, Great Basin and Mojave Desert. The park also includes mountainous terrain with unique rock formations including plateaus, natural arches and other geologic wonders including sandstone buttes and mesas. Over 1,000 species of plants and 250 species of mammals, reptiles, amphibians, fish and birds inhabit the park, most of which thrive in the arid climate (NPS, 2014d).

In 1909, President William Howard Taft designated some of the land within the current park as Mukuntuweap National Monument. In 1919, the US Congress changed the boundaries of the original designation and officially renamed the area Zion National Park. In 1937 and 1956, park boundaries were expanded to include the Kolob Canyons, one of the most recognizable land areas of the park. According to the Annual Park Recreation Visitation report (NPS, 2014e), initial visitors in 1919 numbered 1,814. By 1975, over one million people were visiting the park annually. In 2013, the number of visitors exceeded 2.8 million. This steady increase in visitors forced park managers to reconsider transportation policy and alter infrastructure. Over the past century, major changes have included the construction of Zion Lodge, a tunnel, visitor centres and museums, additional roadways and parking lots and, more recently, a mandatory shuttle transportation system.

The need to change transportation policy became evident when one of the most popular areas of the park, Zion Canyon, routinely experienced traffic congestion and crowding in the 1990s. Before the shuttle became mandatory, on busy weekends, 11,000 visitors in 5,000 vehicles would crowd into Zion Canyon daily (Clark, 2001), with only 450 parking spaces available, most of them allocated to Zion Lodge. This heavy traffic and lack of parking led to the launch of an alternative transportation system in May 2000. The shuttle was meant to minimize the adverse impacts of visitors and their vehicles, such as traffic jams, overcrowding, illegal parking, horn honking, excessive idling of vehicles and air and noise pollution. Shuttle buses now loop through Zion Canyon and the gateway community of Springdale, which is located immediately adjacent to the southwestern part of the park. During the months of April through October, the shuttle is required for all visitors to Zion Canyon. During the peak season (April through October), the shuttle runs continuous loops through Zion Canyon and Springdale, making nine stops in the park and six in the community (NPS, 2014f). Visitors are required to ride the shuttle through Zion Canyon because of the crowding and parking issues through the main corridor of the park. Shuttle usage is encouraged but not required in Springdale. The Zion National Park mandatory shuttle system consists of a fleet of propane-powered shuttle buses that emit fewer particulates and create less noise than the average automobile. There are 30 propane-powered buses in operation in the park, along with 27 accompanying shuttle trailers (connecting to one of the shuttle buses, producing a double-length bus). During the planning stages of the shuttle system, business owners on Zion Boulevard were invited by park officials to request a shuttle stop in front of their businesses. The use of the shuttle is free both in the town of Springdale and for visitors to the park from 5:45 am to 11 pm, with shuttles making stops every 7 to 10 minutes (Zion Natural History Association, 2014). Visitors can park their vehicles in Springdale and take the shuttle into the park. At full capacity, each double-length shuttle bus can hold a total of 66 people, thereby replacing approximately 25 private automobiles that would otherwise travel through or compete for parking within the park. Approximately 1.5 million visitors rode the shuttle during the first year of operation. In subsequent years, the number of riders has steadily increased and approached the three million mark in 2005. A few minor alterations have been made to the shuttle system to accommodate the summer influx of visitors, with operations now beginning in early April and including additional runs later in the evening on weekends.

To accommodate the ever-increasing number of visitors, many changes have been made to the infrastructure of the park and the surrounding community of Springdale. The town has grown in size since being settled by pioneers in 1862 and now consists of approximately 529 permanent residents (US Census Data, 2010). Because of its close proximity, the town and the park are intimately connected. Springdale offers visitors lodging, dining, banking, grocery shopping, gift and souvenir shopping and other travel-related services for tourists and residents of the town. In the 2012 National Park Visitor Spending Effects Report, the NPS estimated that Zion National Park had over 2.9 million visitors spending more than US $152 million (NPS, 2014c). Visitor spending contributed to 1,854 jobs (many seasonal), US $76.2 million in labour income, US $124.9 million in Gross Domestic Product (GDP) and US $192.7 million in sales to consumers and exports or business-to-business exchanges (NPS, 2014c).

Visitor assessment of the Zion National Park Alternative Transportation System

Mace *et al.* (2013) conducted a longitudinal evaluation of the mandatory shuttle system in Zion National Park. In 2000, 2003 and 2010, shuttle-riding park visitors were asked to rate attributes of the shuttle including crowding, accessibility, freedom, efficiency, preference and overall success, along with experiential park factors such as scenic beauty, naturalness, solitude, tranquillity, air quality and soundscape conditions. In 2000, 280 visitors (68.2 per cent response rate) completed the survey at one of three shuttle stops located in Springdale. In 2003 and 2010, 202 (67.3 per cent) and 209 (69.7 per cent) visitors participated in the study, respectively. Results of this survey suggest that visitors in 2000 initially reported some reservations with the new mandatory shuttle system; however, by 2003, those reservations appeared to diminish. Over the past decade, it was found that all of the shuttle-related variables including overall success, preference, comfort, efficiency, accessibility and freedom all improved. Only crowding on the shuttle appeared to increase with time. The shuttle-related variables that showed the most pronounced improvements were perceived freedom, shuttle accessibility and efficiency.

The perception of park-related variables or overall satisfaction with the park experience including scenic beauty, naturalness, solitude, tranquillity, the soundscape and air quality also showed similar patterns of improvement over the past decade. Specifically, it was found that ratings of scenic beauty, naturalness, solitude and tranquillity significantly improved each year of the study. Participants also reported decreases in noise pollution and improvements in air quality in the park. These results are encouraging and provide empirical evidence of the ameliorating effects of public transit on the environmental impacts of visitors and their vehicles.

This study also provides evidence for the use of alternative transportation on public lands as a way of enhancing the overall park experience for visitors.

Economic assessment of the Zion National Park Alternative Transportation System in the gateway community of Springdale, Utah

Research is needed to determine the possible economic impacts of alternative transportation systems on gateway communities. Specifically, research could address the following questions:

1 How are local business owners affected by a shuttle system?
2 How does this segment, obviously dependent on the money of the tourists, compare to visitors in the perception of a shuttle system?
3 Do businesses benefit from the implementation of a shuttle system?

These questions were addressed by asking local business owners and employees about their perception of the Zion National Park mandatory shuttle system and the impact it was having on the community, parking, vehicular traffic, foot traffic in their business, sales, as well as improvements that could be made to the shuttle system. Respondents were also asked to evaluate the effectiveness of the shuttle system in reducing such problems as air and noise pollution, traffic, parking congestion, and the impact of the shuttle on the environment and park experience.

A total of 59 business owners and employees participated in the study including 49 (68 per cent) residents of the town of Springdale and 10 (32 per cent) from neighbouring towns and cities. The age of the respondents ranged from 18 to 70, with 21 men (35.6 per cent), 37 women (62.7 per cent) and one unstated (1.7 per cent) completing the survey. Open-ended questions assessed impressions of the shuttle service and the effects of the shuttle on specific business attributes. The majority of respondents (86.8 per cent) feel the shuttle system is not a cause of concern to their business, with only a few businesses reporting negative consequences (13.2 per cent). When asked to elaborate on their negative response, business owners (n = 4) commented foot traffic had declined significantly since the installation of the shuttle service, and they would like the shuttle to stop directly in front of their business. The chosen location of shuttle stops can have an impact on where visitors venture in a gateway community, especially when weather conditions (e.g. excessive heat) preclude distant foot travel. Consequently, business owners in gateway communities are concerned visitors may spend significantly less time window shopping, and may spend more time in the park (Turnbull, 2009), further reducing lunch-related sales.

Business owners in gateway communities also have parking and traffic-related concerns, especially when the shuttle system connects the town and park. Table 6.1 summarizes the responses (three options: 'yes', 'no', 'no response') of business owners and employees in Springdale to questions regarding such concerns. About 16.9 per cent (19.6 per cent if 'no responses' are removed from the sample) of

Table 6.1 Responses provided by business owners and employees in the gateway community of Springdale to questions regarding the effects of the shuttle service on parking and traffic.

Survey item	*N*	*Percentage*
Before the installation of the shuttle service was parking a problem near your business?		
No	41	69.5
Yes	10	16.9
No Response	8	13.6
Is parking still a problem since the installation of the shuttle service?		
No	45	76.3
Yes	7	11.9
No Response	7	11.9
Before the installation of the shuttle service was vehicle traffic near your business a concern?		
No	34	57.6
Yes	15	25.4
No Response	10	16.9
Is vehicle traffic still a problem since the installation of the shuttle service?		
No	42	71.2
Yes	6	10.2
No Response	11	18.6

business owners feel vehicular parking near their business was a problem before the shuttle system, with 11.9 per cent (13.5 per cent if 'no responses' are removed) stating parking continues to be a problem following shuttle implementation. Traffic problems are more frequent, with 25.4 per cent (30.6 per cent if 'no responses' are removed) of participants reporting issues near their business prior to shuttle implementation, and 10.2 per cent (12.5 per cent if 'no responses' are removed) continuing to feel traffic is a problem after the transportation system began operation. Overall, of the small number of business owners that felt parking or vehicle traffic was a problem prior to the shuttle system, very few continued to view these traffic issues as problematic post-shuttle. Furthermore, as illustrated by these results, mandatory park shuttle systems can have positive impacts by reducing parking and traffic issues in gateway communities.

While a couple of business owners reported being adversely affected due to a reduction in foot traffic, the majority (51.9 per cent) indicated an increase in foot traffic following the installation of the shuttle system. The increase in foot traffic for the majority of business owners in town also translates into an increase in sales and revenue. In fact, 63 per cent of businesses report an increase in sales after the shuttle system began operation.

Finally, business owners were asked to openly comment on how the shuttle system had affected their businesses. Thirty two (54 per cent) business owners felt the shuttle system affected their business in a positive manner. Of those 32, eight made comments about how the shuttle had positively impacted their business, park, and/or town. Six commented on how the shuttle system had improved the overall park experience and three others felt the shuttle was an added convenience to the community, providing tourists greater access to businesses and other offerings in the town. Conversely, 10 (17 per cent) business owners felt the shuttle system had a negative effect on their businesses, with three making comments about how the shuttle system did little to help their businesses. An additional two people believed the shuttle was forcing people into the park without encouraging tourists to stop in town and patronize their businesses. Finally, two others felt the shuttle system encouraged people to bag their lunches and supplies rather than support the restaurants and other accommodations available in town. Overall, however, based on the quantitative data and qualitative comments, the majority of business owners felt the shuttle system had a positive impact on their businesses, the park experience and the community.

Business owners and employees were asked to rate shuttle- and park-related variables using a 5-point Likert scale with 1 representing a low or negative response and 5 representing a high or positive response. Mean and standard deviation for these items are included in Table 6.2. These items are identical to those used in the

Table 6.2 Ratings of shuttle- and park-related variables as provided by business owners and employees on a 5-point Likert scale (i.e. 1 = low or negative response; 5 = high or positive response). For each variable, mean and standard deviation are shown.

Survey item	*M*	*SD*
Shuttle-related variables		
Crowding	2.82	1.17
Comfort	3.24	1.09
Efficiency	4.04	.89
Accessibility	4.22	.99

(*continued*)

Table 6.2 Ratings of shuttle- and park-related variables as provided by business owners and employees on a 5-point Likert scale (i.e. 1 = low or negative response; 5 = high or positive response). For each variable, mean and standard deviation are shown *(continued)*.

Survey item	*M*	*SD*
Freedom	3.25	1.20
Shuttle preference	3.88	1.05
Overall shuttle success	4.02	.92
Park-related variables		
Scenic beauty	4.25	.87
Naturalness	4.31	.84
Solitude	3.16	1.25
Tranquillity	3.32	1.22

visitor assessment survey conducted by Mace *et al.* (2013). Of note, all of the shuttle and park related variables were rated above the midpoint of the scale, indicating a general positive rating. In addition, business owners feel the shuttle is efficient, accessible, successful, and positively impacts the scenic beauty and naturalness of the park. Based on these results, business owners and employees feel the shuttle is positively impacting the overall park experience and enhancing the surrounding landscape. It is important to point out the ratings of business owners essentially parallel the positive visitor response to the shuttle and park experience over the past decade (Mace *et al.*, 2013).

Key findings and areas for future research

In 2000, Zion National Park and the gateway community of Springdale, Utah worked in partnership to institute a mandatory alternative transportation system. The shuttle system was installed with the specific purpose of managing the sheer number of visitors to the park and to reduce the adverse impacts on the park and community. During the peak summer months, visitors vied for limited space in parking lots and roadways, causing damage to natural resources, frightening wildlife and impacting the overall park experience. The results of visitor and business surveys suggest the mandatory shuttle system has reduced many of the environmental issues that plagued the park and the gateway community. Furthermore, the alternative shuttle system has positively impacted the visitor experience, while maintaining, and in some cases improving, the economic vitality of Springdale's businesses. The success of the shuttle system in Zion and Springdale is a result of park officials and community leaders working together for the past 20 years to develop and maintain a popular and effective shuttle system. The transportation system planning and implementation process has even been recognized by the National Parks and Conservation Association, which awarded the town of Springdale the prestigious National Parks Achievement Award (Clarke, 2001). The relationship between gateway communities such as of Springdale, Utah and public land managers is essential to the planning, design, implementation and management of alternative transportation systems. The on-going success of the mandatory alternative transportation system in Zion National Park and Springdale should serve as a model for other public land management agencies and gateway communities considering new or improved transportation systems throughout the United States and around the world.

The results of this experience also have broader implications for the preservation of social and cultural identity of gateway communities, and environmental resources of surrounding areas. Specifically, responses from the business owners and longitudinal visitor surveys (Mace *et al.*, 2013) suggest that alternative transportation systems can both maintain current levels of economic well-being and satisfaction with the park experience, and significantly improve them over time. Future research should continue investigating the long-term impact of mandatory and voluntary alternative transportation systems on these outcomes in other parks and public lands around the world. Research should also focus on regional, cultural and environmental differences that may exist in each community that could potentially impact the perceived acceptability and usability of each type of alternative transportation system. Additional attention should also be given to creating better indicators of economic well-being, social and cultural outcomes, and satisfaction levels with the shuttle and park experience (e.g. Lawson *et al.*, 2009). Despite potential contextual limitations of the current research, the results are promising and suggest that alternative transportation systems can assist in the preservation of natural resources, and enhance the cultural and economic well-being of surrounding gateway communities.

References

Ansson, R. J. (1996) 'Our national parks—overcrowded, underfunded, and besieged with a myriad of vexing problems: How can we best fund our imperiled national park system?', *Journal of Land Use and Environmental Law*, vol 41, no 1, pp. 1–27.

Clarke, W. M. (2001) 'The national park or parking system?', *National Parks*, vol 75, no 7–8, pp. 34–37.

Daigle, J. J. (2008) 'Transportation research needs in national parks: A summary-and exploration of future trends', *The George Wright Forum*, vol 25, no 1, pp. 57–64.

Dilworth, G. and Shafer, S. (2004) 'Visitor perceptions of intelligent transportation systems in a national park', in K. Bricker (ed) *Proceedings of the 2004 Northeastern Recreation Research Symposium (Vol. Gen. Tech Rep. NE-326)*, USDA Forest Service Northeastern Research Station, Newtown Square, PA, pp. 158–163.

Eaton, B. and Holding, D. (1996) 'The evaluation of public transport alternatives to the car in British National Parks', *Journal of Transport Geography*, vol 4, no 1, pp. 55–65.

Gallegos, J. (2004) 'Alternative transportation: Who we are', National Park Service, Department of Interior, www.nps.gov/transportation/alt/wwa.htm, accessed 12 September 2005.

Gallegos, J. (2005) 'Alternative transportation', National Park Service, Department of Interior, www.nps.gov/transportation/alt/wwa.htm, accessed 12 September 2005.

Hallo, J. and Manning, R. (2009) 'Transportation and recreation: A case study of visitors driving for pleasure at Acadia National Park', *Journal of Transport Geography*, vol 17, pp. 491–499.

Harrison, G. S. (1975) 'People and parks: Reactions to a system of public transportation in Mt. McKinley National Park, Alaska', *Journal of Leisure Research*, vol 7, no 1, pp. 6–15.

Holding, D. and Kreutner, M. (1998) 'Achieving a balance between "carrots" and "sticks" for traffic in national parks: the Bayerischer Wald project', *Transport Policy*, vol 5, no 3, pp. 175–183.

Howe, J., McMahon, E. and Prospt, L. (1997) *Balancing Nature and Commerce in Gateway Communities*, Island Press, Washington, DC.

Lawson, S., Newman, P., Choi, J., Pettebone, D. and Meldrum, B. (2009) 'Integrated transportation and user capacity research in Yosemite National Park: The numbers game', *Transportation Research Record: Journal of the Transportation Research Board*, vol 2119, pp. 83–91.

Laube, M. M. and Stout, R. (2000) 'Grand Canyon National Park—Assessment of transportation alternatives', *Transportation Research Record*, vol 1735, no 1, 59–69.

Loomis, L. (2002). On H.R. 4622 the 'Gateway Communities Cooperation Act of 2002', Subcommittee on National Parks, Recreation, and Public Lands Committee on Resources, U.S. House of Representatives, pp. 1–7.

Mace, B. L., Bell, P. A., and Loomis, R. J. (2004) 'Visibility and natural quiet in national parks and wilderness areas: psychological considerations', *Environment and Behavior*, vol 36, no 1, 5–31.

Mace, B., Marquit, J. D., and Bates, S. C. (2013) 'Visitor assessment of the mandatory alternative transportation system at Zion National Park', *Environmental Management*, vol 52, no 5, pp. 1271–1285.

Manning, R., Valliere, W., Minteer, B., Wang, B., and Jacobi, C. (2000) 'Crowding in parks and outdoor recreation: A theoretical, empirical, and managerial analysis', *Journal of Park and Recreation Administration*, vol 18, no 4, pp. 57–72.

Miller, C. A. and Wright, G. (1999) 'An assessment of visitor satisfaction with public transportation services at Denali National Park and Preserve', *Park Science*, vol 19, no 2, pp. 18–21.

Morgan, J. N. (1985) 'The impact of travel costs on visits to US national parks: Intermodal shifting among Grand Canyon visitors', *Journal of Travel Research*, vol 24, no 3, pp. 23–28.

NPS (2012) 'Chapter 5. Effects of overflights on wildlife', in *National Park Service Report on Effects of Aircraft Overflights on the National Parks System*, National Park Service, Washington, DC.

NPS (2012) 'Buses and shuttles in national parks', National Park Service, Washington, DC, www.nps.gov/transportation/busses_shuttles.html, accessed 3 June 2012.

NPS (2013) 'Alternative transportation in the parks', National Park Service, Washington, DC, www.nps.gov/transportation/alternative_transportation.html, accessed 2 September 2013.

NPS (2014a) 'Annual Visitation Summary Report for 2013', National Park Service, Washington, DC, https://irma.nps.gov/Stats/SSRSReports/National%20Reports/Annual%20Visitation%20Summary%20Report%20(1979%20-%20Last%20Calendar%20Year), accessed 29 March 2014.

NPS (2014b) 'NPS Transportation', National Park Service, Washington, DC, www.nps.gov/transportation/index.html, accessed 29 March 2014.

NPS (2014c) 2012 'National park visitor spending effects: Economic contributions to local communities, states, and the nation', Natural Resource Report NPS/NRSS/EQD/NRR – 765, pp. i–42, National Park Service, Washington, DC.

NPS (2014d) 'Nature and Science', National Park Service, Washington, DC, www.nps.gov/zion/naturescience/index.htm, accessed 29 March 2014.

NPS (2014e) 'Annual park recreation visitation for Zion National Park', National Park Service, Washington, DC, https://irma.nps.gov/Stats/SSRSReports/Park%20 Specific%20Reports/Annual%20Park%20Recreation%20Visitation%20(1904%20 -%20Last%20Calendar%20Year)?Park=ZION, accessed 29 March 2014.

NPS (2014f) 'Shuttle System', National Park Service, Washington, DC, www.nps.gov/zion /planyourvisit/shuttle-system.htm, accessed 29 March 2014.

Pettengill, P., Manning, R., Anderson, L., Valliere, W. and Reigner, N. (2012) 'Measuring and managing the quality of transportation at Acadia National Park', *Journal of Park and Recreation Administration*, vol 30, no 1, 68–84.

Reeves, R. (2006) *Tackling Traffic: Sustainable Leisure Transportation in National Parks—An Overview of National Park Authority Involvement*, Council of National Parks, London, UK.

Runte, A. (1997) *National Parks: the American Experience, 3rd ed.* University of Nebraska Press, Lincoln, NE.

Safe, Accountable, Flexible, Efficient Transportation Equity Act: A Legacy for Users. (2005) Public Law, 109–59.

Shiftan, Y., Vary, D. and Geyer, D. (2006) 'Demand for park shuttle services—A stated-preference approach', *Journal of Transport Geography*, vol 14, no 1, 52–59.

Transportation Equity Act for the 21st Century (1998). Public Law 105–178.

Transportation Program Accomplishment (2014) www.nps.gov/transportation /transportation_program_accomplishments.html, accessed 29 March 2014.

Turnbull, K. F. (2009) 'Innovative transportation planning partnerships to enhance national parks and gateway communities', National Cooperative Highway Research Program, Transportation Research Board.

Turk, J. T. and Campbell, D. H. (1997) 'Are aquatic resources of the Mt. Zirkel wilderness area affected by acid deposition and what will emissions reductions at the local power plants do?', *USGS Factsheet 043-97*, pp. 1–4.

US Census Data. (2010) Census 2010 Total Population: Springdale, Utah.

White, D. D. (2007) 'An interpretive study of Yosemite National Park visitors' perspectives toward alternative transportation in Yosemite valley', *Environmental Management*, vol 39, no 1, pp. 50–62.

White, D. D., Aquino J., Budruk, M. and Golub, A. (2011) 'Visitors' experiences of traditional and alternative transportation in Yosemite National Park', *Journal of Park and Recreation Administration*, vol 29, pp. 38–57.

Wilson, B. (2000) 'Transit and the park experience: preservation, access, economics and opportunity', *Community Transportation*, vol 18, no 7, pp. 1–8.

Youngs, Y. L., White, D., and Wodrich, J. (2008) 'Transportation systems as cultural landscapes in national parks: the case of Yosemite', *Society & Natural Resources: An International Journal*, vol 21, no 9, pp. 797–811.

Zion Natural History Association (2014) 'Shuttle system offers relaxing and informative way to see Zion Canyon', www.zionpark.org/article_shuttle_08.php, accessed 29 March 2014.

7 Sustainable mobility within natural areas from the perspectives of persons with disabilities

Brent Lovelock

Part of this chapter is a revised version of a paper that first appeared in 2010 in *Tourism Management* (vol 31, no 3, pp. 357–366).

Introduction

On May 29, 2014, conservationists around New Zealand were on tenterhooks as they waited nervously for a decision from the Minister of Conservation, Dr Nick Smith, as to whether or not he was going to approve a controversial access proposal involving the construction of a 44-kilometre monorail through Fiordland National Park, part of the South West New Zealand World Heritage Area. This was the third in a series of proposals that were all designed to improve the flow of visitors between the tourist hub of Queenstown and Milford Sound, a major attraction within Fiordland National Park. All proposals had been bitterly opposed by conservationists on the grounds of the impacts on landscapes and ecosystems, with opponents questioning their environmental sustainability. Proponents, including the developers, defended the 'Fiordland Link Experience's' environmental credentials, and stressed the economic benefits it would bring. But, significantly, proponents of this and similar motorised transport options also identified their potential to provide a nature-based wilderness tourism experience for visitors with mobility problems. Such access developments would 'allow those aged and less physically mobile visitors to experience wilderness landscapes en route to Milford Sound' (Skytrail, 2001). Of course, the irony that inserting a motorised aerial gondola or a monorail through a wilderness area would actually destroy the very wilderness values that visitors may have come to experience was not lost on opponents to these developments.

The arguments around the above proposals that received the greatest emphasis were centred on aspects of sustainability, and particularly environmental sustainability and economic sustainability. The other pillar of sustainability, social sustainability, was relatively under-emphasised, with the exception of the argument concerning the benefits to persons with disabilities, which was the only real aspect of social sustainability that was put forward, and then in a relatively understated way.

This chapter considers the provision of sustainable transport within protected natural areas for persons with disabilities. The chapter first explores the perspectives of persons with disabilities in terms of the type and level of access to natural areas that they desire and believe to be appropriate. This is done by revisiting a study, undertaken in southern New Zealand, that considers the perspectives of disabled persons about the need for, and impact of, motorised transport within remote and wilderness areas. Then the meaning of sustainable tourism and sustainable transport is analysed in terms of how it may relate to disability, access and natural areas, and ultimately how this may help guide decisions in this field.

Access to natural areas for persons with disabilities

Disability is described by the World Health Organisation as 'any restriction or lack (resulting from impairment) of ability to perform an activity in the manner or within the range considered normal for a human being' (WHO, 1980). There are an estimated 650 million disabled people in the world – comprising roughly 8 per cent of the world's population (UNWTO, 2011). In the European Union (EU), 11 per cent of the population is classified as disabled, comprising 50 million people with mobility impairments (Münch and Ulrich, 2011). Such figures are projected to grow – not only with the increase of the world's population, but also with changing demographics. In developed countries, for example, the share of those aged 80 years and over is expected to more than double from 4 per cent in 2010 to 9.4 per cent in 2050 (OECD, 2010). This ageing of the population and the greater longevity of individuals can be expected to lead to increasing numbers of people at older ages with a disability. As Green *et al.* (2011, p. 219) note, 'every person is likely to experience disability of a permanent or transient nature in his or her own lifetime' (2011, p. 219). Thus, an increasingly large part of our society will share many of the access barriers currently faced by persons with disabilities (Var *et al.*, 2011).

Altruistic concerns aside, the tourism industry is slowly awakening to the size of the disabled market and the opportunities it presents to destinations and operators (Var *et al.*, 2011). The disabled community has been described as 'the largest minority group in the world' (Etravelblackboard.com, 2010), and persons with disabilities, their caregivers, families and friends, in the brutal terms of the marketplace, collectively comprise a substantial and largely untapped niche market, potentially worth 'billions of Euros for the tourism industry' (Var *et al.*, 2011, p. 602). This growing interest has been captured in the emergent discourse around 'Accessible Tourism'; but creating accessible tourist destinations is not simply charity: 'it is good business' (UNESCAP, 2007).

What exactly are the access barriers?

The tourism literature is replete with examples of the access problems faced by disabled travellers (e.g. Israeli, 2002; Takeda and Card, 2002; Shaw and Coles, 2004; Daniels *et al.*, 2005; Freeman and Selmi, 2010). The travel experiences of persons with disabilities are still 'highly restricted by physical accessibility barriers, such as: transportation constraints, inaccessible accommodation and tourism sites as well as information barriers' (Puhretmair and Buhalis, 2008, p. 969). Studies reveal lower participation rates in tourism for those with disabilities (e.g. Packer *et al.*, 2002), and this particularly applies to nature-based tourism settings, where people with disabilities, as a whole, are under-represented (Hartmann and Walker, 1988). For example, in the United States of America, while 14.4 per cent of the population was reported to have a mobility-disability, only 2.3 per cent of users of public recreation areas were mobility impaired (Bricker, 1995). In England, people who are without disability are more likely to visit woodlands than their socio-demographic counterparts are (Morris *et al.*, 2013). Not only are the disabled under-represented in terms of access to natural sites, but persons with disabilities are more likely to identify constraints to participation in outdoor recreation than those without disabilities due to inadequate facilities (among other constraints) (Williams *et al.*, 2004). In UK studies, lack of access to sites has been documented as a key barrier by disabled respondents engaging in environmental activities (Countryside Agency, 2005; Burns *et al.*, 2008; Natural England, 2008).

However, addressing access problems for persons with disabilities poses special challenges in natural areas – due to the often difficult terrain and the potential high cost of favouring physical access. The biophysical sensitivity of the environment also poses social and political challenges for the development of 'hardened' routes or motorised access within protected and natural areas, as unmodified nature, or 'wilderness', is a resource which is strongly defended by conservationists and wilderness advocates.

What do legislation and policy tell us?

This raises the question of the role of legislation and policy in guiding access developments within protected and natural areas. The United Nations Convention on the Rights of Persons with Disabilities (2006) outlines that state parties shall take all appropriate steps to ensure that reasonable accommodation is provided; where 'reasonable accommodation' means necessary and appropriate modifications and adjustments not imposing a disproportionate or undue burden (on others). But within most protected area contexts, for example in the national parks of the USA, access for those with disabilities is also governed by the 'reasonable modifications' requirement, which dictates that such modifications and adjustments must not fundamentally alter the nature of the programme or services provided. This has been challenged, however, under the provisions of the powerful Americans with Disabilities Act (ADA, 1990). Thus the policy environment is complex, characterised by national or state level law indicating a 'right' of access to nature for persons with disabilities, but in effect, park system or individual park level policy often serve to protect natural areas from the impacts of such access developments. No clear guidance is provided, this even being the case for World Heritage Sites, where there is no overarching guideline or policy for addressing the 'wilderness' (interpreted loosely) access needs of visitors with disabilities.

A number of landmark cases have highlighted the difficulties faced by protected area managers in addressing the dual mandate of providing universal access while addressing the preservation of natural resources. One such case concerns the Boundary Waters Canoe Area, in Minnesota (USA), where in 1977, in the debate prior to the passing of the Boundary Waters Canoe Area Wilderness Act, a US Senator for Minnesota stated publicly that reducing motorised access in the area would discriminate against 'the handicapped, the elderly, and women' (Lais, 1995, p. 26). Wilderness advocates responded that allowing motorised access would deny everyone, including those with disabilities, a true wilderness experience (Bricker, 1995). In a more recent and ongoing case, a group of disabled plaintiffs seeking to use all-terrain vehicles (ATVs) in an area of the Adirondack State Park (New York State, USA) that is closed to motorised vehicles took their argument to court. The judge ruled in favour of the disabled group, noting that the remaining wilderness portions of the park should not be available only to the able-bodied (Skidmore, n.d.).

What do persons with disabilities desire in the way of access?

One major problem is that there is a paucity of data on the wishes and needs of persons with disabilities with respect to access (Münch and Ulrich, 2011). This 'invisibility' of disabled people in many areas of environmental concern has also been pointed out by Fenney and Snell (2011). Such invisibility appears to particularly apply to the issue of access to natural areas, as empirical research into the perceptions, attitudes and behaviours of people with disabilities around these contexts has been sparse. However, the few studies that do exist, point to strong similarities with the able-bodied, with nature featuring highly on the preferred holiday itineraries of those with mobility-disabilities (Ray and Ryder, 2003). For example, in a survey of motivations and features important for travel, the top six for those with mobility-disabilities were: to learn about nature; landscape and wildlife photography; mountains; lakes and streams; wilderness/undisturbed nature, trees and wild flowers; and tropical forests (Ray and Ryder, 2003, p. 64). Generally, persons with disabilities visit wilderness for the same reasons as other visitors (Bricker, 1995; Brown *et al.*, 1999).

However, it has been argued that the realities and limitations of being disabled may make contact with nature 'just that much more precious' for persons with disabilities (McAvoy, 1996 in Jaquette, 2005, p. 9). Heintzmann (2014) in a review article of the benefits of visiting natural areas also stresses the spiritual benefits to people with disabilities. Additionally, more prosaic benefits are also available, with studies demonstrating that persons with disabilities who visit wilderness settings transfer the outcomes they gain in wilderness into their daily lives (McAvoy *et al.*, 2006). In a qualitative study in the UK, Burns *et al.* (2013) found that access to the outdoors afforded disabled participants the opportunity to escape and challenge what they see at times as a restricted and routinised life. This is emphasised in this quote from one of the participants in the study: 'There's so much denied to us that you get to a level where things that are natural become more important. There's so much else you can't do. You live a very restricted life, getting out is necessary' (Burns *et al.*, 2013, p. 1068). That study also found that constructions of risk are used to reinforce the message that disabled people should not be in natural areas (Burns *et al.*, 2013). The authors suggest that this is an example of what Thomas (2007 in Burns *et al.*, 2013) terms 'psycho-emotional disablism'.

Figure 7.1 Aerial gondola in Wulingyuan World Heritage Area, China: an example of how transportation systems may enable visitors to reach otherwise difficult to access sites (source: photo by Brent Lovelock).

However, such disablism, whether socially or physically constructed, can potentially be mitigated through enhancing built access to and within natural areas (Figure 7.1). Perhaps surprisingly, the few studies into whether or not persons with disabilities actually want such enhanced access reveal some ambiguity. A National Council on Disability study of wilderness users with disabilities addressed this question. The study's respondents were split over whether or not current access to wilderness was sufficient, but the majority of respondents were not supportive of increasing access inside wilderness areas, and three-quarters of respondents were against allowing mechanised devices in legally recognised wilderness areas, with only 21 per cent favouring the use of all-terrain vehicles (Bricker, 1995).

Analysis of disabled people's perceptions of access to nature in New Zealand

Given the lack of data noted above, the growth of nature-based tourism, and increasing pressure for access to protected areas, (plus the urgency of a number of 'live' proposals to enhance motorised visitor access to protected areas in the south of New Zealand), a clear need was identified to generate data to help inform decisions around such access proposals. As so little is known about what persons with mobility impairments actually want in the way of access (the only comparable study was that of the National Council on Disability, undertaken in the United States 14 years previously, and with actual users of wilderness), a new study was conceived and undertaken in the South Island of New Zealand. This study has been reported on in full previously in Lovelock (2010), but is revisited here as the only extant research that specifically focuses on the expressed wishes of persons with disabilities with regard to access to natural and protected areas.

The survey questionnaire

A survey questionnaire was developed with the aim of canvassing respondents' (both disabled and enabled) perspectives on access to natural areas. The questionnaire comprised sections that addressed: (i) respondents' mobility; (ii) their views on access to New Zealand's backcountry; and (iii) their socio-demographics.

The mobility section assessed respondents' health, mobility impairment, use of assistive devices and accessibility problems encountered while travelling. The level of mobility was mainly based upon a self-assessment of seven items representing common day-to-day activities utilising the NHANES III (National Health and Nutrition Examination Survey) scale, which ranges from 'unable to do' to 'no difficulty'. This scale is widely utilised in cross-population studies of mobility (e.g. Iburg *et al.*, 2001).

The section focusing on access to the New Zealand backcountry had two questions about current and past recreational use of the outdoors. Respondents were then asked about their position regarding the development of further motorised access to New Zealand's remote backcountry. The next question asked respondents to indicate how they feel the development of further motorised access may affect various aspects of the remote backcountry environment, utilising six items to represent these aspects, rated on a five-point scale from 'harmful' to 'beneficial'. A further question was asked about specific forms of transport development for the remote backcountry, considering nine items that were rated on a five-point scale from 'disapprove' to 'approve'. Two more questions asked if there should be further opportunities to allow those with mobility problems to access the backcountry in New Zealand, and if, given the opportunity, respondents would participate in some form of motorised access to a remote/wilderness area of a New Zealand national park.

There were three sampling frames identified for the survey: organisations with a high proportion of members with mobility impairments and located in two southern regions; tourists in the tourist destination of Queenstown staying in hotels and/or participating in tours that cater for persons with or without disabilities; residents of the city of Dunedin – including persons with disabilities and persons without disabilities. For the first sampling frame, questionnaires were distributed to all aged and disability-related organisations in the region that indicated a willingness to participate. This included organisations such as Parafed and CCS Disability Action that specialise in individuals affected by, among others, multiple sclerosis, stroke, arthritis, amputation, muscular dystrophy and polio. These organisations distributed the questionnaire to their members through newsletters and meetings. Three retirement homes and two age-concern organisations were also included. For the tourist sampling frame, questionnaires were supplied to four tour companies who cater to the senior market and/or tourists with disabilities, and to a sample of hotels that cater to the disabled and non-disabled market. For the resident population sampling frame, questionnaires were distributed to the Dunedin urban population, using a stratified random sampling procedure based upon socio-economic indicator data. The total number of returned usable questionnaires was 431.

Findings

A summary of the findings is presented here – and readers are urged to visit Lovelock (2010) for the full analysis of data. Older age groups predominated in the sample, with over half the sample coming from the 65+ age group. Respondents self-rated their level of mobility impairment, with a high percentage (64.5 per cent) having a long-standing illness, disability or infirmity. Over half the sample considered that their mobility was impaired for everyday activities. One third of the sample used assistive devices in their everyday lives. There were some differences in terms of the level of self-reported impairment for everyday activities compared with outdoor activities (e.g. bush walks, wildlife viewing, hiking), with a much higher percentage feeling that they were 'highly impaired' for the outdoor activities. Using the NHANES III scale, there were significant differences between the scores of the self-described mobility-impaired and other respondents: the former, as expected, demonstrating lower scores and greater difficulties with daily activities. The above analysis, together with the level of self-reported accessibility problems encountered when travelling in general, suggested that two distinct groups could be identified in the sample and that self-described mobility impairment was an appropriate variable to divide the sample into these groups for further analysis: 'impaired' and 'not-impaired'.

Overall, slightly more participants approved than disapproved of the development of further motorised access to New Zealand's remote backcountry and national parks by, for example, monorail, gondola, 4WD, aircraft or jet boat, with a substantial number (one-third) feeling neutral on the matter (Table 7.1). When the sample is considered in two groups, impaired and not-impaired, the former comes out more strongly in favour of further motorised access development, and the latter against such development.

It is important to note that 'impairment' is not the only variable associated with attitude to the development of motorised access, and that other socio-demographic variables may come into play. Associations were found, for example, with education: respondents with higher levels of education tending to be opposed to the further development of motorised access. Occupation was also associated with attitude towards the development of motorised access: those in managerial and professional occupations being opposed to development, while those in trades occupations being in favour. Associations were also found with both past level of recreational use and current level of recreational use. Those with high levels of both past and current recreational use of the outdoors tended not to favour the further development of motorised access, and conversely those who never used or use the outdoors for recreation were more in favour of further access development.

Table 7.1 Approval for the development of further motorised access by impaired, not-impaired and all respondents.

Further develop motorised access		*Impaired*	*Not-impaired*	*All respondents*
Disapprove	Frequency	54	64	118
	%	24.9	36.4	30.0
Neutral	Frequency	68	64	132
	%	31.3	36.4	33.6
Approve	Frequency	95	48	143
	%	43.8	27.3	36.4
Total		217	176	393

Source: Lovelock (2010, p. 363)

Overall, participants felt that the development of motorised access would be beneficial to the visitor experience and to the tourism industry, but harmful to natural landscapes, wildlife and the feeling of solitude, and neutral in terms of the impact upon the wilderness experience (Figure 7.2). The marginally higher mean scores for the impaired respondents suggested that compared with the not-impaired group, they felt the impacts of developing motorised access would be less harmful and more beneficial.

Participants expressed different levels of approval or disapproval for various modes of transport development within remote backcountry and wilderness. These ranged from the simple development of walking tracks and wheelchair tracks through to aerial cableways, scenic flights and other motorised forms of transport (Figure 7.3). Overall, participants found walking tracks and wheelchair tracks to be acceptable forms of transport development. Respondents were more 'neutral' in their feelings towards scenic flightseeing, gondola/aerial cableway, roads, jet boat and monorail. However, motorbikes/ATVs and 4WDs were considered to be less acceptable transport modes. The impaired group scored all transport modes marginally higher (greater approval) than the not-impaired group (apart from walking tracks).

Overall, the majority of respondents (69.2 per cent) agreed in general with the development of further opportunities to allow those with mobility-impairments to access New Zealand's backcountry and national parks (Table 7.2). When the two groups (impaired and not-impaired) were compared, there was no significant difference in their responses to this item.

Overall, just over one-half of participants (57.1 per cent) indicated that they would, if given the opportunity, participate in some (non-specified) form of motorised access to a remote/wilderness area of a New Zealand national park.

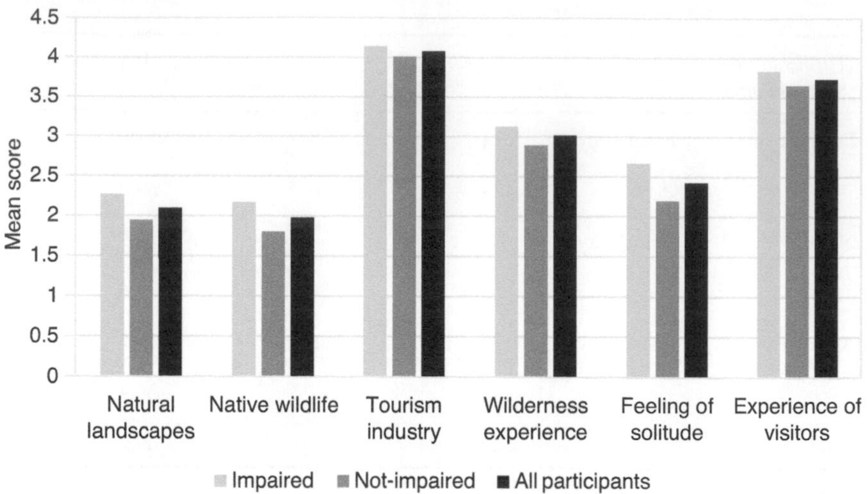

Figure 7.2 How development of motorised access may impact upon natural and experiential factors (1 = harmful; 5 = beneficial) (source: Lovelock, 2010, p. 363).

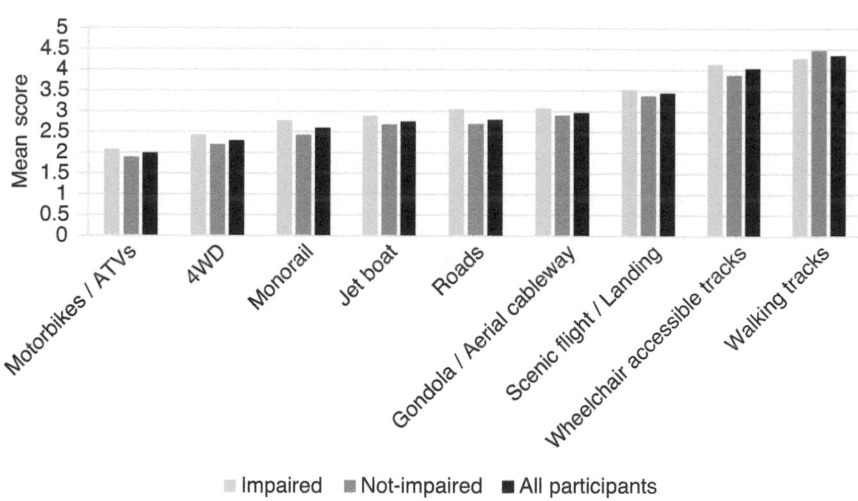

Figure 7.3 Approval for different types of transport development in wilderness (1 = disapprove; 5 = approve) (source: Lovelock, 2010, p. 363).

When the two groups (impaired and not-impaired) were compared, a higher percentage of the impaired group indicated that they would participate. For both groups, however, a significant portion would not participate in such an activity.

Table 7.2 Approval for the development of further motorised access for the mobility-impaired by impaired, not-impaired and all respondents.

Develop motorised access for mobility-impaired		*Impaired*	*Not-impaired*	*All respondents*
Agree	Frequency	154	116	270
	%	72.6	65.2	69.2
Disagree	Frequency	58	62	120
	%	27.4	34.8	30.8
Total		212	178	390

Source: Lovelock (2010, p. 364)

Discussion

The needs, wants and desires of persons with mobility-related disabilities in relation to accessing protected and natural areas are virtually unknown. This study is one of a meagre few that have broached this topic, therefore representing a source of empirical data that can contribute to the debate around the development of sustainable transport options in natural areas. Yau *et al.* (2004, p. 946) note that 'travel in a context designed primarily for people without disabilities poses unique challenges', and this is especially true for natural and protected areas, with their landscape and environmental challenges.

The study presented in this chapter reveals attitudes towards the further development of motorised access in remote natural landscapes. While the study supports the view of Yau *et al.* (2004) that people with disabilities generally have the same needs and desires for tourism as others, significant differences are observed between these groups. The group with mobility-disabilities expressed a stronger desire for enhanced access in such environments. Furthermore, they viewed the potential impacts of such development on the natural landscapes, wildlife and feeling of solitude as being more benign and less harmful than did the able-bodied respondents in the study. Despite these differences, however, less than half of the mobility-disabled sample approved of further motorised access, and those with mobility-disabilities still considered the impacts of such development to be harmful. Additionally, while those with mobility-disabilities revealed higher levels of approval for four out of nine different access/transport development/modes, they still disapproved of most types of development, with strong support only shown for walking tracks and wheelchair tracks.

The findings of this study challenge the simple assumption of proponents of enhanced motorised access to natural areas that the 'aged and disabled' desire such development. The data does tentatively support suggestions that people with disabilities desire the same 'pristine experience, untainted by disability accommodations' as those without disabilities (McAvoy, 1996 in Jaquette, 2005, p. 9). Respondents in this study do not support large-scale mechanised transport developments, such as gondolas, monorails and roads within wilderness areas.

To date, the debate around the sustainability of such transport developments has typically focused on their environmental or economic aspects, with little or no emphasis placed upon their social sustainability. This limited view of sustainability is aptly demonstrated by the fact that the Minister's decision regarding the proposal for a monorail through Fiordland National Park and World Heritage Area (see Introduction) was negative, and that conservationists throughout New Zealand celebrated. More specifically, in his decision, the Minister stated that 'this proposal does not stand up either *economically* or *environmentally*... The door is still open but [further, new] proposals will need to be both *environmentally* sustainable and *economically* viable' (Smith 2014).

This blinkeredness may relate to our astoundingly poor understanding of socially sustainable tourism. Especially so, considering that sustainability has been the defining paradigm for tourism for the last 25 years, and that the concept of sustainable tourism development has become almost universally accepted as a desirable approach (Sharpley, 2003). To date, the limited research on socially sustainable tourism has focused on local participation in tourism planning, benefit sharing among the host community, and tourism workers' rights, with little focus on access and inclusiveness for the actual or potential visitor. Perhaps this aspect of socially sustainable tourism has become partially hidden within a research agenda that is now beginning to espouse 'social tourism', 'inclusive tourism', 'accessible tourism' and 'tourism for all'.

Griffin and Stacey (2011), leading the foray into this aspect of socially sustainable tourism, advocate a socially inclusive 'tourism for all'. They point out that holidaying is a socially exclusionist activity, and that illness and disability (physiological and psychological), mobility and access may act as constraints. Their 'tourism for all', 'sits comfortably' within the concept of sustainable tourism that includes 'socially motivated practices to make holidays accessible for all' (Griffin and Stacey, 2011, p. 435; see also Darcy *et al.*, 2010). The links between tourism for all and sustainability have been recognised at the European Union level, where 'making holidays available to all' is explicitly identified as one of the eight key sustainability challenges facing the tourism sector with the challenge of physical disability having been highlighted (among others) (EC, 2007 in Griffin and Stacey, 2011).

However, how does this conceptualisation of sustainability fit with the management of natural and protected areas and the development of sustainable transport? Teasing out Dillard *et al.*'s (2009) definition of social sustainability, socially sustainable tourism could be defined as 'the tourism processes that generate social health and wellbeing now and in the future'. Thus, truly sustainable tourism transport options would need to aspire to this criterion. Of course, in protected natural areas, it could be argued that the environmental aspect of sustainability should be paramount. This is reflected in Eagles *et al.*'s (2001, p. 10) description of sustainable tourism in protected areas, which although incorporates 'provid[ing] people with the ability to learn, experience and appreciate the natural and cultural heritage of the site', also embraces a wider set of criteria that largely focus on environmental outcomes. Similarly, while Gladwell *et al.* (2013, p. 6) argue for the health and wellbeing benefits to both the disabled and able bodied of a more inclusive access to nature, this does come with a caveat that the pressure of development 'would potentially destroy the natural environment that elicits these health benefits'.

Such preservationist prescriptions for our protected areas, epitomised in works such as Joseph Sax's (1981) *Mountains Without Handrails*, provide little comfort for those with disabilities who may wish to experience wilderness, in ways that are not considered appropriate or what Sax terms 'ethical'. It seems that this group is only being catered for in a de facto way through more general motorised access developments that are demanded by and cater to the (generally able-bodied) modern seekers of wilderness, many of whom may come from emerging markets that may be non-traditional (or non-ethical) in how they interact with wilderness. Perhaps we need to realise that, especially within our larger protected areas, there could be space for all seekers of wilderness, and that this is not necessarily restricted to those heading off with a backpack, boots and map. What is needed is the application of system and site planning which ensure that the infrastructure needed to cater for disabled visitors to protected areas does not overly detract from the experiences of other users.

This chapter has presented data on the perceptions, attitudes and needs of persons with disabilities regarding transport options for natural and protected areas. Persons with disabilities are socially and physically excluded from many aspects of life that others enjoy, and this particularly applies to access to natural areas. It behoves us to work with the disabled community and other socially excluded groups to explore their needs and desires for access to such resources. The fact that the findings of our work may sometimes favour some enhanced motorised access does not mean that this will always be the case, for every development proposal in every national park. Ultimately, the knowledge acquired is valuable for planners and decision makers when it comes to designing socially (and economically and environmentally) sustainable transport options for our natural and protected areas.

Acknowledgements

The author is very grateful to Gerald Sides for research assistance.

References

ADA (1990) Americans with Disabilities Act of 1990, Pub. L. No. 101–336, 104 Stat. 328.

Bricker, J. (1995) 'Wheelchair accessibility in wilderness areas: The nexus between the ADA and the Wilderness Act', *Environmental Law*, vol. *25*, no 4, pp. 1243–1270.

Brown, T.J., Kaplan, R., and Quaderer, G. (1999) 'Beyond accessibility: Preference for natural areas', *Therapeutic Recreation Journal*, vol 33, no 3, pp. 209–221.

Burns, N., Watson, N. and Paterson, K. (2013) 'Risky bodies in risky spaces: disabled people's pursuit of outdoor leisure', *Disability & Society*, vol 28, no 8, pp. 1059–1073.

Countryside Agency (2005) *By All Reasonable Means: Inclusive Access to the Outdoors for Disabled People*, Cheltenham, Countryside Agency.

Daniels, M. J., Drogin Rodgers, E. B. and Wiggins, B. P. (2005) '"Travel Tales": an interpretive analysis of constraints and negotiations to pleasure travel as experienced by persons with physical disabilities', *Tourism Management*, vol 26, no 6, pp. 919–930.

Darcy, S., Cameron, B. and Pegg, S. (2010) 'Accessible tourism and sustainability: a discussion and case study', *Journal of Sustainable Tourism*, vol 18, no 4, pp. 515–537.

Dillard, J., Dujon, V. and King, M.C. (2009) 'Introduction', In J. Dillard, V. Dujon and M.C. King (eds.) *Understanding the Social Dimensions of Sustainability*, Taylor and Francis, New York.

Eagles, P. F. J., Bowman, M. E. and Chang-Hung Tao, T. (2001) *Guidelines for Tourism in Parks and Protected Areas of East Asia*, IUCN – The World Conservation Union, Gland.

Etravelblackboard.com (2010) 'Australia misses the plane on accessible tourism', 12 October 2010, www.etravelblackboard.com/article/109845/australia-misses-the-plane-on-accessible-tourism, accessed 12 November 2011.

Freeman, I. and Selmi, N. (2010) 'French versus Canadian tourism: Response to the disabled', *Journal of Travel Research*, vol 49, no 4, pp. 471–485.

Gladwell, V. F., Brown, D. K., Wood, C., Sandercock, G. R. and Barton, J. L. (2013) 'The great outdoors: how a green exercise environment can benefit all', *Extreme Physiology & Medicine*, vol 2, no 3, www.extremephysiolmed.com/content/2/1/3, accessed 5 May 2014.

Green, R. J. (2011) 'An introductory theoretical and methodological framework for a Universal Mobility Index (UMI) to quantify, compare, and longitudinally track equity of access across the built environment', *Journal of Disability Policy Studies*, vol 21, no 4, pp. 219–229.

Griffin, K. and Stacey, J. (2011) 'Towards a "tourism for all" policy for Ireland: achieving real sustainability in Irish tourism', *Current Issues in Tourism*, vol 14, no 5, pp. 431–444.

Hartmann, L. A. and Walker, P. J. (1988) 'Outdoor recreation participation by disabled people', in *Outdoor Recreation Benchmark 1988: Proceedings of the National Outdoor Recreation Forum, Tampa, Florida, Jan 13–14, 1988* (pp. 105–127), General Technical Report SE-52. USDA, Forest Service, Southeastern Forest Experimental Station.

Heintzman, P. (2014) 'Nature-based recreation, spirituality and persons with disabilities', *Journal of Disability & Religion*, vol 18, no 1, pp. 97–116.

Iburg, K. M., Salomon, J. A., Tandon, A. and Murray, C. J. L. (2001) 'Cross-population comparability of self-reported and physician-assessed mobility levels: Evidence from the Third National Health and Nutrition Examination Survey', Global Programme on Evidence for Health Policy Discussion Paper No. 14. N.p, World Health Organisation, Geneva, Switzerland.

Israeli, A. (2002) 'A preliminary investigation of the importance of site accessibility factors for disabled tourists', *Journal of Travel Research*, vol 41, pp. 101–104.

Jaquette, S. (2005) 'Maimed away from the earth: Disability and wilderness', *Ecotone*, Spring, pp. 8–11.

Lais, G. (1995) 'Wilderness inquiry', *International Journal of Wilderness*, vol 1, no 2, pp. 26–29.

Lovelock, B.A. (2010) 'Planes, trains and wheelchairs in the bush: Attitudes of people with mobility disabilities to enhanced motorised access in remote natural settings', *Tourism Management*, vol 31, pp. 357–366.

McAvoy, L. (2001) 'Outdoor for everyone: Opportunities that include people with disabilities', *Parks and Recreation*, vol 36, no 8, pp. 24–36.

Münch, H. and Ulrich, R. (2011) 'Inclusive tourism', in A. Papathanassis (ed) *The Long Tail of Tourism: Holiday Niches and their Impact on Mainstream Tourism*, Gabler Verlag, Springer Fachmedien, Wiesbaden, Germany.

Natural England (2008) *A Sense of Freedom: The Experiences of Disabled People in the Natural Environment*, Natural England, Peterborough, UK.

OECD (2010) Health Statistics and Indicators, Organisation for Economic Cooperation and Development, Paris, France, www.oecd.org/health/healthdata, accessed 5 June 2014.

Packer, T. L., McKercher, B. and Yau, M.K. (2002) 'Understanding the complex interplay between tourism, disability and environmental contexts', *Disability and Rehabilitation*, vol 29, no 4, pp. 281–292.

Pühretmair, F. and Buhalis, D. (2008) 'Accessible tourism introduction to the special thematic session', in K. Miesenberger, J. Klaus, W. Zagler and A. Karshmer (eds) *Computers Helping People with Special Needs: 11th International Conference ICCHP 2008 Proceedings*, Springer-Verlag, Berlin, pp. 969–972.

Ray, N. M. and Ryder, M. E. (2003) '"Ebilities" tourism: An exploratory discussion of the travel needs and motivators of the mobility-disabled', *Tourism Management*, vol 24, no 1, pp. 57–72.

Sax. J. L (1980) *Mountains Without Handrails: Reflections on the National Parks.* University of Michigan Press, Ann Arbor, MI.

Shaw, G. and Coles, T. (2004) 'Disability, holiday making and the tourism industry in the UK: A preliminary survey', *Tourism Management*, vol 25, no 3, pp. 397–403.

Skidmore, M. (n.d). 'Disabled rights to the wilderness: Whose waterfall is it anyway?' http://law.fordham.edu/publications/articles/700flspub226.pdf, accessed 11 November 2007.

Skytrail (2001). Proposal for gondola. Pamphlet. Np.

Smith, N. (2014) 'Fiordland Link monorail declined,' www.beehive.govt.nz/release/fiordland-link-monorail-declined, accessed 5 June 2014.

Takeda, K. and Card, J. A. (2002) 'U.S. tour operators and travel agencies: Barriers encountered when providing package tours to people who have difficulty walking', *Journal of Travel & Tourism Marketing*, vol 12, no 1, pp. 47.

United Nations (2006). Convention on the rights of persons with disabilities, www.un.org/disabilities/default.asp?navid=12&pid=150, accessed 22 November 2007.

UNESCAP (2003) *Barrier-free Tourism for People with Disabilities in the Asian and Pacific region*, United Nations Economic and Social Commission for Asia and the Pacific, New York.

UNESCAP (2007) 'Promoting tourism for people with disabilities', (Press release 22.11.2007), United Nations Economic and Social Commission for Asia and the Pacific, www.scoop.co.nz/stories/wo0711/s00956.htm, accessed 27 November 2007.

Var, T., Yesiltas, M., Yayli, A. and Öztürk, Y. (2011) 'A study on the travel patterns of physically disabled people', *Asia Pacific Journal of Tourism Research*, vol 16, no 6, pp. 599–618.

WHO (1980) 'International Classification of Impairments, Disabilities, and Handicaps: A manual of classification relating to the consequences of disease', Published in accordance with resolution WHA29.35 of the Twenty-ninth World Health Assembly, May 1976, World Health Organization, Geneva, Switzerland.

Yau, M. K., McKercher, B. and Packer, T. L. (2004) 'Travelling with a disability: More than an access issue', *Annals of Tourism Research*, vol 31, no 4, pp. 946–960.

Part III

Practices – Experiences around the world

8 Managing sustainable mobility in natural areas

The case of South Tyrol (Italy)

Anna Scuttari and Maria Della Lucia

This chapter is a revised version of a paper that first appeared in 2013 in the *Journal of Sustainable Tourism* (vol 21, no 4, pp. 614–637).

Introduction

The 'tourist transport system' (Page, 1994, 2005; Hall, 2005; Duval, 2007) is a complex system of nodes and interconnections which both allows the physical movement of visitors across space and over time, and gives them their travelling experience (Page, 1994). Although the relationship between tourism and transport is acknowledged to be crucial (Kaul, 1985), until recently managerial studies kept these fields separate (Hall, 1999; Lumsdon and Page, 2004) because transport was considered to be an accessibility issue (Prideaux, 2000) rather than a component of the tourism experience (Schiefelbusch *et al.*, 2007).

The inter-sectoral application of the sustainability paradigm has pushed managerial tourism and transport studies towards convergence, in an attempt to balance the environmental, social and economic impacts of transports (McIntosh, 1990; Hall, 1999; Gössling, 2002; Dickinson and Robbins, 2008), and to create more ethical and eco-friendly consumption patterns, in transport as elsewhere (Yeoman, 2005; Kelly *et al.*, 2007). This process of convergence towards sustainable tourism mobility creates opportunities for resource conservation, local development, well-being and competiveness (Pechlaner *et al.*, 2012c), but is challenging to implement in practice due to the complexity of the system itself and the multidimensional nature and scale of its impacts. The complexity of the system is principally a result of its open nature – its dependence on market sources – but also of the non-linear interactions between system components – small interventions in traffic management can have a big influence on tourism and vice versa (Cilliers, 1998; Salmon *et al.*, 2012). The most significant impacts of transport – locally and globally – are: a) environmental – energy and space consumption, greenhouse gas emissions, localized pollution and noise (Peeters *et al.*, 2004); b) social – stress, anxiety and fear of travelling (McIntosh, 1990) and the competition between residents and guests using public transport and/or road systems; c) economic – the construction, maintenance and operational costs of infrastructure and means of transport and the monetization of environmental and social impacts (Alpine Convention, 2007).

Whether or not the environmental, social and economic impacts of transport can be mitigated depends on the vision of the public and private networks engaged in an area's development, and their ability to act collaboratively. In comparison with pristine lands (Dickinson and Robbins, 2007, 2008) – where conservation/environmental issues prevail, natural areas where people live and work are difficult to manage because their socio-economic development, often driven by tourism, must be reconciled with the protection of natural resources and the provision of satisfying tourist experiences. Their developmental trajectories depend on how growth (economic) and regulation (environmental and social) issues are combined (Weaver, 2012). Participatory territorial planning is crucial to foster the coordination of policies in different sectors (Duval, 2007; Bramwell, 2011; Bramwell and Lane, 2011) and to design alternative and shared traffic management strategies by involving different interest groups (Gronau and Kagermeier, 2007). However, participatory planning encounters obstacles in multi-stakeholder contexts as inter-sectoral coordination is challenged by stakeholders' diverse backgrounds, aims, roles, competences and unevenly distributed power relations.

Effective tourist mobility analyses provide policymakers or destination managers with a more precise picture of the tourist transport system. Acquiring accurate data is the first step to activating participatory processes aimed at supporting the planning of integrated tourism and mobility management policies/strategies, which may strengthen sustainable development (Guiver *et al.*, 2007; Kelly *et al.*, 2007). 'Carrot' (or pull) measures (Holding and Kreutner, 1998; Cullinane and Cullinane, 1999) – incentives to use sustainable transport (e.g. enhanced public transport, low-impact mobility rental, promotion of zero impact holidays, mobility cards, etc.) – are usually preferred by tourists, local stakeholders and politicians. Yet they can only achieve a significant modal shift towards sustainable means of transport if combined with 'stick' (or push) measures (Stradling *et al.*, 2000; Holding, 2001) – disincentives on the use of private vehicles (e.g. traffic restrictions, access tolls and parking fees). Successful combinations of measures are place-specific and should result from ad hoc systematic tourist mobility analyses and a weighted balance between stakeholder interests. However, both data collection and wide stakeholder engagement are still rare when it comes to practical cases of policy planning. Moreover, the traffic management measures that are implemented are generally not combined with awareness-raising campaigns on sustainable transport use.

This chapter contributes to filling the data-management gap on tourism mobility in two ways: providing an innovative tourism traffic analysis and an exploratory evaluation of the effects of traffic management measures on tourism flows. The estimation and analysis of tourism traffic was carried out in South Tyrol, an Alpine destination at the forefront of destination management and tourism sustainability (Brida *et al.*, 2014), awarded for being Italy's greenest region (Fondazione Impresa, 2013) and experimenting with pioneering sustainable mobility initiatives. The tourism traffic analysis was used to estimate the volumes and environmental impacts of inbound private transport tourism and same-day visit traffic in South Tyrol. The exploratory evaluation of the effects of traffic management measures on tourist demand was tested in two South Tyrolean pilot projects – the 'Alpine Pearls' and 'Alpe di Siusi' initiatives. Additional details and critical analyses of the case are provided by Scuttari *et al.* (2013, 2014) who focused on the integration of mobility into destination planning according to sustainable development principles and on tools for estimating tourist transport impact and increasing sustainable mobility.

Study area

South Tyrol is an autonomous province located in north-eastern Italy that covers an area of 7,400 square kilometres, has a population of over 500,000 inhabitants and is composed of 116 municipalities (ASTAT, 2013). This Alpine region was chosen as a case study to investigate the management of sustainable tourist mobility because it successfully combines the protection of natural resources, the provision of quality recreation and socio-economic development. In fact, three elements characterise the territory. First, about half (40 per cent) of the province's surface area is protected at the European, Italian and local level, and the Dolomites, which partly fall within South Tyrol, have recently been inducted onto the UNESCO World Natural Heritage List. Second, the province is among the most competitive tourism regions in Italy (Cracolici *et al.*, 2006) with 5.3 million arrivals, 27.6 million overnights and over 8 per cent of the GDP coming from the tourism sector (ASTAT, 2009a). Third, South Tyrol has a special status within the Italian law, this allowing it greater autonomy in formulating tourism and mobility policies. Its Destination Management Organization system is a pioneering Italian example of effective public-private destination governance based on a three-tier structure – provincial (Alto Adige Südtirol Marketing), inter-municipal (Tourist Consortia) and municipal (Tourism Associations) (Pechlaner *et al.*, 2012b, 2012d) – which places sustainability at the heart of the provincial vision and mission. This focus on sustainability has resulted in South Tyrol ranking first in Italy according to the Green Economy Index (Fondazione Impresa, 2013). The combination of these unique features has allowed South Tyrol to pioneer both sustainable tourism and mobility policies through an integrated public transport system (ASTAT, 2014; Pechlaner *et al.*, 2012a, 2013) including high capacity cableways (Brida *et al.*, 2014), significant provincial investment in transport infrastructure, a well-developed networks of trails and cycling paths, and several incentives to the use of public transit (e.g. multi-day passes).

Despite all efforts, however, 89 per cent of incoming tourists still access South Tyrol by means of private transport and 55.7 per cent use it during their holiday (ASTAT, 2009b). To overcome this critical situation, some pioneering initiatives in traffic management have been tested in some South Tyrolean municipalities. Among these, the 'Alpine Pearls' project (www.alpine-pearls.com) and the 'Alpe di Siusi' initiative (www.alpedisiusi.net) are particularly interesting as they rely on 'carrot and stick' measures, respectively. The effects of traffic management measures on tourism flows were explored in two pilot cases – the seven South Tyrolean 'Alpine Pearls' (Moso in Passiria-Moos in Passeier, Nova Levante-Welschnofen, Nova Ponente-Deutschnofen, Tires-Tiers, Racines-Ratschings, Cornedo-Karneid and Funes-Villnoess) and the 'Alpe di Siusi' initiative involving a single area in the Castelrotto-Kastelruth municipality (Figure 8.1). The South Tyrolean 'Alpine Pearls'

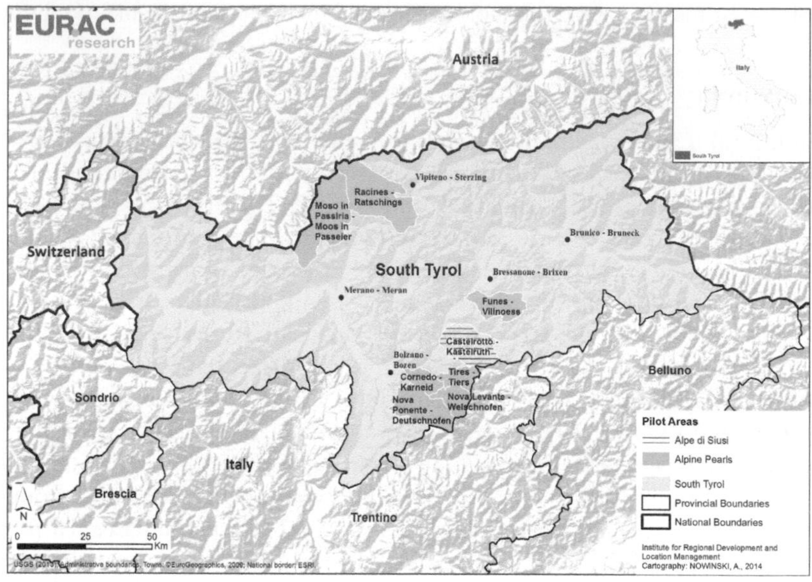

Figure 8.1 South Tyrol in north-eastern Italy and areas of implementation of the 'Alpine Pearls' project and 'Alpe di Siusi' initiative.

belong to a network of 29 Alpine locations in six Alpine countries that promote sustainable mobility as a core component of the tourism offer by implementing mainly 'carrot' measures, such as car-free access incentives and a range of local alternatives to private cars, from ordinary public transport to amusing, innovative e-vehicles. By contrast, the 'Alpe di Siusi' initiative employs 'stick' measures to promote sustainable mobility, restricting traffic flows to one of the largest European plateaus in the high season, and providing alternative bus and cableway services.

Methodology

Tourism-related traffic (V_T) was assumed to be a portion of total inbound traffic (V_{IT}), that is the daily inbound movement of light vehicles (i.e. motorbikes, cars and small vans) crossing the border to South Tyrol. V_T was estimated by breaking V_{IT} down into its components on a daily scale. Components were estimated using tourism and traffic databases from official national or provincial institutions reporting secondary data on tourism (tourists and same-day visits, inbound and outbound, in all accommodation facilities) and on traffic flows (crossing the border to South Tyrol). Outbound tourism flows were included because they contribute to V_T, as they generate inbound traffic when South Tyrolean tourists travel back to the province.

The analysis of V_{IT} was based on a linear, deterministic model with constant coefficients and resulted from the application of the following formula:

$$V_{IT} = \beta_0 + \beta_1 * \Sigma_T x_T + e_i \qquad (1)$$

where

V_{IT} = total inbound traffic volumes

β_0 = non-tourism traffic

$\beta_1 = \frac{n}{ot}$, where β_1 is a conversion parameter of tourism-related flows into traffic flows based on the composition of the tourist nucleus (*n*) and the percentage of own transportation inbound arrivals (*ot*)

x_T = tourist flows (arrivals)

T = different tourism types (tourists and same-day visits, inbound and outbound)

$\beta_T * \Sigma_T x_T$ = tourism-related traffic V_T

e_i = residual component

The estimation of V_{IT} components was based on two reasonable assumptions concerning β coefficients. First, the non-tourism traffic (β_0) was assumed to be a constant twice as intense on working days as during holidays because of the higher number of commuters during weekdays (this assumption was based on an analysis of low-intensity tourism periods in November 2007). Second, the conversion of tourist flows (x_T) into tourism-related traffic flows (V_T) was assumed to be possible by applying a parameter (β_1) based on: a) the composition of the tourist nucleus (*n*), that is the average number of people using one vehicle; b) the percentage of own transportation inbound arrivals (*ot*). These parameters are place-specific and sector-specific (official sample survey of the ASTAT on tourist expenditure of inbound tourists) and they were applied to all tourist flows (i.e. inbound and outbound tourists and same-day visits), as no equivalent information is available on same-day visits and outbound flows.

Starting from these assumptions, the breakdown of inbound traffic volumes (V_{IT}) followed two steps:

- the identification and quantification of daily V_{IT} components (both tourism-related and not) and the conversion of annual data into daily data through appropriate distribution if daily data were not available;
- the progressive removal from the V_{IT} of the components thus obtained, under the condition of non-negativity of the residual component.

Eight components of V_{IT} were identified sequentially (Table 8.1): one refers to non-tourism traffic (β_0) and seven to tourism traffic (V_T). Among the tourism-related components, three refer to inbound tourism-related phenomena – inbound tourism (V_{TTI} and V_{TTSI}) and inbound same-day visits (V_{TEI}) – and four refer to outbound tourism-related phenomena – outbound tourism (V_{TTO} and V_{TTSO}), outbound same-day visits (V_{TEO}) and same-day tourist visits outside the province (V_{TET}). The residual component (e_i) was assumed to refer to potential errors in the estimation of inbound and outbound same-day visits, since this was the only component based on an annual estimate through sample surveys. Therefore, e_i was added to the known inbound (V_{TEI} + e_i*) and outbound (V_{TEO} + e_i**) same-day visit components in a measure proportional to the incidence of each of these components on the annual estimate of same-day visits (e_i = e_i*+ e_i**).

Table 8.1 Components of V_{IT}.

Type of flow		*Components*	*Definition*	*Sources*
Non-tourism $ß_0$		$ß_0$	Traffic generated by non-tourism activities mainly composed of commuter flows	Census: Traffic flows on Brenner Motorway by Brenner Motorway S.p.A, Traffic flows on state and province roads by ASTAT; visitor movement to accommodation facilities by ISTAT [National Institute of Statistics]
Tourism V_T	Inbound tourism-related traffic	V_{TTI}	Traffic generated by inbound tourist arrivals in hotels, complementary and free accommodation facilities	Census: Visitor movement to accommodation facilities by ISTAT Sample surveys: Italian tourists' Journeys and Holidays by ISTAT, International Tourism by Bank of Italy, Tourist expenditure by ASTAT
		V_{TTSI}	Traffic generated by inbound tourist flows not officially registered by accommodation facilities (hidden)	Same sources as for TTI component
		$V_{TEI} + e_i^\star$	Traffic generated by inbound same-day visitor flows	Sample surveys: Italian tourists' Journeys and Holidays by ISTAT, International Tourism by Bank of Italy, Tourist expenditure by ASTAT

Type of flow	*Components*	*Definition*	*Sources*
Outbound tourism-related traffic	V_{TTO}	Traffic generated by outbound tourist flows, registered on return to the province	Census: Visitor movement to accommodation facilities by ISTAT Sample surveys: Italian tourists' Journeys and Holidays by ISTAT, International Tourism by Bank of Italy, Tourist expenditure by ASTAT
	V_{TTSO}	Traffic generated by outbound tourist flows, not officially registered by accommodation facilities (hidden), registered on return to the province	Same sources as for TTO component
	$V_{TEO} + c_i$**	Traffic generated by outbound same-day visitor flows registered on return to the province	Sample surveys: Italian tourists' Journeys and Holidays by ISTAT, International Tourism by Bank of Italy, Tourist expenditure by ASTAT
	V_{TET}	Traffic generated by inbound tourists on same-day trips outside the province, registered on return to the province	Census: Visitor movement to accommodation facilities by ISTAT Sample surveys: Tourist expenditure by ASTAT

Source: our elaboration of Scuttari *et al.*, 2013

The environmental impacts of tourism traffic volumes in terms of emissions and energy consumption were estimated starting from inbound tourism-related traffic (Table 8.1). The availability of this data led to a bottom-up approach being chosen for the environmental impact assessment (Becken and Patterson, 2006), rather than a top-down one, based on macroeconomic input-output tables (Filimonau *et al.*, 2011).

The estimation was obtained by applying a general equation (2), according to which the total impact (I_e) of a transport activity (TR) depends on the volume of this activity (V_{TR}) and on a coefficient (β_{TR}), which is the impact factor specific for each unit of activity (Peeters *et al.*, 2004). The impact factors per unit of activity ($ß_{TR}$) were derived from the literature (UITP, 2003; Peeters *et al.*, 2007) and the volumes of activity (V_{TR}) from our previous analysis on inbound tourism-related traffic. The calculation of emissions and energy consumption through (2) had previously required an equivalence relation of inbound tourism-related traffic to the corresponding volume of transport activity (V_{TR}). This is because impact factors ($ß_{TR}$) use person per kilometre covered (p/km) as the unit of activity, while inbound tourism-related flows are expressed in vehicles. Moreover, as there was no data on the distance in kilometres (*d*) covered from tourists' home regions to South Tyrol, the total impact I_e of inbound tourism-related phenomena was expressed per kilometre covered to access the destination, and in absolute terms for internal mobility.

$$I_e = \Sigma_{TR} (ß_{TR} * V_{TR}) \qquad (2)$$

where

$$V_{TR} = \frac{V_{TTI} + V_{TTSI} + V_{TEI} + e_i^{\star}}{n} * d$$

n = tourist nucleus
d = distance covered

The exploratory analysis of the impact of traffic management measures on tourist flows focuses on the quantitative effects on tourist demand and introduces two main assumptions. First, the proxy used to assess the impact of traffic management measures on tourism demand is the official time series of arrivals before and after their introduction. A cross-check of the arrival series with accommodation facility series (beds) for the same periods of observation was introduced to verify whether or not marketing effects (i.e. bed capacity) had interfered with policy effects (i.e. traffic management measures). Second, each pilot case is assumed to adopt only one type of measure, the prevailing one. However, all locations are using a mix of measures, which have simultaneous effects that cannot be measured separately at this stage. Most of the 'Alpine Pearls' carrot measures were introduced in 2006, while the 'Alpe di Siusi' initiative introduced stick measures in 2003. The variation in the mean yearly percentage of arrivals and bed capacity before and after the introduction of the measures was calculated between 1999–2006 and 2006–2013 for the 'Alpine Pearls', and between 1993–2003 and 2003–2013 for the 'Alpe di Siusi' initiative. The village of Moso in Passiria-Moos in Passeier was excluded from the analysis, as it was included in the project only in 2010.

Results

Tourism-related traffic (V_T) accounts for more than half (51 per cent) of the almost 14 million light vehicles that entered South Tyrol in the tourist year 2007/2008 (Figure 8.2). Its main component is inbound tourism-related traffic,

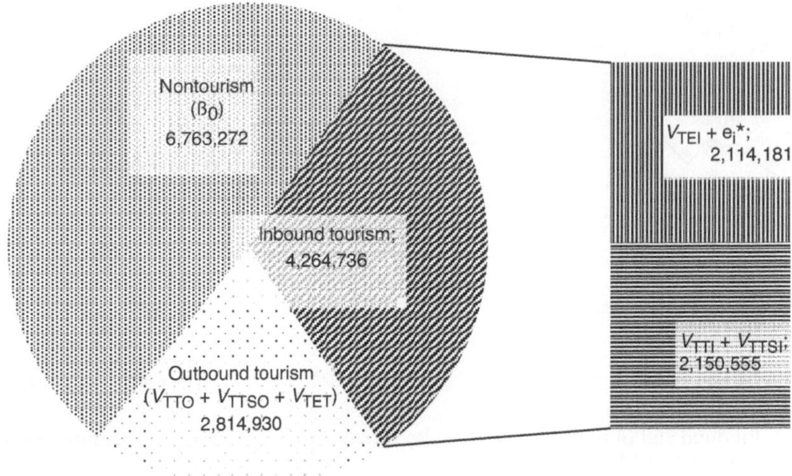

Figure 8.2 Components of total inbound traffic (V_{IT}) (source: ASTAT, own elaboration).

which represents almost one-third (31 per cent) of total inbound traffic (V_{IT},) and is the second largest component after non-tourism traffic (β_0). The number of inbound tourism vehicles (V_{TTI} and V_{TTSI}) and same-day visit vehicles (V_{TEI} + e_i★), into which it is divided, is almost equivalent. By contrast, outbound same-day visit vehicles (V_{TEO} + e_i★★) contribute more than tourism (V_{TTO}, V_{TTSO} and V_{TET}) to outbound tourism-related traffic, which accounts for only one-fifth (20 per cent) of V_{IT}.

The monthly percentage distribution of tourism-related traffic (V_T) shows a summer peak for all components (Figure 8.3). In particular, 65 per cent of both inbound and outbound same-day visit traffic (V_{TEI} + e_i★ and V_{TEO} + e_i★★) and 79 per cent of same-day visits associated with inbound tourists travelling outside the province daily (V_{TET}) occur in the summer season. Concerning tourism, about two-thirds of outbound tourism (V_{TTO} + V_{TTSO}) and over 60 per cent of inbound tourism (V_{TTI} + V_{TTSI}) occur between May and October. The peak month is August for all traffic flows, except for those related to outbound tourism (V_{TTO} + V_{TTSO}), which is higher in June. Finally, a daily analysis shows the relation between visitor movements (arrivals, departures) and traffic flows at weekends and in working days as well as during holidays. Same-day visits are concentrated in the weekends, while peak tourist flows mainly occur on Saturdays and returning outbound flows on Sundays. Traffic flows are, of course, influenced by holidays in both South Tyrol and tourists' countries of origin, while single-peaks are evident during German holidays (e.g. Corpus Christi), for example.

In the tourist year 2007/2008, the 4.2 million vehicles (an average of more than 36,000 vehicles per day) accessing South Tyrol for tourism and same-day visits consumed 50 million Mj of energy per kilometre covered and emitted over 1,600

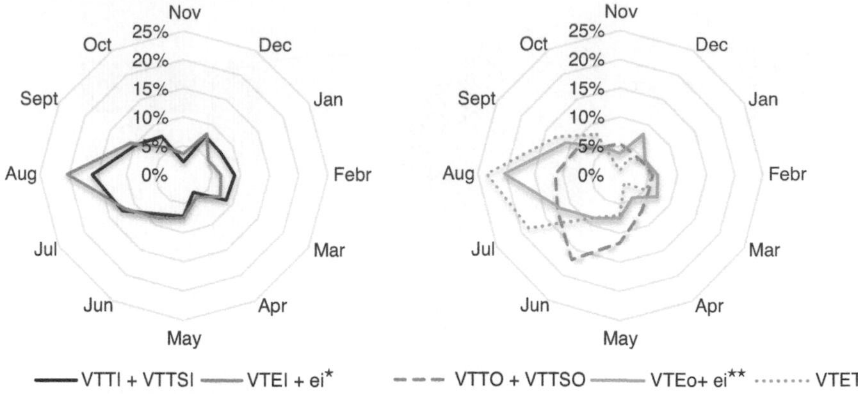

Figure 8.3 Inbound and outbound tourism-related traffic: monthly percentage composition (source: ASTAT, own elaboration).

tons of CO_2, 1,751 tons of CO_2 equivalent, 282 kg of particulates and 6,267 kg of nitrogen oxide per kilometre covered. The movement of vehicles for internal mobility consumed 694 million Mj of energy and produced 29,313 tons of CO_2, 30,790 tons of CO_2 equivalent, 4,959 kg of particulates and 110,200 kg of nitrogen oxides.

In the six municipalities of the South Tyrolean 'Alpine Pearls' where 'carrot' measures were introduced, the overall trend in arrivals is positive for all municipalities, but variations in arrivals before and after traffic management measures, and the comparison of arrivals with average provincial values show contrasting results. One group of municipalities (Cornedo-Karneid, Funes-Villnoess and Tires-Tiers) shows a higher variation in arrivals after traffic management than before, and this is even more significant than the variation at the province-scale, except for Cornedo-Karneid. This could suggest a positive relationship between traffic management and tourist flows. Conversely, a second group of municipalities (Nova Levante-Welschnofen, Nova Ponente-Deutschnofen and Racines-Ratschings) reveals the opposite trend: the variation in arrivals is lower after traffic management than before. In the Castelrotto-Kastelruth municipality, where private motorised traffic has been restricted (though the tourist area can be reached by bus and cableway), increases in arrivals were recorded both before and after the introduction of 'stick' measures (Table 8.2). The higher increases took place after the restrictions, in line with the general provincial arrival trend. Marketing policies do not seem to influence arrival trends in either the South Tyrolean 'Alpine Pearls' or the 'Alpe di Siusi'. In the municipalities showing higher arrival variations after traffic management than before, the corresponding mean yearly variations of bed capacity are negative or slightly positive. Moreover, they are always lower than the corresponding arrival increase.

Table 8.2 Mean yearly percentage variations of arrivals in pilot areas.

	Mean yearly % variations before traffic management		*Mean yearly % variations after traffic management*	
	Arrivals	*Bed capacity*	*Arrivals*	*Bed capacity*
'Alpe di Siusi' initiative	*(1993/2003)*		*(2003/2013)*	
Castelrotto-Kastelruth	+3.4%	−0.2%	+3.7%	+1.5%
South Tyrol	**+2.3%**	**−0.4%**	**+2.6%**	**+0.4%**
South Tyrolean 'Alpine Pearls'	*(1999/2006)*		*(2006/2013)*	
Cornedo-Karneid	+2.0%	−1.5%	+2.2%	−2.9%
Funes-Villnoess	+1.6%	−0.9%	+4.5%	−0.4%
Nova Levante-Welschnofen	+4.7%	−0.3%	+0.7%	−0.6%
Nova Ponente-Deutschnofen	+6.7%	+2.6%	+0.9%	+0.4%
Racines-Ratschings	+7.8%	+2.9%	+3.6%	+1.1%
Tires-Tiers	+1.7%	+1.3%	+3.9%	−0.5%
South Tyrol	**+3.5%**	**+0.4%**	**+2.6%**	**+0.3%**

Source: our elaboration

Discussion

The South Tyrol case study provides interesting insights into how data management can support effective decision-making processes and initiatives in the tourism transport domain. It addresses both the volumes and environmental impacts of inbound private transport generated by tourism and same-day visit traffic in destinations, and the effects of impact mitigation policies (traffic management measures) on tourist flows.

Tourism-related inbound traffic in South Tyrol represents more than half of total inbound traffic. Inbound tourism-related traffic represents the main share of tourism-related traffic and is concentrated in August. Inbound same-day visits contribute to one-half of inbound tourism-related traffic and are more concentrated in the summer season (mainly August) than is tourism as a whole. Outbound tourism-related traffic is less significant, and occurs mainly in June. Daily distributions show a concentration of flows during weekends and over holiday periods.

The methodology used for tourism traffic analysis – a deterministic model with constant coefficients – is not novel per se, but its novel application to tourism and transport generates a synergy, which could allow the integrated management of the two domains. Moreover, the halfway stage – the transformation of tourism-related vehicles into people – allows widely neglected but valuable phenomena such as same-day visits and hidden tourism to be addressed, both in terms of magnitude and behaviour. The main weakness of the model is that accurate and effective analyses require a wide, complete, up-to-date and place-specific range of data on traffic, tourism and visitor travel habits derived from reliable surveys/sources. Moreover, the model does not go into details about tourists' behaviour (e.g. speed, fleet composition, driving style), this preventing a more accurate estimation of emissions.

The pioneer tourism traffic management initiatives introduced across the province were intended to reduce car-based mobility and to encourage the use of public transport. The analysis of the relationship between traffic management measures and tourist flows in the two pilot cases showed that the measures adopted do not reduce tourist demand. They do, however, significantly reduce the environmental impact of tourism-related traffic. The arrival trend remained positive in all the pilot municipalities after the introduction of traffic management measures and in four out of seven cases growth trends increased, in line with what shown by the Austrian 'Alpine Pearl' Werfenweng (Federal Ministry of Agriculture, Forestry, Environment and Water Management *et al.*, 2006). The exclusion of possible marketing effects on arrivals due to an increase in capacity supports this result.

Despite these encouraging results, however, this analysis is still preliminary and exploratory. First, the method used to address the effects of traffic management measures on tourist flows does not consider the many short, medium and long-term factors which may influence demand (e.g. seasonality, accommodation capacity, quality of accommodation, environmental certifications, thematic packages, image of the destination). Second, the proxy used to measure these effects refers to entire municipalities, while in some cases (e.g. 'Alpe di Siusi' initiative) traffic management measures were only introduced in some parts of the municipalities. Third, especially for the 'Alpine Pearls', the observation timeframe may be too short to reveal significant changes. Fourth, the traffic measures have not been widely promoted, and especially not by the 'Alpine Pearls', where sustainable mobility options are still scarcely visible on tourism websites. Further, larger studies are necessary to evaluate both the quantitative impacts of traffic management measures on tourist flows (e.g. reduction of traffic volumes and environmental impact, increase in public transport use) and their qualitative impact on tourist mobility behaviour (e.g. modal shifts, travel habits, consumption and spending patterns).

This case study, in combination with others, may help overcome institutions' and tourist operators' considerable scepticism about the effects of tourist traffic management and encourage a systematic use of tourist traffic analyses, including environmental impact studies, as tools to monitor and manage the implementation of traffic management policies and initiatives.

Conclusion

South Tyrol is proactively tackling emerging tourism challenges like the growing scarcity of natural resources, biodiversity loss and poor climate change awareness

(Weaver, 2012). Its autonomous status and high public expenditure on services compared to the national average (Brida *et al.*, 2014) have allowed this province to formulate innovative tourism and transport policies, to pioneer initiatives in traffic management and to test innovative green policies. The latter seem to be making sustainability a societal norm in South Tyrol as the province has endeavoured to follow an 'incremental development path' (Weaver, 2012). This regulation-driven process is designed to foster sustainable tourism practices – both niche and mass – through a sustainable integrated strategy of spatial differentiation, which makes wide use of tourism mobility / traffic regulation initiatives. In protected areas, regulations aim to preserve the natural and cultural assets that provide a unique sense of place, quality of life, and perceived competitive advantage. The tourism development strategy of the interprovincial Dolomites UNESCO World Heritage Site (Elmi and Wagner, 2013), for example, includes a mix of sustainable mobility / traffic measures and tourist and stakeholder awareness-raising campaigns, which enhance the experiential component of tourist journeys. Interprovincial cooperation allows South Tyrol to share best practices with the neighbouring Alpine provinces (Belluno, Pordenone, Trento, Udine) involved in the management of the Site.

This incremental development path towards sustainability has contributed to South Tyrol's position at the forefront of tourism and mobility / transport policy-making. However, private vehicle tourism-related traffic is still significant both in terms of volume and environmental impact, as this and other studies demonstrate. Both guests and residents face physical and psychological barriers to shifting from private to public means of transport, while the Alpine landscape, with its lower density and steep terrain, implies 'last-mile' problems when using traditional public transport services (Teal and Becker, 2011). Low awareness and cognitive dissonance in both tourism demand (guests) and supply (tourism operators) further inhibit a shift towards public transport (Van Exel and Rietveld, 2009; Higham *et al.*, 2013).

While the integration of tourism management and traffic analysis unveils great opportunities for the preservation of natural resources and recreational experiences, it also raises a number of relevant questions. How can the awareness of the importance of encouraging sustainable behaviour in the tourism mobility field be increased on both the demand and supply side? How can inter-sectorial coordination and the integration of tourism and traffic policies be encouraged? How can the physical and psychological barriers associated with shifting to public means of transport be overcome? How should/could green strategies help overcome the science-policy and the attitude-behaviour gaps? These and other issues can be investigated in future research by taking a territorial-tourist governance perspective.

Acknowledgements

The authors would like to acknowledge support received from Dr Ludwig Castlunger (Provincial Statistics Institute of Bolzano), Professor Pier Luigi Novi Inverardi (University of Trento) and Dr Marco Giuliani (University of Bologna/ EURAC research) in the development of the model.

References

Alpine Convention (2007) 'The real costs of transport in transalpine corridors', EURAC, Bolzano.

ASTAT (2009a) 'Il conto satellite del turismo per l'Alto Adige – 2005' [The tourism satellite account in South Tyrol – 2005], Istituto Provinciale di Statistica, Bolzano, Italy, www.provinz.bz.it/astat/it/service/845.asp, accessed 31 March 2014.

ASTAT (2009b) 'Profilo dei turisti in Alto Adige. Anno turistico 2007/08' [Tourists' profile in South Tyrol. Tourist year 2007/2008], Istituto Provinciale di Statistica, Bolzano, Italy, www.provinz.bz.it/astat/it/mobilita-turismo/turismo.asp, accessed 31 March 2014.

ASTAT (2013) 'Annuario statistico della provincia di Bolzano – 2013' [Statistical yearbook of the Autonomous Province of Bolzano – 2013], Istituto Provinciale di Statistica, Bolzano, Italy, www.provincia.bz.it/astat/download/Jahrbuch_2013.pdf, accessed 31 March 2014.

ASTAT (2014) 'Soddisfazione degli utenti del trasporto pubblico – Indagine multiscopo sulle famiglie 2013' [Satisfaction of public transport users – Multi-purpose survey to families 2013], Istituto Provinciale di Statistica, Bolzano, Italy, www.provinz.bz.it/astat/it/256.asp?News_action=4&News_article_id=370799, accessed 31 March 2014.

Becken, S. and Patterson, M. (2006) 'Measuring national carbon dioxide emissions from tourism as a key step towards achieving sustainable tourism', *Journal of Sustainable Tourism*, vol 14, no 4, pp. 323–338.

Bramwell, B. (2011) 'Governance, the state and sustainable tourism', *Journal of Sustainable Tourism*, vol 19, no 4–5, pp. 459–477.

Bramwell, B. and Lane, B. (2011) 'Critical research on the governance of tourism and sustainability', *Journal of Sustainable Tourism*, vol 19, no 4–5, pp. 411–421.

Brida, J. G., Deidda, M. and Pulina, M. (2014) 'Tourism and transport systems in mountain environments: analysis of the economic efficiency of cableways in South Tyrol', *Journal of Transport Geography*, vol 36, pp. 1–11.

Business Location Südtirol (2012) 'Alto Adige. La Green Region d'Italia' [Italy's Green Region. South Tyrol]. www.bls.info/upload/file/Alto_Adige._La_Green_Region_d_Italia_web[1].pdf, accessed 31 March 2014.

Cilliers, P. (1998) *Complexity and Postmodernism: Understanding Complex Systems*, Routledge, London.

Cracolici, F., Nijkamp, P. and Rietveld, P. (2006) 'Assessment of tourist competitiveness by analysing destination efficiency', Tinbergen Institute Discussion Paper, 097/2, University of Amsterdam, Amsterdam.

Cullinane, S. and Cullinane, K. (1999) 'Attitudes towards traffic problems and public transport in the Dartmoor and Lake District National Parks', *Journal of Transport Geography*, vol 7, pp. 79–87.

Dickinson, J. E. and Robbins, D. (2007) 'Using the car in a fragile rural tourist destination: A social representations perspective', *Journal of Transport Geography*, vol. 15, pp. 116–126.

Dickinson, J. E. and Robbins, D. (2008) 'Representations of tourism transport problems in a rural destination', *Tourism Management*, vol 29, pp. 1110–1121.

Duval, D. T. (2007) *Transport and Tourism. Modes, Networks and Flows*, Channel View Publications, Clevedon [u.a.].

Elmi, M. and Wagner, M. (2013) 'Turismo sostenibile nelle Dolomiti. Una strategia per il Bene Patrimonio Mondiale UNESCO' [Sustainable Tourism in the Dolomites. A strategy for the UNESCO World Heritage Site], EURAC, Bolzano, Italy.

Federal Ministry of Agriculture, Forestry, Environment and Water Management; Federal Ministry for Transport, Innovation and Technology; and Federal Ministry of Economics and Labour. (2006) *Environmentally friendly travelling in Europe, challenges and innovations facing environment, transport and tourism*, BMLFUW, Wien.

Filimonau V., Dickinson J. E., Robbins D. and Reddy, M. V. (2011) 'A critical review of methods for tourism climate change appraisal', *Journal of Sustainable Tourism*, vol 19, no 3, pp. 301–324.

Fondazione Impresa (2013) 'Indice di Green Economy 2013' [Green Economy Index 2013]. www.fondazioneimpresa.it/wp-content/uploads/2013/12/Indice-di-green-economy-2013.pdf, accessed 31 March 2014.

Gössling, S. (2002) 'Global environmental consequences of tourism', *Global Environmental Change*, vol 12, pp. 283–302.

Gronau, W. and Kagermeier, A. (2007) 'Key factors for successful leisure and tourism public transport provision', *Journal of Transport Geography*, vol 15, pp. 127–135.

Guiver, J., Lumsdon, L. and Weston, R. (2007) 'Traffic reduction at visitor attractions: the case of Hadrian's Wall', *Journal of Transport Geography*, vol 16, pp. 142–150.

Hall, C. M., (2005) *Tourism: Rethinking the Social Science of Mobility*, Prentice Hall, Harlow.

Hall, D. R. (1999) 'Conceptualizing tourism transport: inequality and externality issues', *Journal of Transport Geography*, vol 7, pp. 181–188.

Higham, J., Cohen, S. A., Peeters, P. and Gössling, S. (2013) 'Psychological and behavioural approaches to understanding and governing sustainable mobility', *Journal of Sustainable Tourism*, vol 21, no 7, pp. 949–967.

Holding, D. M. (2001) 'The Sanfte Mobilitaet project: Achieving reduced car-dependence in European resort areas', *Tourism Management*, vol 22, pp. 411–417.

Holding, D. M. and Kreutner, M. (1998) 'Achieving a balance between "carrots" and "sticks" for traffic in national parks: The Bayerischer Wald project', *Transport Policy*, vol 5, pp. 175–183.

Kaul, R. N. (1985) *Dynamics of Tourism: A Trilogy (Vol. 111) Transportation and Marketing*, Sterling Publishers, New Delhi.

Kelly, J., Haider, W., Williams, P. W. and Englund, K. (2007) 'Stated preferences of tourists for ecoefficient destinations planning options', *Tourism Management*, vol 28, pp. 377–390.

Lumsdon, L. and Page, S. (2004) 'Progress in transport and tourism research', in L. Lumsdon, S. Page, (eds) *Tourism and Transport: Issues and Agenda for the New Millennium*, Elsevier, Oxford.

McIntosh, I. B. (1990) 'The stress of modern travel', *Travel Medicine International*, vol 8, pp. 118–21.

Page, S. (1994) *Transport for Tourism*, Routledge, London.

Page, S. (2005) *Transport and Tourism: Global Perspectives*, Pearson Education, Harlow.

Pechlaner, H., Bonelli, A., Scuttari, A. and Martini, M. (2012a) 'Customer satisfaction analysis on regional rail transport in South Tyrol', EURAC, Bolzano, http://interregiorail.eu, accessed 31 March 2014.

Pechlaner, H., Herntrei, M., Pichler, S. and Volgger, M. (2012b) 'From destination management towards governance of regional innovation systems – the case of South Tyrol, Italy', *Tourism Review*, vol 67, no 2, pp. 22–33.

Pechlaner, H., Pichler, S. and Herntrei, M. (2012c) 'From mobility space towards experience space: Implications for the competitiveness of destinations'. *Tourism Review*, vol 67, no 2, pp. 34–44.

Pechlaner, H., Volgger, M. and Herntrei, M. (2012d) 'Destination management organizations as interface between destination governance and corporate governance', *Anatolia: An International Journal of Tourism and Hospitality Research*, vol 23, no 2, pp. 151–168.

Pechlaner, H., Scuttari, A., Martini, M. and Bonelli, A. (2013) 'Analisi della soddisfazione del trasporto su gomma' [Customer satisfaction analysis on regional bus transport], EURAC, Bolzano, http://agenzia-mobilita.bz.it/it/amministrazione/485.asp, accessed 31 March 2014.

Peeters, P. M., van Egmond, T. and Visser, N. (2004) 'European tourism, transport and environment. Second draft deliverable 1 for the DG-ENTR MusTT project',

NHTV CSTT, Breda, The Netherlands, www.cstt.nl/userdata/documents/appendix_deliverable_1_subject_matter_review_30082004.pdf, accessed 31 March 2014.

Peeters, P., Szimba, E. and Duijnisveld, M. (2007) 'Major environmental impacts of European tourist transport', *Journal of Transport Geography*, vol 15, pp. 83–93.

Prideaux, B. (2000) 'The role of the transport system in destination development', *Tourism Management*, vol 21, no 1, pp. 225–241.

Salmon, P. M., McClure, R. and Stanton N. A. (2012) 'Road transport in drift? Applying contemporary systems thinking to road safety', *Safety Science*, vol 50, 1829–1838.

Schiefelbusch, M., Jain, A., Schäfer, T. and Müller, D. (2007) 'Transport and tourism: Roadmap to integrated planning developing and assessing integrated travel chains', *Journal of Transport Geography*, vol 15, pp. 94–103.

Scuttari, A., Della Lucia, M. and Martini, U. (2013) 'Integrated planning for sustainable tourism and mobility. A tourism traffic analysis in Italy's South Tyrol region', *Journal of Sustainable Tourism*, vol 21, no 4, pp. 614–637.

Scuttari, A., Della Lucia, M. and Martini, U. (2014) 'La mobilità sostenibile tra destination management e mobility management. Un'analisi esplorativa in Alto Adige' ['Sustainable mobility between destination management and mobility management. An exploratory analysis in South Tyrol region'], *Mercati e Competitivitá*, vol 1, pp. 125–151.

Stradling, S. G., Meadows, M. L. and Beatty, S. (2000) 'Helping drivers out of their cars. Integrating transport policy and social psychology for sustainable change', *Transport Policy*, vol. 7, pp. 207–215.

Teal, R. F. and Becker, A. J. (2011) 'Business strategies and technology for access by transit in lower density environments', *Research in Transportation Business & Management*, vol 2, pp. 57–64.

UITP (2003) *Billete al futuro. Las 3 paradas de la movilidad sostenible* [A ticket to the future. The 3 stops of sustainable mobility], Unión Internacional de Transportes Públicos, Bruxelles.

Van Exel, N. J. A. and Rietveld, P. (2009) 'Could you have made this trip by another mode? An investigation of perceived travel possibilities of car and train travellers on the main travel corridors to the city of Amsterdam, The Netherlands', *Transportation Research Part A: Policy and Practice*, vol 43, pp. 374–385.

Weaver, D. B. (2012) 'Organic, incremental and induced paths to sustainable mass tourism convergence', *Tourism Management*, vol 33, no 5, pp. 1030–1037.

Yeoman, I. (2005) 'Tomorrow's world – Consumer and tourist', *VisitScotland*, vol 1, no 2, pp. 1–31.

9 On-demand transport systems to remote natural areas

The Swiss case of Bus Alpin and AlpenTaxi

Roger Sonderegger and Widar von Arx

Introduction

The network of public transport in Switzerland is regarded as one of the best in the world. Almost all inhabitants are provided with a good public transport service. The Swiss law, in fact, requires that public transport be provided to all settlements with a permanent population of over 100 and an hourly service be provided to settlements over 500 inhabitants. The main reasons for the success of this public transport system are synchronised timetables for trains and buses, a dense and integrated network, the validity of one ticket for one journey (even if various transport companies are involved) and modest prices for customers (e.g. popular flat rate). Swiss policymakers and the Swiss population have backed these assets in many referenda and also ensured there is a solid financial basis for them. The outstanding quality of this system results in a record high usage of public transport in Switzerland as the modal share of public transport is nearly 28 per cent and only 65 per cent of all kilometres are covered by private transportation (ARE and Swiss Statistics, 2012). The average Swiss inhabitant covers a distance of over 2,000 kilometres per year by train and about 50 per cent of all inhabitants of Switzerland are owners of a public transport pass (VöV, 2014; Litra, 2014).

The limits of this high quality transport system become visible in tourism regions in remote areas. In fact, as transport for tourism and leisure is not subsidised by the Swiss government, there are many attractive places that are not served by public transport, especially in mountain regions. Nonetheless, the users of public transport represent a key visitor group for Alpine tourism regions, this meaning that remote regions without access to the public transport system face the danger of losing a big market share. Further, areas lacking adequate public transportation often suffer from a high presence of motorised traffic, which causes noise, pollution and uncontrolled parking, all contributing to a reduction in the quality of tourism.

Two services in Swiss mountain regions tackle these issues successfully: Bus Alpin and AlpenTaxi. They both offer an on-demand transport service for the 'last mile' between locations served by traditional public transport and tourist attractions. Both services are driven by the idea of fostering public transport and improving accessibility in the Alps. While Bus Alpin essentially provides new services similar to traditional public transport, AlpenTaxi is mostly based on already existing services.

This chapter will discuss the two services in detail, presenting their concepts and business models, analysing their effects on the environment (e.g. reduction of CO_2 emissions), society (e.g. passengers transported) and the economy, investigating the main reasons of their success and eventually discussing the possibility to transfer the concepts of the two initiatives to other contexts. In particular, the chapter attempts to answer the following questions:

1 Why are such mobility services important for the destinations they serve?
2 How are they organised and financed?
3 How successful are they, i.e. how many passengers are attracted over the year?
4 What makes these two concepts successful?
5 Do they replace car trips, or do they mostly generate new trips?
6 How much CO_2 is cut by Bus Alpin and AlpenTaxi during a year?
7 Are there relevant effects on traffic levels in the areas served by these transport options?

Bus Alpin and AlpenTaxi are portrayed based on the information available on their respective webpages, and on other public documents. Additional information on the two services was obtained from telephone interviews with the founders and managers, as well as further written commentaries by them (Bernhard *et al.*, 2008; Bernhard, 2012; Mountain Wilderness, 2012; Bernhard, 2014). The quality of the information provided for both services is certified by the national management of the two services. As only little information is available for AlpenTaxi regarding the number of passengers and modal shift from private vehicles, some rough estimates were obtained through an analysis of the visits to AlpenTaxi's website.

Study area

Switzerland is a country with a strong connection to the Alps (Figure 9.1). Not only do they cover almost two-thirds of the national surface, but they also play an important role in the national identity. Many of the eight million inhabitants have their own second home in the Alps. Alpine tourism has a very long and strong tradition in Switzerland, and it is still a major industry today. With €33.5 billion and 215,000 employees, tourism is responsible for around 5 per cent of the Swiss economy (GDP €599 billion, 4 million employees) (Swiss Statistics, 2014; and Swiss Tourism Association, 2014). The most important sector within the tourism industry is accommodation, providing about half of all jobs and half of all turnovers. As all major conurbations in Switzerland (including Zurich, Basel, Geneva, Lausanne and Bern) are within a 90-minute drive from the Alps, the connection by public transport very good, and overnight prices very high, it is attractive for most Swiss people to visit the Alps or the Jura mountains in a day trip. Foreigners, however, tend to stay longer because of longer travel times: in fact, more than half of all overnight stays in Switzerland are realised by foreigners.

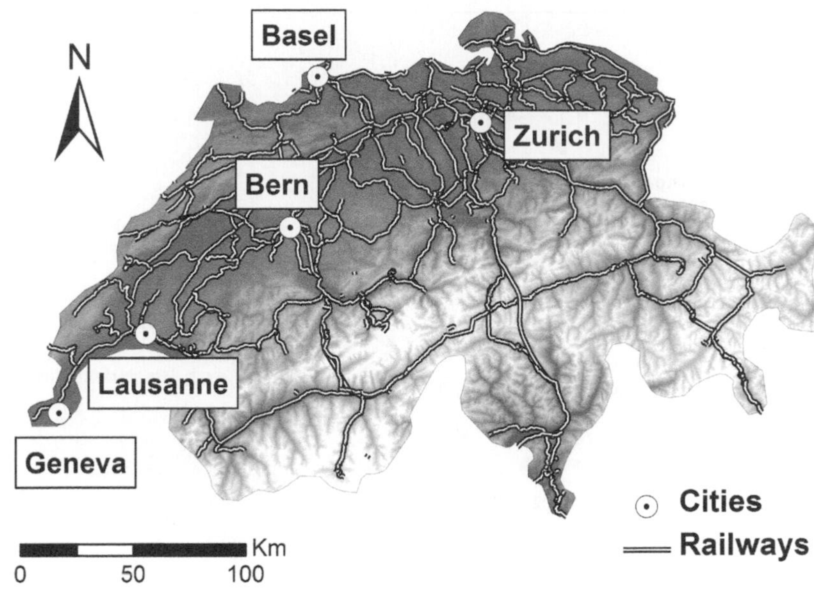

Figure 9.1 While Switzerland is a mountainous country, its widespread public transit network (the map shows railways) allows residents and visitors to easily reach most of the territory.

Within the Swiss mountains, however, only some transit corridors and areas of mass tourism are well connected to the major conurbations. Bus Alpin and AlpenTaxi serve tourism destinations located in remote mountain regions characterised by little infrastructures and almost no coverage of public transport. However, given their specific characteristics (i.e. Bus Alpin needs some minimum ridership in order to work, whereas AlpenTaxi can easily respond to low demands), the two initiatives generally serve different locations.

Bus Alpin and AlpenTaxi

Bus Alpin

Bus Alpin is a network of public transport in the Swiss Alps.[1] It consists of several small lines in various regions, serving mainly tourism purposes. Based on a successful pilot scheme in 2006/2007, Bus Alpin has by now become a stable service in many mountain valleys, which would not be served by public transport otherwise. In most cases, it connects the last point served by regular public transport with tourism attractions, such as a trailhead, an outlook or a restaurant. Today, Bus Alpin is active in nine regions across the Swiss Alps and has three lines in the Jura Mountains (Figure 9.2).[2] Bus Alpin services are more active in summer than in winter (i.e. only

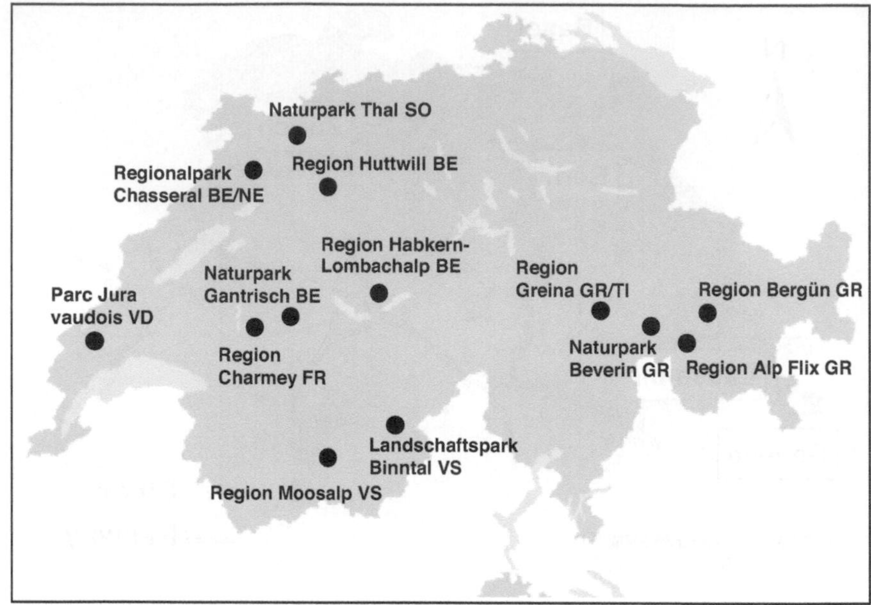

Figure 9.2 The local networks of Bus Alpin (source: Adapted from Bernhard [2014, p. 1]. Base map: © http://d-maps.com/carte.php?num_car=24779&lang=en).

four services operated in the winter season) and partially use private roads or roads that are banned to private cars.

Each of the bus lines is operated by a local network of actors. There are major differences in the organisation of these networks, but they typically include the local government, a regional nature park management (if existent), and the destination management organisation. For the driving service itself, the local network usually employs a professional bus company based in the region (e.g. a local division of the nationwide known post bus company 'Postauto'). Most regional organisations collaborate with various sponsors to cover the operation costs.

As of 2007, a national network called Alpentäler-Bus[3] coordinated the participating regions, offering marketing and communication services, exchange of know-how (especially on organisational and financial issues, sometimes on legal issues) and ensuring a co-funding by the public administration for the one year pilot scheme. Additionally, the national network has organised sponsoring on a national level to support the idea.

The services offered in each region vary a lot, depending on the demand and the financial background of the local organisation. In the Greina region, for example, a total of eight daily services operate on a regular and scheduled basis in three different valleys. In the upper part of the Binntal, in contrast, there are only three services a day, and only on demand. Reservation is compulsory in both regions. Often, these bus services can also be booked on a private basis (e.g. group transport) and most services are now integrated into the national timetable for public transport.

The operation costs of Bus Alpin are between Swiss Francs (CHF) 30,000 and 70,000 (€29,000 and 67,000) per year. However, the percentage of the costs covered by fares varies between 30 and 90 per cent. As these bus services cover leisure purposes and are therefore not eligible for public funding, some alternative funding must be obtained. These extra funds come from local and national sponsors, both of which are indispensable for every single line within the Bus Alpin network. Many local and some regional governments pay substantial subsidies to Bus Alpin services, arguing that they are of public interest due to the value added to and the jobs created within a region, especially in the tourism sector. The costs for the national network are covered by its members, i.e. the local networks. Each member pays between CHF 2,000 and 4,000 (€1,900 and 3,800) per year, resulting in a budget of about CHF 30,000 (€29,000) per year.

AlpenTaxi

Similar to Bus Alpin, the project AlpenTaxi responds to an existing issue in leisure traffic in the mountains, namely how to enable alpinists and hikers to undertake their activities using public transport. These visitor groups may choose AlpenTaxi because they do not own a car, because of restrictions on private motorised access to specific roads or because they are interested in linear hikes (i.e. the start and the end of the hike do not coincide). AlpenTaxi is especially relevant on private roads and where a four-wheel drive vehicle is needed.

An AlpenTaxi service might be a classic taxi, a bus on demand or more special transport modes, such as cableways. Among the 1,774 cableways listed in the AlpenTaxi website (www.alpentaxi.ch), most are only available on demand and often operated by farmers or restaurant owners. In other words, AlpenTaxi, though based on already existing services, never competes with traditional public transport: it only covers niches that would otherwise be uncovered.

The national organisation of AlpenTaxi is rather simple. It is based on an online high quality map where all the taxi services and cableways as well as the road- and rail-based public transportation networks are visible (see www.alpentaxi.ch). Through a click on the item of a local taxi company or a cableway, further information including contact details, and possible hiking and climbing options appear.

Any transport mode that complies with the requirements of AlpenTaxi is eligible to appear on the website for free. Founded as a common means for communication by alpinists and first published as a printed brochure, AlpenTaxi now mainly works as an online platform, which is linked to the Swiss Alpine Club, Schweiz Mobil (i.e. the Swiss portal for slow mobility) and Bergportal (i.e. an information platform for alpinists). It is run by the environmental NGO Mountain Wilderness.

AlpenTaxi today is a project with little financial issues as all operation costs are covered by fares and there is no constant need for sponsoring or public funding.

In order to stimulate its growth and allow the construction of a new webpage in 2010, federal public funding was raised by Mountain Wilderness, which operates the information platform on its own account, supported by sponsors.

Comparing Bus Alpin and AlpenTaxi

Bus Alpin, which was awarded various times for its contribution to sustainable transport, can be considered a substantial innovation in leisure transport in the past ten years, as shown by the considerable share of people that shifted to public transport after its introduction. In a poll among passengers carried out in 2007, some 30 per cent indicated they would have come by car or driven elsewhere, had Bus Alpin not existed. According to Bernhard's (2014) calculations, an added value of CHF one to two million in the relevant regions and a reduction of around 100 tons of CO_2 are directly caused by Bus Alpin. Table 9.1 shows the yearly number of passengers in the various regions in the period 2006–2013.

The homepage www.busalpin.ch was visited by over 20,000 people in 2011 and Bus Alpin produced over 100,000 information flyers for local promotions.

Table 9.1 Number of passengers on Bus Alpin in the various regions between 2006 and 2013.

Region	*2006*	*2007*	*2008*	*2009*	*2010*	*2011*	*2012*	*2013*
Alp Flix GR				2,406	2,211	2,199	2,537	3,003
Bergün GR							1,319	1,625
Binntal VS	6,735	5,688	6,400	10,263	11,726	10,200	11,135	10,056
Chasseral BE/NE			600	760	1,043	1,704	1,825	1,731
Gantrisch BE	151	273	363	302	389	291	247	611
Greina/Bleniotal GR/TI	6,433	6,682	7,306	8,564	7,662	7,363	5,440	9,134
Habkern-Lombachalp BE								2,500
Huttwil BE							245	247
Jura vaudois VD								*
Moosalp VS		*	*	*	*	*	*	*
Thal SO				865	811	935	603	733
TOTAL	13,319	12,643	14,669	23,160	23,842	22,692	23,351	29,640

*= flat-rate rides only, in addition to regular post bus line (cannot be compared to the other Bus Alpin passenger numbers)
Source: Bernhard (2014, p. 2)

Regarding quality management and promotion, the Bus Alpin office and the local networks define a checklist of tasks such as timetable planning or budget control.

The main success factors of Bus Alpin may be described as follows:

- Services are targeted to the existing demand (e.g. tourist attraction not served by public transport and sometimes not even reachable by private motorised vehicle)
- A local organisation board ensures long-term local engagement for the projects
- The local financial support is of mixed origin, mostly with long-term partners
- A stable national network ensures an intense exchange of know-how and offers a broad communication through strong partners (e.g. WWF, VCS,[4] SchweizMobil)
- A strong national brand allows for national communication through the media
- Pilot schemes with intense testing made the product mature for multiplication
- Thanks to the project character in the pilot scheme, national public funding could be raised
- Promotion of all Bus Alpin services helps cross-selling
- All local opinion leaders in tourism and transport planning were involved in an early phase
- Operation costs are generally low due to part-time job models for drivers

Certainly, the combination of local engagement with national know-how and support can be considered one of the key factors for the success of Bus Alpin. In each of the 12 regions where it is adopted, Bus Alpin has represented a feasible solution for an existing problem of leisure traffic. A national network helps tackle management issues, convince regional and national sponsors, raise public funds and gain public visibility.

Similar to Bus Alpin, AlpenTaxi meets a pre-existing transport demand in the Alps, and it is also a very successful example to learn from. Unfortunately, to date, there is only limited information available on the effects of AlpenTaxi, particularly regarding the number of passengers, their motives to use the service, the reduction of CO_2, etc. In fact, it would be a great empirical effort to investigate these issues as all services of AlpenTaxi have a great variety of users, and it is unknown if they are on board due to the information provided by AlpenTaxi, or due to other information or motives. Nonetheless, the number of visits to the webpage of AlpenTaxi gives at least an idea of the size of the initiative. After the launch of the new webpage in 2010, the number of single visitors to the webpage has grown quickly. Starting from around 420 visits per month in 2011, it has grown to more than 650 in 2013, corresponding to a growth of around 75 per cent in two years. The average number of visits per month amounts to 560 over the period 2011–2013. Peaks in access are observed in winter and in summer, supporting the idea that the main user groups consist of hikers, alpinists and backcountry skiers. Although the number of visitors remains largely unknown, the effects and the potential of AlpenTaxi might at least be illustrated by the following example.

For a hiking trip on the popular route 'Via Alta della Vallemaggia', the hamlet of Daghéi is a good starting point. As there is no public transport from Brione (village served by scheduled buses) to Daghéi, excursionists would be forced to use the private vehicle for the whole journey between their place of residence and Daghei. However, the presence of AlpenTaxi between Brione and Daghéi allows them to choose the public transportation. Considering two people coming from Zurich, this would imply a potential cut of around 30 litres of fuel, equivalent to about 60 kilograms of CO_2. This shows that the real potential of AlpenTaxi lies in allowing people who traditionally rely on the private vehicle to use the public transport. Considering that the AlpenTaxi's website receives around 500 visits per month and assuming that 20 per cent (i.e. 100) of those who visit the website actually use AlpenTaxi, the potential reduction in car use would be around 20,000 kilometres per month, or 240,000 kilometres per year, which correspond to 36 tons of CO_2. Since the visits to the webpage of AlpenTaxi are increasing, this potential is also likely to grow further in the future.

The main success factors of AlpenTaxi are the following:

- Local services covered by local organisations ensure a constant adaptation to the demand
- A national information platform provides easy access to information
- The target group is environmentally sensitive, well informed and financially strong
- All services are free of charge for providers and end users
- The quality of the information is guaranteed by an active NGO

Table 9.2 summarises the main characteristics of Bus Alpin and AlpenTaxi.

Comparing their goals and approaches, Bus Alpin and AlpenTaxi could also be regarded as two competing concepts, though AlpenTaxi promotes the available Bus Alpin services. The main difference between the two concepts certainly lies in the depth of their collaboration with the networks of local agencies. While AlpenTaxi merely displays good quality information through the web and through partners, Bus Alpin collaborates closely with the local staff (marketing and communication with the media are managed through Samuel Bernhard, founder of the Bus Alpin network).

The large-scale national initiative SchweizMobil (www.schweizmobil.ch), which was founded first to promote hiking and biking for leisure and tourism, combines and integrates almost all relevant information for sustainable mobility in Switzerland, and can therefore compete with AlpenTaxi. Technically, SchweizMobil is made on a very high level, providing suggestions for excursions, integration of the timetables for public transport, the official Swiss topographic maps, booking options for hotels, GPS tracks, etc. Almost all the information provided on the AlpenTaxi's website can also be found on the Schweizmobil's one, leading to serious questions about the future of AlpenTaxi.

Apart from their not-for-profit character and their similar goals, however, Bus Alpin, AlpenTaxi and SchweizMobil share a feature that has been crucial for their success: they all rely on a strong partnership between big and established companies,

Table 9.2 Main characteristics of Bus Alpin and AlpenTaxi.

	Bus Alpin	*Alpen Taxi*
Extension	12 regions	300 lines
Passengers/year	25,000–30,000	Unknown
Fare (single trip)	CHF 2–10 (€ 1.90–9.60)	Open, depending on distance/service
Costs covered by fares	30–90%, depending on region	100%
Modal shift from car	Ca. 30%	Unknown
Reduction of CO_2	Ca. 100 tons/year	Ca. 36 tons/year (estimation)
Financial scheme	Public, local and national sponsors	Fares and Mountain Wilderness
Year of foundation	2006	1996
National partners	SAB, VCS, SAC, Postauto[4]	None, only for communication
National coordinator	Samuel Bernhard	Mountain Wilderness

Source: Bernhard *et al.* (2008), Bernhard (2012), Mountain Wilderness (2012)

and semi-professional or even volunteered local initiatives. Just like the Swiss hiking trail network could never be maintained without volunteer work by regional associations, many small schemes in the area of semi-public transport could not be kept running without the help of volunteers or part-time working staff, especially considering the high wages paid in Switzerland.

There are several tourism hot spots around the world that are afflicted by transport-related issues (e.g. traffic congestion), which may seriously affect the quality of the recreational experience. While parking management or road pricing may offer adequate solutions to such issues, the effectiveness of these measures can be enhanced through combination with innovative services such as Bus Alpin and AlpenTaxi (e.g. Val Genova,[5] Italy).

What can we learn from these successful examples for application to other contexts? Certainly, there are great differences between places, but there are some general principles that can be derived from Bus Alpin and AlpenTaxi, and drive the implementation of similar services elsewhere.

1 Projects for new services work better when originating from an existing problem and a real demand.
2 A combination with a national brand and a national marketing and fundraising campaign seems very promising for other contexts and other countries, too.

Collaboration with a strong national partner or various partners is highly recommended.

3 Considerable ridership can only be achieved with stable services over many years, since the increase in ridership takes time. Ideally, the organisation and the management in charge of the service should not change.
4 A continuous and high service quality is indispensable.
5 A pilot scheme can help avoid mistakes and raise funds.
6 A sound financial basis cannot rely on fares only, but must include public funds and sponsors. Raising public and private funds would be easier for a set of similar projects than it is for a single project.
7 Low operation costs are crucial. They can be achieved mainly by using rolling stock of existing services (e.g. school bus, public transport) and by employing non-professional drivers.
8 Any new service should avoid competition with existing schemes, especially with public transport.
9 Building up a network with existing organisations helps promote a new service. Partners must represent the target group as precisely as possible.[6]
10 An intense stakeholder inclusion policy, starting in a very early project phase, is important for acceptance of new solutions.
11 A monitoring system is needed in every project in order to measure its success (at least a systematic monitoring of all passengers).
12 A combination with traffic management measures (e.g. parking fees, ban on private cars) is recommendable.
13 A technically easy and transparent mode to reserve and pay these services is advisable (e.g. applications for smartphones).

Both projects make a real contribution to sustainable transportation in the Alps. They have shown that small projects on a local scale might have a great impact if well integrated into a national framework. With a yearly ridership of several tens of thousands, they are highly capable of reducing noise, pollution and CO_2, and improving the safety and the leisure quality in the regions served. Furthermore, Bus Alpin and AlpenTaxi induce a series of indirect synergy effects such as an increased use of public transport (due to the increase in transport quality), a decrease in car ownership, and even positive effects on regional economies. As all services are managed by local organisations and people, the added value remains within the peripheral region and opportunities are offered to local people for complementing their traditional incomes (e.g. agriculture, hotel businesses).

Conclusion

In remote tourism areas not served by public transport, services such as Bus Alpin or AlpenTaxi may enable visitors to reach popular spots (e.g. trailhead, outlook) without the need of a private vehicle, thus contributing to the preservation of natural resources and the quality of the recreational experience. Ideally, such services can be combined with other measures, such as road pricing or parking management, to maximise modal shift from private vehicles. The basic conditions for establishing Bus Alpin or AlpenTaxi services are found in many places in the Alps and beyond: popular tourist attractions, basic organisational structures, a clear commitment to sustainable tourism, lack of coverage by public and/or private transport and potential sponsors at the local level. While for a scheduled Bus Alpin service to be effective, a regular demand should be well documented before the start, AlpenTaxi could start from very basic conditions (e.g. low documented demand, availability of a private cableway). Socio-economic conditions may vary considerably across countries, but other Alpine countries such as Austria, Germany or Italy would have the potential to reproduce similar brands.

A more serious challenge instead might be the creation of a national – or at least regional – network, which incorporates a large body of information about transportation options. The wealth of information provided by SchweizMobil in Switzerland reaches far beyond Bus Alpin and AlpenTaxi, and provides an international benchmark for country-level information on alternative mobility. In the future, services like Bus Alpin and AlpenTaxi and online platforms like SchweizMobil should be increasingly integrated to give tourists worldwide the possibility to enjoy car-free vacations in remote natural contexts.

Notes

1 All information in this chapter taken from Bernhard *et al.* (2008), Bernhard (2012), Bernhard (2013) and SAB (2009).

2 The 12 regions with Bus Alpin are as follows (* stands for Jura region): Alp Flix region, Bergün region, Beverin Regional Natural Park, Binntal Regional Nature Park, Chasseral Regional Natural Park*, Gantrisch Regional Natural Park, Greina region, Habkern-Lombachalp region, Huttwil region, Jura Vaudois Regional Natural Park*, Moosalp region, Thal Regional Natural Park*.

3 Alpentäler is a German name, which was not understood in the French, Italian and Romansh speaking parts of Switzerland and therefore needed to be replaced. In 2008, a new organisation was founded under the name 'IG Bus Alpin', which was later renamed 'Bus Alpin' – a Romansh (or French) speaking name with an intercultural touch.

4 Schweizerische Arbeitsgemeinschaft für die Berggebiete SAB (NGO for rural development), VCS Verkehrs-Club der Schweiz (NGO for sustainable transport), Schweizer Alpen-Club SAC (Swiss Alpine Club), and PostAuto Schweiz AG (Swiss Post Bus company).

5 In the Val Genova (located in the Parco Naturale Adamello Brenta, Italy), parking fees are used to co-finance a new transport service consisting of a train on rubber wheels and a small bus.

6 In the case of AlpenTaxi and Bus Alpin, the partners were SBB (Swiss Federal Railways), Schweiz Mobil (an information platform for human powered mobility), SAC (Swiss Alpine Club), Bergportal (an information platform for alpinists) and the national network of Swiss parks.

References

ARE and Swiss Statistics (2012) 'Swiss Microcensus on Mobility and Transport', Office for Spatial Development and Swiss Statistics, Bern/Neuchâtel.

Bernhard, S. *et al.* (2008) 'Alpentäler-Bus / Bus Alpin, Final Report (short version)', www.busalpin.ch.

Bernhard, S. (2012) 'Dokumentation Bus Alpin: Februar 2012, Zürich', www.busalpin.ch, accessed 10 September 2014.

Bernhard, S. (2013) 'Information for the media', provided on 13 December 2013, www.busalpin.ch, accessed 10 September 2014.

Bernhard, S. (2014) 'Bus alpin baut Angebot weiter aus', unpublished document.

Litra (2014) 'Litra Verkehrszahlen 2013', www.litra.ch/de/Zahlen-und-Fakten/LITRA-Verkehrszahlen.

Mountain Wilderness (2012) 'AlpenTaxi.ch – Dein Weg zum Berg', Mountain Wilderness, www.AlpenTaxi.ch.

SAB (2009) 'Verkehrsmanagement in Schweizer Berggemeinden mit touristischen Ausflugszielen', www.sab.ch.

Solèr, R., Sonderegger, R., von Arx, W. and Cebulla, L. (2014) 'Sanfte Mobilität für Ihre Gäste', Lucerne School of Business, Lucerne, Switzerland.

Swiss Statistics (2014) 'Gross domestic product', www.bfs.admin.ch, accessed 21 October 2014.

Swiss Tourism Association (2013) 'Swiss tourism in numbers', www.stv.ch, accessed 21 October 2014.

VöV (2014) 'Fakten und Argumente zum öffentlichen Verkehr der Schweiz', Edition 2012, Berne, Switzerland.

10 Reducing visitor car use while securing economic benefits in protected areas

Application of a market segmentation approach in the Lake District National Park (UK)

Davina Stanford

This chapter is a revised version of a paper that first appeared in 2014 in the *Journal of Sustainable Tourism* (vol 22, no 4, pp. 666–683).

Introduction

The imperative to reduce carbon emissions in all areas of human activity is well understood from an environmental point of view. In economic terms, the Stern Review (Stern, 2007) warns that the cost of actions to ensure that the worst impacts of climate change are avoided might be around 1 per cent of global GDP, whereas the cost associated with inaction is 'equivalent to losing at least 5 per cent of global GDP each year, now and forever'. In the UK, for example, the transport sector currently accounts for around 24 per cent of domestic emissions of carbon dioxide (CO_2) and will be required to play a full role in reducing these emissions and contribute to meeting overall economy-wide targets. Emissions from road traffic (80 per cent of which is car transport) dominate the domestic transport sector and accounted for 92 per cent of the total in 2006 with leisure trips accounting for 14 per cent of the estimated CO_2 emissions from all modes of passenger transport by journey purpose based on the UK 2002/2006 average (Department for Transport, 2008).

In rural areas, cars can be a threat to tourism as the visual, noise and atmospheric pollution associated with car traffic may negatively affect the environment and the recreational experience of visitors. Further, increasing traffic congestion calls for the construction of new or wider roads as well as larger parking sites, which lead to greater erosion on nearby footpaths (Sharpley and Sharpley 1997). Visitor surveys undertaken by the English National Park Authorities (NPAs) and tourism bodies demonstrate that traffic and congestion are considered a threat to the special qualities of national parks in general and damage the visitor experience (English National Parks Authorities Association, 2007). Local communities are also heavily affected by tourism transport (Jurowski *et al.*, 1997; Lindberg and Johnson, 1997). Nevertheless, visitors to rural areas are typically car-based (Dickinson and Dickinson, 2006) and reducing car use is particularly difficult due to lack of economic resources to manage alternative transportation system, but also lack of detailed knowledge about visitor preferences towards transportation options.

Dickinson and Dickinson (2006) have highlighted the inability of existing sustainable transport studies in exploring visitor preferences as they identify factors of success or failure through quantitative surveys based on standard attitude behaviour models. They observe that 'traditional transport attitude and behaviour studies, while useful for establishing baseline information and trends, do little to further our understanding of the social realities that underpin people's attitudes towards transport and tourism and their decisions about transport behaviour' (Dickinson and Dickinson, 2006, p. 193). Anable (2005) also identifies limitations in the field and writes that in travel research methodology the combination of instrumental, situational and psychological factors affecting travel choice and illustrating differences for distinct groups of people is overlooked.

A body of work is developing that seeks to understand travel behaviour through segmenting the market. Market segmentation is a technique used by marketers to identify the specific characteristics of different customer groups and then use those insights to meet customer needs. For example, Dallen (2007) used market segmentation to understand the attitudes of tourists and the local community towards using the Looe Valley Branch Railway Line in Cornwall (UK). Notable is the approach proposed by Anable (2005), who used multidimensional attitude statements to segment a population of day travellers into potential mode switchers using cluster analysis. Six distinct psychographic groups were extracted, each with varying degrees of mode switching potential. In both cases, the importance of the segmentation approach is emphasised to highlight the unique combination of preferences and attitudes and the complexities and diversity of different groups that need to be understood in order to optimise the chance of influencing travel behaviour.

A DEFRA report (2008) argues the need to segment audiences and to tailor messages accordingly. Interventions need to start from an understanding of current lifestyles (and life stages) for different population groups, even if the longer-term aspiration is to bring about a fundamental shift in that lifestyle or a particular behaviour. More recently, the UK Department for Transport (2011) has used a market segmentation approach to provide a framework for local authorities and other planning organisations to help develop effective, targeted sustainable transport initiatives taking into account the nature of the local population.

Travel behaviour in particular is supposed to largely neglect the influence of environmental information, with a range of other factors overriding environmental concerns. Coulter *et al.* (2007) suggest that many barriers thwart changes in travel behaviour and that car travel is viewed as essential and necessary. Habitual behaviour such as car use limits people's propensity to consider other forms of transport and there are perceived disincentives for switching to alternative forms of transport. Hence, it is reasonable to assume that we will have to carefully account for these factors to achieve even a relatively modest impact on people's travel behaviour.

However, actions to induce changes in travel behaviour cannot be separated from the goal of maintaining economic benefits for the destination. Ultimately, it would not be beneficial to a destination to target a market segment on the basis of its willingness to change the environmental behaviour, if that market segment is one of the least beneficial for the destination in economic terms. An adequate balance between economic and environmental concerns needs to be sought because too often being environmentally friendly is associated with sacrifice and a loss in economic benefits (Moeller *et al.*, 2011). Yet this requires cost-benefit analyses to be less biased towards economic aspects, and better account for other aspects such as the environment (Northcote and Macbeth, 2006). Research is therefore needed to identify market segments that are most amenable to a positive behavioural change, while guaranteeing a positive economic contribution to the destination.

Researchers have previously aimed to measure the environmental impact of tourism in relation to economic gain or yield (Gössling *et al.*, 2005; Becken and Simmons, 2008; Moeller *et al.*, 2011). Gössling *et al.* (2005), for example, took an eco-efficiency approach, which uses life cycle assessment concepts to measure environmental damage (estimated through carbon dioxide equivalent emissions) per unit of value generation (computed through actual visitor spend). Moeller *et al.* (2011) sought to identify market segments that are environmentally friendly and have high expenditures by asking about general travel behaviour, specific travel behaviour on the last trip (including spend), general attitudes towards the environment (using the New Ecological Paradigm scale proposed by Dunlap *et al.*, 2000) and socio-demographic information. Respondents' total spend per day was used as the dependent variable and as the indicator of economic contribution.

This chapter presents a study aimed at understanding how best to reduce the transport-related environmental burden of protected area visitors while maintaining economic benefits. The specific objectives of the study are to test attitudes towards behavioural change, based on a social psychological framework, and to identify market segments that might demonstrate a high propensity towards both a positive behavioural change and a high contribution to the destination in economic terms. The hypothesis being that, if market segmentation is not a useful approach in achieving this, there would be little difference in the preferences and behaviours of the identified market segments. Ajzen's theory of planned behaviour (Ajzen, 1988) was used as the conceptual framework to analyse attitudes towards behavioural change with regard to tourism transport choices according to market segments. The study was conducted in the Lake District National Park (UK) as part of a research project commissioned by Natural England,[1] and Friends of the Lake District and Cumbria Tourism, to explore ways of reducing the transport-related environmental burden of visitors to Cumbria (for full details, see Stanford, 2014). In this context, the term environmental burden includes issues such as air quality, visual and noise impact, as well as the broader issue of carbon emissions.

Study area

The Lake District National Park, England's largest national park (2,292km^2), is located in Cumbria, in the northwest of England (Figure 10.1). Consistent with its name, the park has over 14 lakes (Windermere being the largest one at 14.8km^2), which formed due to the combined effect of high rainfall, deep glacial valleys and relatively impermeable volcanic rocks. The county of Cumbria, in which

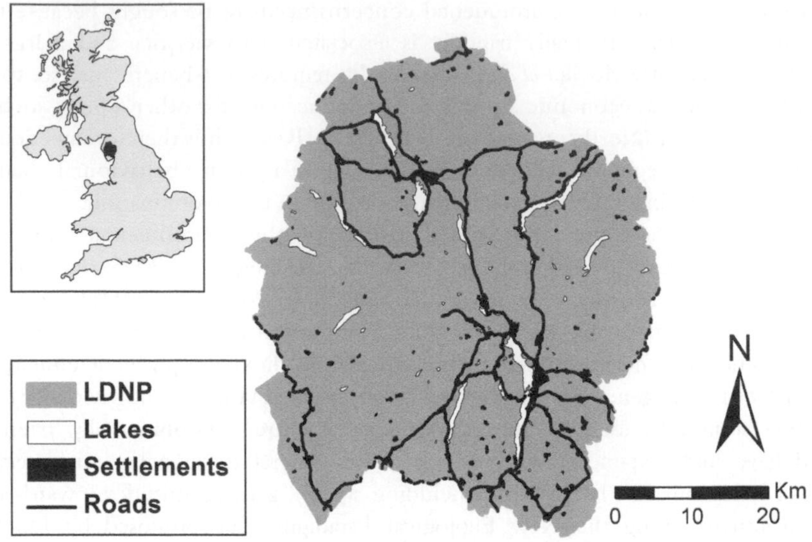

Figure 10.1 The Lake District National Park (LDNP) in Cumbria (northwestern England).

the Lake District is situated, is a rural and mountainous region, which covers an area of 6,768km² with a population of just under 500,000. The county has above average employment in agriculture, tourism-related activities and construction (Cumbria County Council, 2014). According to Cumbria Tourism's website (Cumbria Tourism 2013), in 2013 Cumbria and the Lake District received around 40 million visitors (34.2 million day trippers and 5.4 million overnight visitors) who brought in UK £2.2 billion to the local economy and provided the equivalent of 32,805 full-time jobs.

Given the economic significance of the tourism industry to Cumbria and the amount of visitors, it is worrying in environmental terms that 85 per cent of visitors use a motorised vehicle (car, van, motorbike or motor-home) to arrive, and 80 per cent of tourists use cars and other types of motorised vehicles to travel around the destination (Cumbria Tourism, 2006). In such a context, even a small reduction in private vehicle reliance could result in a significant reduction in CO_2 emissions. Moreover, the quality of life of residents could enormously benefit from a reduction in the visual and aural pollution associated with traffic congestion on the roads and in car parks.

Method

A survey was developed that sought to understand statements regarding attitudes to car use in the Lake District. The survey also gathered information on visitor demographics and other details such as type of accommodation. The results were then mapped against the known spend for each of the market segments used for Cumbria Tourism's staying visitors and the results for each stage combined. Cumbria Tourism has a bespoke market segmentation based on an analysis of people's reported activities as visitors to the region split into day and staying visitors (Cumbria Tourism, 2006). The segmentation provides information on demographics, accommodation, activities, experiences and motivation. The present study, however, is only based on staying visitors.

The survey was based on Ajzen's framework and asked respondents to consider their next trip to Cumbria. Ajzen's theory of planned behaviour (1988) is a social-psychological framework, which is often used in social sciences to predict behaviour. The theory is one of the most frequently cited and reliable ways to predict human behaviour (Ajzen, 2011). The theory proposes that the immediate determinant of an individual's behaviour is largely influenced by his or her intentions to perform or not that behaviour and his or her perceived control over that behaviour. Intentions are the product of three constructs, summarised as follows:

- Personal attitudes towards performing the behaviour – the individual's beliefs that a given action will produce positive or negative outcomes.
- Subjective norms – a person's belief that specific individuals or groups think he or she should or should not perform the behaviour (this includes individuals such as parents, spouse, children, friends and managers).
- Perceived behavioural control – the individual's perception of the difficulty of performing the behaviour reflecting both past experience and anticipated obstacles.

The three constructs themselves are linked to underlying beliefs: behavioural beliefs for attitudes; normative beliefs for subjective norms and control beliefs for perceived behavioural control.

The theory has been applied in a tourism context to explore the social influences predisposing visitors to engage in specified behaviours in socio-cultural contexts (Goh, 2010) and pro-environmental behaviours (Stanford, 2006; Powell and Ham, 2008; Ham *et al.*, 2009; Reigner and Lawson, 2009; Ong and Musa, 2011; Serenari *et al.*, 2012). It has also been used to identify travel behaviour segments (Anable 2005). In some of these examples, the theory has been adapted and simplified. For example, Reigner and Lawson (2009) adapted the construct measurements to better fit with the behaviour of interest in their study and to more directly address the management interests of the national park in which the study took place. Similarly, other studies (e.g. Sparks and Shepherd, 1992; Cheung *et al.*, 1999; Ong and Musa, 2011) have found that measuring attitudes, subjective norms and perceived behavioural control helps predict intention and behaviour. Based on the need to simplify the theory for the context of the Lake District National Park and following the example of these previous examples, Figure 10.2 shows how the theory was operationalised for this study.

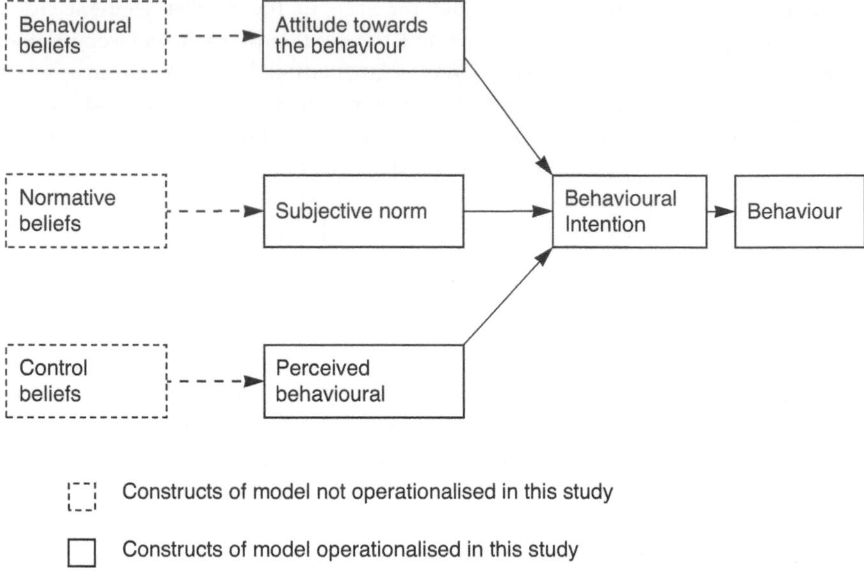

Figure 10.2 Layout showing how the Ajzen's theory of planned behaviour was operationalised for this study (source: Ajzen, 1988).

Respondents were asked to agree or disagree with statements reflecting: their personal attitude to reduce car use, the influence of others on the decision to reduce car use (subjective norm), the perceived difficulty in reducing car use (perceived behavioural control) and the intention to reduce car use (Table 10.1).

The survey was conducted with a nationally and regionally representative sample of 2,000 adults in Great Britain with face-to-face interviews conducted at home by Ipsos MORI interviewers using the Computer Assisted Personal Interviewing system (CAPI). This method allows a representative national sample to be obtained and, if compared to online surveys, does not exclude respondents without access to the Internet.

The responses provided in the survey were then grouped based on Cumbria's six visitor segments (Table 10.2) to demonstrate which segments had highest economic contribution to the destination.

Results

The sample consisted of 390 completed questionnaires. In gender terms, respondents split almost equally with 53 per cent males and 47 per cent females. Almost a quarter of all respondents (23 per cent) were aged 18–34, 42 per cent of the total

Table 10.1 Application of Ajzen's Theory in the present study. Respondents were asked whether they agreed or disagreed with four statements reflecting their attitude, subjective norms, perceived behavioural control and intention.

Ajzen's Theory	*Do you agree or disagree with the following statements?*
Attitude	I think I should use my car less when I visit the Lake District
Subjective norm	People who are important to me think I should use my car less when I visit the Lake District
Perceived behavioural control	It would be very difficult for me to use my car less when I next visit the Lake District
Intention	I intend to use my car less when I next visit the Lake District

Table 10.2 Market segments of visitors to Cumbria with specification of their share of total visitors and their average daily spend.

Market segment	*Share of total visitors (%)*	*Average daily spend per person (£)*
New Explorers	22	40.43
Wilderness Couples	17	37.85
Frequent Adventurous Independents	16	35.54
Old Scenery Watchers	15	34.14
Cultured Families	14	31.02
Familiar Families	16	30.13

Source: Cumbria Tourist Board, 2006

were between 35 and 54 and the remaining 35 per cent were over 55. Regarding the type of accommodation, 42 per cent stayed in serviced accommodations (e.g. hotels, guesthouses, B&Bs), 40 per cent stayed in non-serviced accommodations (e.g. camping, caravanning) and 18 per cent gave other answers (e.g. cannot remember). The majority travelled as a family (36 per cent), while 33 per cent travelled as a couple, 25 per cent with friends and 6 per cent alone or with other kinds of groups. Four-fifths of the interviewees (81 per cent) had a car. These socio-demographic data are consistent with those collected through a Cumbria visitor survey.

The rates of agreement with the four statements listed in Table 10.1 are shown in Figure 10.3. On average, 43 per cent of respondents feel they should use their car less, and 40 per cent state that they also intend to do so. This intention is a little higher than might be expected as typically this questioning framework implies a drop between the stated preference (i.e. those who think they should) and the actual intention to perform a behaviour. This could perhaps be the result of social desirability bias, whereby respondents give answers that are not necessarily truthful but are the answers which they think are most socially acceptable. Most importantly perhaps is the result showing that the majority of respondents (60 per cent) think it would be difficult for them to reduce their car use on their next visit. The influence of others is less important: not even 30 per cent of respondents indicate that they agree with this statement.

Table 10.3 summarises the percentages of respondents who agreed with each of the statements cross-matched with Cumbria's visitor segments. New Explorers and Familiar Families show more agreement than average with statements regarding attitudes towards positive behavioural change. They are more likely than average to agree that they intend to reduce their car use (45 per cent and 48 per cent, respectively) and are less likely to perceive this as difficult (both 55 per cent). Frequent Adventurous Independents demonstrate a similar pattern, but the small sample size of this segment may raise issues of reliability. Wilderness Couples show the least agreement with the statements regarding attitudes, significant others and intentions (37 per cent, 22 per cent and 30 per cent, respectively).

Table 10.4 summarises findings from the combination of attitudinal declarations collected through the survey and expected economic contribution as specified in data from the Cumbria Tourist Board (2006). New Explorers and Frequent Adventurous Independents show both a high propensity to behavioural change

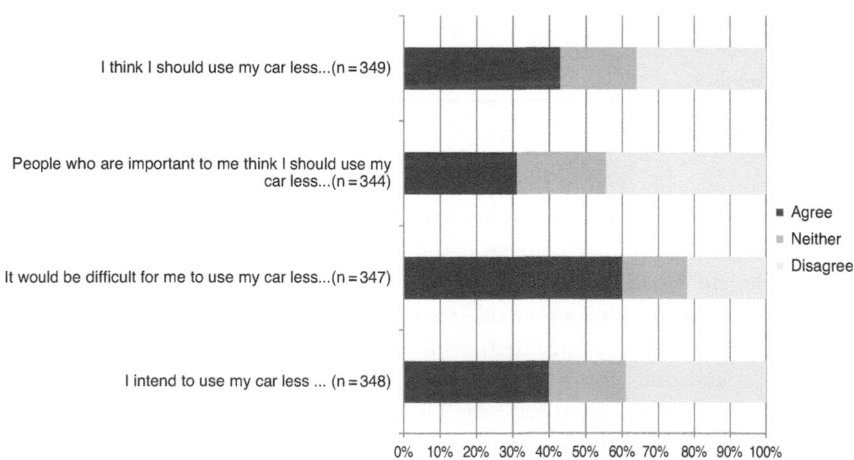

Figure 10.3 Rates of agreement with the statements presented in the questionnaires.

Table 10.3 Percentages of respondents (within the entire sample and within each of the Cumbria's visitor segments) who agreed with the four statements presented in the survey.

	I should…	*Other people think I should…*	*It would be difficult to…*	*I intend to…*
	…reduce my car use on my next visit to the Lake District			
Average				
%	*43*	*28*	*60*	*40*
Cross-matched segments	*%*			
New Explorers (n = 34)	56	31	55	45
Old Scenery Watchers (n = 85)	37	31	63	37
Familiar Families (n = 90)	41	24	55	48
Wilderness Couples (n = 94)	37	22	61	30
Cultured Families (n = 64)	42	38	58	41
Frequent Adventurous Independents (n = 21)	65	29	74	45

and high expected spend, therefore representing higher yield segments. Wilderness Couples also have a high economic impact, but may be the most resistant to changing their travel behaviour and present a greater challenge in terms of all round yield. Other segments are all expected to have a minor economic impact, which makes them less attractive even if they show a significant willingness to change their behaviour.

Discussion

Ajzen (1988) suggests that three main factors influence behaviour: personal attitude towards the action (how people feel personally about doing something), subjective norm (the influence of significant others or society) and perceived behavioural control (in this case the ease or difficultly of performing a certain action). In other words, the factors that influence behaviour are essentially personal belief on the one hand and social influences on the other: these will ultimately be moderated by a person's perception of how easy or difficult undertaking a certain action will be.

Table 10.4 Summary of willingness to behavioural change, propensity to spending and proposed action for each of the visitor segments.

Segment	*Willingness to change*	*Spend*	*Action for segment*
New Explorers	Easier segment to influence	High spend	Encourage segment
Old Scenery Watchers	Most resistant to change	Low spend	Less beneficial segment
Familiar Families	Easier segment to influence	Low spend	Increase economic value of segment
Wilderness Couples	Most resistant to change	High spend	Change travel behaviour of segment
Cultured Families	Average resistance to change	Low spend	Less beneficial segment
Frequent Adventurous Independents	Easier segment to influence	High spend	Encourage segment

It is interesting to observe that in the case of Lake District National Park, though the majority of respondents know they should reduce their car use, they also share a perception that a reduction in car use will be difficult. In some ways, this is encouraging, as at least the underlying values which drive behaviour are aligned, for the majority, with responsible action in environmental terms. It does however provide challenges for destination planners who are expected to ensure that alternative transport is both relatively easy to use and perceived as such. Clearly, some segments perceive change to be more difficult than others, but this seems to be associated with the kind of activity performed by each segment. For example, wilderness lovers may be more reluctant to leave their car as they generally travel at times when no alternative transportation would be available and reach locations not served by alternative transportation. Similarly, elderly segments might find it harder to shift from car to public transport due to their limited physical mobility.

The adopted approach and the results obtained prove that it is possible to identify market segments that both have a greater propensity for a positive behaviour change (in this case a reduction in private car use) and guarantee a higher economic contribution to a destination. Returning to the concept of yield, there are clear implications here for administrators and park managers who should attract segments that would ensure a greater environmental and economic contribution. In the case of the Lake District National Park, for example, encouraging New Explorers and Frequent Adventurous Independents segments looks like a win-win solution: one that is consistent with the concept of sustainable tourism.

However, the method does have limitations. The Ipsos survey, for example, can be criticised for focusing on 'professional respondents' rather than a representative sample of people. Furthermore, for some of the market segments (e.g. Frequent Adventurous Independents), the number of observations seems too low to provide statistically robust estimates. The greatest weakness of the method, however, undoubtedly lies in the impossibility to control the discrepancy between statements and actual intentions: respondents may answer questions based on what is socially desirable rather than what they really think or do (Phillips, 1976; Nachimas and Nachimas, 1981; Fisher, 1993; Jones, 1996; Singleton and Straits, 1999; Ballantyne and Hughes, 2006). This issue, which is common to all approaches based on stated preferences, may in fact impair the reliability of the results, by providing an overly optimistic vision (i.e. respondents may look more willing to change behaviour than they really are) that does not correspond to reality. Hence, the results should be interpreted more as an indication of which segments are more willing to change behaviour rather than a precise estimate of how willing a given segment is to change behaviour.

Conclusion

This research adds to the literature on behaviour change in an environmental context by addressing some of the shortcomings of studies exploring sustainable transport behaviour. In particular, the proposed approach has attempted to go beyond mere baseline data to provide a more rounded understanding of behaviour by applying Ajzen's theory of planned behaviour and adding an additional layer of depth through a combination of market segmentation and the concept of yield. The study recognises that a shift from the private vehicle to alternative transportation is particularly difficult in natural settings for various reasons (e.g. need to carry bulky gear, scarce autonomy allowed by public transit systems), but that some visitor segments may be more willing than others to undertake such a shift. Yet adopting policies that target only these segments may be detrimental for a destination if the same segments do not guarantee sufficient economic revenues.

Instead, this study emphasises that identifying visitors who are more easily influenced to reduce their car use and have a higher economic impact on a destination is a key element of transport behaviour management and subsequently sustainable tourism. Combining the attributes of market segments for both environmental and economic benefits helps us attract tourists who tread lightly and pay their way, therefore guaranteeing the preservation of our natural lands.

Note

1 Natural England is an executive non-departmental public body responsible to the Secretary of State for Environment, Food and Rural Affairs with a purpose to protect and improve England's natural environment and encourage people to enjoy and get involved in their surroundings.

Acknowledgements

Many thanks to Cumbria Tourism, Friends of the Lake District and Natural England who funded the research; to Dr Xavier Font at Leeds Beckett University for commenting on early drafts; and to colleagues at TEAM Tourism who collaborated on the project.

References

Ajzen, I. (1988) *Attitudes, Personality and Behaviour*, Open University Press, Milton Keynes, UK.

Ajzen, I. (2011) 'The theory of planned behaviour: Reactions and reflections', *Psychology & Health*, vol 26, no 9, pp. 1113–1127.

Anable, J. (2005) '"Complacent Car Addicts" or "Aspiring Environmentalists"? Identifying travel behaviour segments using attitude theory', *Transport Policy*, vol 12, no 1, pp. 65–78.

Ballantyne, R. and Hughes, K. (2006) 'Using front-end and formative evaluation to design and test persuasive bird feeding warning signs', *Tourism Management*, vol 27, no 2, pp. 235–246.

Becken, S. and Simmons, D. (2008) 'Using the concept of yield to assess the sustainability of different tourist types', *Ecological Economics*, vol 67, no 3, pp. 420–429.

Coulter, A., Clegg, S., Lyons, G., Chatterton, T. and Musselwhite, C. B. A. (2007) 'Exploring public attitudes to personal carbon dioxide emission information', Department for Transport, London, UK.

Cumbria County Council (2014) 'Cumbria's Economic Ambition: Cumbria County Council's Role in Delivering a Thriving Economy', Cumbria County Council, Cumbria, UK, www.cumbria.gov.uk/elibrary/Content/Internet/538/755/1929/6478/4133311542.pdf, accessed 24 October 2014.

Cumbria Tourism (2006) 'Cumbria Visitor Survey'.

Cumbria Tourism (2013) 'Economic Impact of Tourism – Visitor Volume and Value 2013', www.cumbriatourism.org/research/surveys-data.aspx, accessed 19 June 2013.

Dallen, J. (2007) 'Sustainable transport, market segmentation and tourism: The Looe Valley Branch Line Railway, Cornwall, UK', *Journal of Sustainable Tourism*, vol 15, no 2, pp.180–199.

DEFRA (2008). 'A Framework for Pro-Environmental Behaviours', Department for Environment Food & Rural Affairs, UK.

Department for Transport (2008) 'Carbon pathways analysis: Informing development of a carbon reduction strategy for the transport sector', London, UK.

Department for Transport (2011) 'Climate change and transport choices segmentation model: A framework for reducing CO_2 emissions from personal travel', London, UK.

Dickinson, J. E. and Dickinson, J. A. (2006) 'Local transport and social representations: Challenging the assumptions for sustainable tourism', *Journal of Sustainable Tourism*, vol 14, no 2, pp. 192–208.

Dunlap, R. E. and Van Liere, K. D. (1978) 'The New Environmental Paradigm: A proposed measuring instrument and preliminary results', *Journal of Environmental Education*, vol 9, no 4, pp.10–19.

ENPAA (2007) 'Transport Position Statement', English National Parks Authorities Association, UK.

Fisher, R. (1993) 'Social desirability bias and the validity of indirect questioning', *Journal of Consumer Research*, vol 20, no 2, pp. 303–326.

Goh, E. (2010) 'Understanding the heritage tourist market segment', *International Journal of Leisure and Tourism Marketing*, vol 1, no 3, pp. 257–270.

Gössling, S., Peeters, P., Ceron, J. P., Dubois, G., Patterson, T. and Richardson, R. (2005) 'The eco-efficiency of tourism', *Ecological Economics*, vol 54, no 4, pp. 417–434.

Ham, S., Brown, T. J., Curtis, J., Weiler, B., Hughes, M. and Poll., M (2009) 'Promoting persuasion in protected areas: A guide for managers who want to use strategic communication to influence visitor behaviour', CRC for Sustainable Tourism Pty Ltd, Australia.

IPCC (2013) 'Summary for Policymakers', in T. F. Stocker, D. Qin, G.-K. Plattner, M. Tignor, S. K. Allen, J. Boschung, A. Nauels, Y. Xia, V. Bex and P. M. Midgley (eds) *Climate Change 2013: The Physical Science Basis. Contribution of Working Group I to the Fifth Assessment Report of the Intergovernmental Panel on Climate Change*, Cambridge University Press, Cambridge, United Kingdom and New York, NY, USA.

Jones, R. (1996) *Research Methods in the Social and Behavioural Sciences*, Sinauer Associates, Sunderland, MA.

Jurowski, C., Uysal, M. and Williams, D. R. (1997) 'A theoretical analysis of host community resident reactions to tourism', *Journal of Travel Research*, vol 36, no 2, pp. 3–11.

Lindberg, K. and Johnson, R. L. (1997) 'Modeling resident attitudes towards tourism', *Annals of Tourism Research*, vol 24, no 2, pp.402–424.

Moeller, T., Dolnicar, S. and Leisch, F. (2011) 'The sustainability–profitability trade-off in tourism: can it be overcome?', *Journal of Sustainable Tourism*, vol 19, no 2, pp. 155–169.

Northcote, J. and Macbeth, J. (2006) 'Conceptualizing yield: sustainable tourism management', *Annals of Tourism Research*, vol 33, no 1, pp. 199–220.

Ong, T., and Musa, G. (2011) 'An examination of recreational divers' underwater behaviour by attitude–behaviour theories', *Current Issues in Tourism*, vol 14, no 8, pp. 779–795.

Perdue, R. R., Long, P. T. and Allen, L. (1990) 'Resident support for tourism development', *Annals of Tourism Research*, vol 17, no 4, pp. 586–599.

Phillips, B. (1976) *Social Research: Strategy and Techniques*, Macmillan Publishing Co., Inc., New York, NY.

Powell, R. and Ham, S. (2008) 'Can ecotourism interpretation really lead to pro-conservation knowledge, attitudes and behaviour? Evidence from the Galapagos Islands', *Journal of Sustainable Tourism*, vol 16, no 4, pp. 467–489.

Reigner, N. and Lawson, S. (2009) 'Improving the efficacy of visitor education in Haleakalā National Park using the theory of planned behavior', *Journal of Interpretation Research*, vol 14, no 2, pp. 21–45.

Serenari, C., Leung, Y., Attarian, A. and Franck, C. (2012) 'Understanding environmentally significant behavior among whitewater rafting and trekking guides in the Garhwal Himalaya, India', *Journal of Sustainable Tourism*, vol 20, no 5, pp. 757–772.

Sharpley, R. and Sharpley, J. (1997) *Rural Tourism: An Introduction*, International Thomson Business Press, London, UK.

Singleton, R. and Straits, B. (1999). *Approaches to Social Research*, Oxford University Press, Oxford, UK.

Stanford, D. (2006). *Responsible tourism, responsible tourists: What makes a responsible tourist in New Zealand?* (Unpublished doctoral dissertation). Victoria University of Wellington, Wellington, New Zealand.

Stanford, D. (2014) 'Reducing visitor car use in a protected area: a market segmentation approach to achieving behaviour change', *Journal of Sustainable Tourism*, vol 22, no 4, pp. 666–683.

Stern, N. (2007) *The Economics of Climate Change: The Stern Review*, Cambridge University Press, Cambridge, UK.

11 Cycle tourism development in parks

The experience of the Peak District National Park (UK)

Richard Weston, Nick Davies and Jo Guiver

Introduction

The popularity of cycle tourism has to some extent followed that of the bicycle itself. Right after the development of the safety bicycle[1] at the end of the nineteenth century, leisure cycling spread rapidly with many benefiting from this low-cost form of access to the nearby countryside. Cyclists also became more organised: in the UK the Cycle Tourism Club formed in 1878 and saw cycle tourists traveling further afield, often using the railways, as bicycles could easily be transported in guards' vans (Dickinson and Lumsdon, 2010). In the early part of the twentieth century, the bicycle became a common form of transport for working class men as mass production reduced prices and cycling became more egalitarian (Pooley *et al.*, 2013). The inter-war and immediate post Second World War years were 'La Belle Époque' for cycle tourism. During a visit to London in 1934, Lee observed that 'on fine Sunday mornings, while horses rested, Putney High Street filled up with bicycles – buxom girls in white shorts chased by puffing young men, old straw-hatted gents in blazers, whole families on tandems carrying their babies in baskets, and all heading for the open country' (Lee, 1992, p. 247).

Cycling generally declined in popularity from the middle of the twentieth century, both as a means of transport and a leisure activity, with the increase in private car ownership. By the end of the century, apart from a few notable exceptions, it had become a niche leisure activity (Dickinson and Lumsdon, 2010; Pooley *et al.*, 2013). Yet cycling remains popular in some parts of Denmark, Germany and the Netherlands as a transport mode, and in some countries, including Austria and France, for holiday purposes (Lumsdon *et al.*, 2012). Over recent decades, for example, Austrian regional and local authorities have invested in the Danube cycle route (Donauradweg), which is now one of the most popular in Europe (Dickinson and Lumsdon, 2010).

Cycle tourism can be divided into three main categories: cycle touring/cycle holidays, holiday cycling and day cycling. These are defined by the importance of cycling in the trip, the duration of the trip and, to some extent, the cycling proficiency required. The first of these categories is the one most often identified with cycle tourism though it attracts often the smallest group. It is normally undertaken by more experienced cyclists who like holidays where cycling is the principal activity. The main distinction between cycle touring and cycle holiday is that the former implies travelling from place-to-place, often following long-distance multi-day linear or (more rarely) circular routes, whereas the latter revolves around a single base from where a day's cycling normally begins and ends, sometimes with the support of public transport. The second category typically represents a greater share of cyclists, though this can vary depending on the location and the type of cycle route. Here cycling is one of a number of activities undertaken while on holiday and is therefore not the primary motivation for destination choice. The level of cycling experience required can vary within this group, but people doing holiday cycling are typically less experienced and prefer traffic-free cycle routes. The third category groups less experienced cyclists travelling from home to enjoy an easy day of cycling often in the company of friends and family. This is in almost all circumstances the largest demand segment and, given the generally low level of cycling experience, is drawn to traffic-free routes or quiet roads (Downward and Lumsdon, 2001).

This distinction is important as it impacts strongly on the choice of destination for cycling. Studies suggest that cycle tourists are generally motivated by pleasant surroundings, such as open countryside and wildlife, while enjoying mild exercise (Lumsdon *et al.*, 2012). Converted disused railway-lines are popular as they offer moderate slopes and a variety of vistas from cuttings, embankments and viaducts, and are generally constructed as greenways exclusively for non-motorised users, cyclists, walkers, horse riders and disabled users. Infrastructure is the key factor in encouraging cycle tourism: consistent investment pays long-term dividends. This was very well understood in Spain, for example, where the Spanish greenways, or Vias Verdes,[2] program was started in 1993 by the Spanish Ministry of Public Works, Transport and Environment, in partnership with the two state railway companies (the former RENFE and FEVE, now integrated in ADIF) and the Ferrocarriles de Vía Estrecha (narrow gauge railways). Together they created the Spanish Railways Foundation (FFE), which would be responsible for the development of the program. An inventory of the disused railway lines, buildings, bridges and viaducts identified over 7,500 kilometres of disused railway lines and almost 1,000 stations. Using the greenway concept developed in the UK and USA as a benchmark, in 20 years the FFE in partnership with local and regional organisations created over 100 greenways, totalling around 2,000 kilometres of traffic-free routes, which are used to promote active tourism and a healthy lifestyle for the local population. Additionally, over 70 railway stations have been refurbished providing accommodation, refreshments, bike rentals and other cultural facilities.

In Switzerland, a national network for non-motorised traffic has been established promoting active forms of travel for leisure and tourism. The Veloland network, which saw 3.3 million users in 1999 (only one year after it was launched), was later renamed SchweizMobil and broadened its target market to include hiking, mountain biking, skating and canoeing in addition to cycling (Lumsdon *et al.*, 2012). One of the main reasons for its success has been the cooperation between stakeholders, including federal departments, cantonal offices, local authorities, the Principality of Liechtenstein and the various non-motorised traffic specialist organisations, such as the Cycling in Switzerland Foundation. The creation of a network of national and regional routes, bicycle rental schemes and accommodation providers as well as the work with public transport companies have also encouraged multi-modal travel with bicycles carried on trains, buses and boats.

This chapter explores the benefits of developing cycle tourism in national parks as a strategy for improving access to and mobility within parks while mitigating some of the impacts normally associated with leisure travel, such as traffic congestion, atmospheric pollution and noise (Mundet and Coenders, 2010). In the United Kingdom (UK), the establishment of national parks was originally driven by the two potentially conflicting objectives of conserving (and enhancing) natural beauty, wildlife and cultural heritage, and promoting opportunities for the understanding and enjoyment of the special qualities of natural areas. Public access and landscape preservation are not always complementary activities: when conflict does arise, there is a general principle that conservation takes priority. The 1995 Environment Act further added to the duties of national parks, requiring them to seek to foster the economic and social well-being of local communities within the parks, expanding areas of potential conflict. Among other protected areas, the Peak District National Park (PDNP) has a history of promoting cycling and has recently been successful in obtaining funding to further strengthen its position as a popular cycling destination. This is the context that will be analysed in this chapter.

Study area

The Peak District National Park is located at the geographic centre of England and was the first UK national park to be officially designated in 1951. It covers around 1,438 square kilometres of mostly upland areas and can be divided into three areas: the White Peak, Dark Peak and South West Peak. The White Peak's limestone plateaus and rolling dales are home to the park's main settlements. The area is mainly grassland used for dairy farming, with some broadleaved woodland cover and small chalk stream and rivers. The Dark Peak, which is much less populated than the White Peak, presents grit stone outcrops, upland heath and bogs that are more suited to hill farming, with some grouse shooting occurring on the uplands. The area also forms the southern end of the Pennine Mountains and many of the local valleys have been flooded to create reservoirs supplying water to the surrounding urban areas. The South West Peak, which is also sparsely populated, is a mixture of upland moor and lowland pasture, with mixed stock farming use.

The PDNP lies predominantly within the county of Derbyshire, but also covers parts of Staffordshire, and Cheshire. It is surrounded by a number of industrial towns and cities, including Manchester, Sheffield, Stoke-on-Trent, Derby and Nottingham and as a result over 16 million people live within one hour's drive of the park's edge (Figure 11.1). The park attracts over ten million visitors annually with 85 per cent of them arriving by car, creating over four million car journeys every year.[3]

The PDNP has a resident population of around 38,000. Almost 90 per cent of local inhabitants have access to a car and the average car ownership is 1.6 per

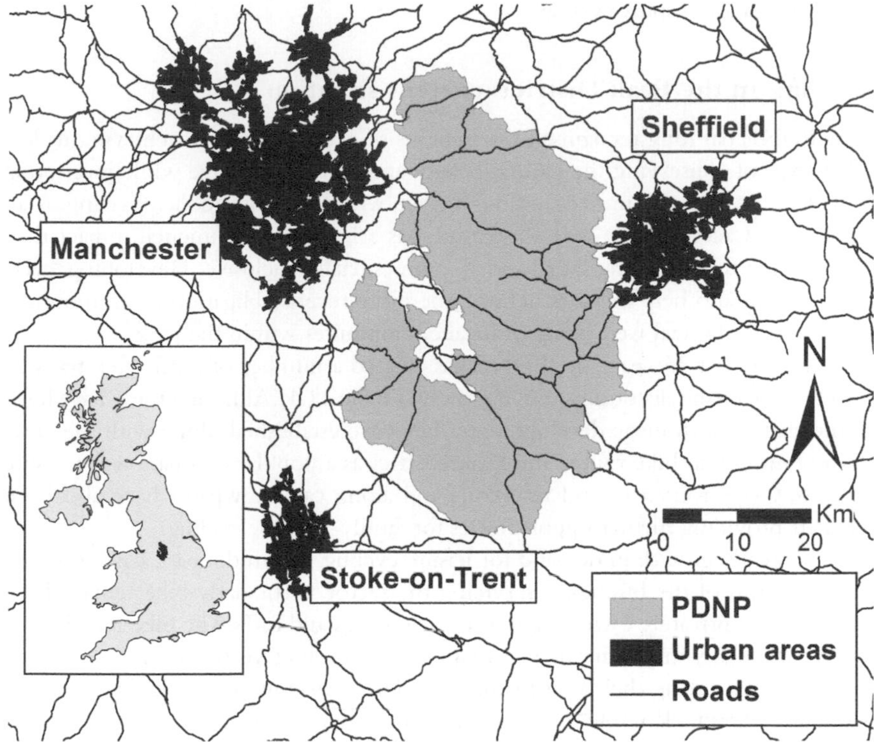

Figure 11.1 The Peak District National Park (PDNP), located at the geographic centre of England, is surrounded by various cities including Manchester, Sheffield and Stoke-on-Trent.

household, compared to an average of 75 per cent and 1.2 respectively for the rest of the country. Recent years have seen the average age of the park's population rising, which is possibly one of the causes of increased car ownership. The combined impact of private car use by residents and visitors to the park has made traffic flows almost double over the last 30 years and as a result there is limited capacity for further growth, either in traffic levels or car parking.

The PDNP is crossed east-west by rail lines connecting the west coast and midland main lines, which provide fast access to London: two hours from Manchester and one and a half hours from Derby. Many of the cities surrounding the park will have their rail connections improved as part of the 'Northern Hub' development, strengthening access and shortening journey times.

Beside the traditional tourism directed to the honeypot sites, such as the market towns of Ashbourne and Bakewell, there is a long history of developing and promoting active tourism in the park. Recreational walkers constitute the backbone of this, but both cycling and rock climbing are popular in the park. Cycling especially has enjoyed increase in popularity following the recent successes of national teams and individuals in sport cycling.

Cycling in the Peak District: current situation

The PDNP has long recognised the benefits of cycle tourism in achieving its key objectives of conserving the natural beauty and cultural heritage while promoting opportunities for the understanding and enjoyment of the public. Cycling as an activity and access mode is less intrusive, has a lower environmental impact than most other forms of transport and is more socially inclusive. Developing cycle tourism can also help the park achieve the more recent obligation to promote the economic and social well-being of local communities within the park.

Over the last 35 years, the PDNP has created a number of traffic-free trails by mostly converting decommissioned railways (Table 11.1). Although these have been somewhat opportunistic developments, they have established, along with the creation of mountain bike routes, the Peak District as a popular destination for leisure cyclists. Cycle routes around reservoirs and along canal towpaths have also been created, providing further opportunities for family-friendly cycling.

The steady growth in demand for leisure cycling within the park has supported an expansion of the businesses servicing the sector, particularly bike rentals (both publicly and privately owned), but also cycle shops and cafés. The bike rentals have been particularly important in encouraging 'non-cycling' visitors to try out cycling as an activity during their visit to the park. The availability of traffic-free trails has of course played a key role in this, as safety is an important issue for inexperienced cyclists (Pooley *et al.*, 2013).

Over the past decade, the PDNP has also developed an electric bike network covering much of the southern and central areas of the park to foster cycling among inexperienced visitors. The network provides charging points at a number of popular attractions, such as cafés, as well as in the market towns, which act as

Table 11.1 Traffic-free trails in the Peak District National Park.

Name	*Length*	*Established*
The Manifold Trail	14 kilometres	1937
The Tissington Trail	21 kilometres	1971
The High Peak Trail	28 kilometres	1971
The Monsal Trail	14 kilometres	1981
The Transpennine Trail	26 kilometres (within the PDNP)	2001
The Thornhill Trail	2.4 kilometres	1994

hubs. The PDNP also has a wealth of bridleways,[4] green lanes[5] and quiet lanes suitable for mountain bikers, and three challenging road climbs popular among sport or club cyclists, one of which featured in the 2014 Tour de France Grand Départ. These elements provide visitors with opportunities to leave their cars at home or holiday accommodation and explore the park by bike.

As suggested earlier, the development of the traffic-free trails within the PDNP has been opportunistic as it largely occurred where existing infrastructure had become redundant and when funding was available. While this has delivered a number of popular trails, they are often disconnected, both from each other and from urban areas. The lack of connection between trails has reduced the opportunity to create circular rides, especially for families and less experienced cyclists, and in some cases has encouraged the illicit use of footpaths leading to conflict with other users.[6] Even where the trails are in close proximity to each other, there is a general lack of signposts that thwarts visitors willing to undertake circular rides using quieter public roads.

One of the main disadvantages of utilising disused railway lines for cycle trails is that they can create a 'corridor mentality'. In fact, while segregation was important for safety reasons when in use as an operational railway, it often reduces the economic and social benefits to communities near the cycle trails, as cyclists are either unsure of the facilities available nearby or discouraged from visiting them as they have to leave the trail often using relatively busy public roads. This in turn has had an impact on the awareness of some associated service providers, and has resulted, for example, in relatively few taking up the 'Cyclists Welcome' accreditation within the PDNP, despite its popularity as a cycling destination.

Urban areas in and on the edge of the park could act as hubs encouraging the use of public transport, but the poor connection to these areas encourages the use of private cars to access the park and travel within it, conceivably increasing traffic levels rather than reducing them.

All of the above issues, combined with the poor connection of the park with surrounding urban areas, have created a situation where around 85 per cent of visitors arrive by car. This clearly generates congestion and conflict between cyclists and drivers on some routes at peak hours, especially where trails enter urban areas. Cycling could be a popular activity in the shoulder months, the early and late parts of the holiday season, but the lack of awareness within the local business community of the cycle tourism market has resulted in a highly peaked season, with low off-season demand.

Cycling in the Peak District: future perspectives

The PDNP was keen to build on its existing cycle tourism product. In 2009, the park received a grant of £2.5 million from Cycling England[7] and the Department for Transport to fund a project called 'Pedal Peak District'. The project included both hard and soft measures including reopening of four tunnels along the Monsal Trail to create a continuous traffic-free route of 14 kilometres, cycle training and bike maintenance. To ensure that the momentum was maintained, the PDNP began to develop a cycling strategy in early 2013 and held the Peak District Cycle Summit, inviting participants from local authorities (including those in neighbouring areas), national government (Department for Transport), cycling organisations and other third sector groups. This inclusive forum recognised that not only did the authority need to achieve a consensus supporting future plans but that the opportunities for funding were likely to be diverse (e.g. facilities designed primarily for utility cycling could also be used in a leisure context and vice versa). The key objective of the summit was to develop a cycle strategy for the wider PDNP that could link the park to the surrounding areas. Hence, two main tasks were set: to map all the gaps in the network, both within and outside the park, and to develop ideas that would improve cycling within the park, such as road crossings, cycle storage, etc. The delegates were then asked to rank these ideas according to two criteria: deliverability and impact. This process helped prioritise ideas for a new funding application.

The Wider Peak District Cycle Strategy (WPDCS) identified five areas where cycling could benefit the Wider Peak District and the park itself:

- Economic: the initial investment in cycling encourages visitors to spend more in the local economy.
- Health and Well-being: in addition to the individual physical and mental health benefits from cycling, there are wider health benefits from reduced vehicle emissions and noise pollution.
- Community: additional leisure opportunities, increased travel sustainability and more socially equitable access to employment and other facilities.
- Transport: improved opportunities for both the local communities and visitors to travel on foot and by bike while reducing congestion in some of the busiest areas in the park.
- Personal discovery, fun and development: opportunities to discover the surrounding countryside in safety, particularly areas that would not normally be accessible.

The strategy creates a hierarchy of main, secondary and complementary routes. The main routes will form the backbone of the network connecting to surrounding towns and cities, while the secondary routes will connect market towns, railway stations, residential areas and key attractions to the main network. The complementary routes will support the main and secondary networks with connections to other places. The development of the main and secondary routes will be prioritised, while complementary routes will be realised as opportunities arise.

In particular, the WPDCS sets out four themes in order to achieve the park's ambition:

1. Increase the network of routes,
2. Support cyclist-friendly infrastructure to stimulate the cycling economy,
3. Promote the Peak District cycle experience, and
4. Develop sustainable transport packages.

In 2013, the PDNP, together with four local authorities, submitted a bid for £5 million, with £2.5 million match funding, to the Department for Transport's

'Linking Communities – Grants to support cycling in National Parks' initiative. The 'Pedal Peak Phase II – Moving Up A Gear' is a combination of both hard and soft measures to develop cycling. Four new cycle routes will be created and a Cycle Friendly Places Grant Fund, open to community organisations as well as local businesses, will be established to improve provision for recreational cycling in and around the park. The new routes will begin the process of connecting some of the existing routes, both to each other and to urban areas, providing direct cycle access to Sheffield and Stoke-on-Trent on a combination of on-road and traffic-free routes, and direct rail links to Manchester, Sheffield and Derby. These improvements will allow 3.5 million people direct rail or cycle access to the park.

Discussion

The recent cycling success of UK athletes, both in the Olympics and the Tour de France, has stimulated a resurgence of interest in cycling, especially for leisure purposes. Besides that, visitors to parks are getting more aware of the environmental impacts of tourism, this boosting the demand for lower impact activities. The combination of an increased interest in active tourism (driven by a desire to increase personal health and well-being) and the growing interest in the natural environment suggests that it is a good time to be encouraging further development of cycle tourism within the PDNP.

While many of the current trails are at capacity, particularly during peak times, considerable potential exists away from these periods. There are also opportunities to add new trails to the existing offer creating a wider network, including connections to the neighbouring urban areas around Manchester and South Yorkshire. The expansion of the network would not only relieve congestion on the busiest trails, but has the potential to mitigate car traffic by encouraging some visitors to use the bicycle to reach the park. The expanded network would also allow multi-modal journeys, with visitors able to travel by train and bike, and would increase economic and social benefits.

However, it is important to maintain a reasonable balance between an increase in demand and the availability of cycling offers, such as new traffic-free trails and cycle-friendly accommodations. The PDNP is already popular with cyclists, but allowing demand to outstrip supply could increase the areas of conflict and have a longer-term detrimental impact on this popularity, the viability of future development plans and eventually the goodwill of local communities.

The development of cycle trails and other forms of countryside access in the UK are often confronted with resistance from other users, particularly landowners. The majority of land in the UK is privately owned and any new access arrangement has often to be negotiated with the owner and others who may have an interest, such as tenant farmers. Some elements of trail development may require planning consent from local authorities. The development could be open to scrutiny from the local population, as some people are resistant to change, particularly if this is likely to impact on their lives (e.g. noise or disturbance during construction, excessive presence of visitors). There are also special interest groups who may need to be consulted, such as wildlife organisations or sport groups (e.g. fishing or shooting clubs). These can all cause additional delays to the planning and development process, and in some cases prevent it from taking place. However, the PDNP has encouraged these groups to engage with the planning process at a relatively early stage, mitigating some of the risk.

Other regions of the UK and national parks have also recognised the potential economic, social and environmental benefits of cycle tourism: the market is already competitive and is likely to be increasingly so. However, the PDNP has a long history of developing cycle routes and has been actively supported by Derbyshire County Council: the local authority in which most of the park lies. Over the last three decades, a good all round cycling offer, both on and off road for all levels of proficiency, has been developed, contributing to make the PDNP a cycling holiday destination.

In terms of delivering the aims of the WPDCS, the new Pedal Peak project will help fill the gaps in the existing network. This will reduce congestion at some of the current popular access points, particularly in the urban areas, and will create some extended riding possibilities for those who stay longer in the park. Nonetheless, the discontinuity of the new traffic-free sections will remain a barrier to the less experienced cyclists who dislike the considerable slopes and traffic levels of the park's roads.

The improved connections should encourage more cyclists to start and end their journeys in the market towns, as this would raise awareness among the business community about the role of cycle tourism for the local economy. Anyway, the PDNP will need to implement further soft measures, such as encouraging accommodation providers to obtain 'Cyclists Welcome' accreditation, as well as hard measures, such as clear signing to cycle-friendly businesses, to ensure the economic benefits are fully realised.

Increasing the potential for multi-modal journeys, particularly train and bike on the newly formed White Peak Loop, should be another priority to help reduce the number of car-based trips. Nonetheless, at present, this is thwarted by the limited bike load capacity (i.e. two bicycles per train) on some of the older rolling stocks operating into the two market towns that are currently the main access stations for the PDNP. A bus service with the possibility to carry bicycles, such as that operating between the traffic-free sections from Stoke-on-Trent, is also an option, but tourists may have problems accessing this type of service.

Conclusion

The Pedal Peak Phase II project attempts to strike a balance between the often conflicting objectives of the PDNP, namely encouraging public access and preventing the negative environmental impacts of such access. The project has also aimed to promote the economic and social well-being of local communities, and this often means encouraging tourism, as other opportunities for employment are limited. Cycling, along with other forms of active tourism, has a relatively low environmental impact, but can offer significant economic and social benefits if well managed.

The presence around PDNP of large urban conurbations that are home to almost a quarter of the UK's population suggests that access to the park will be an ongoing issue. The initiatives so far must be seen as a work in progress, but useful lessons can already be taken from them. First, long-term investment in traffic-free cycle routes has seen the park develop into a popular cycle tourism destination encouraging economic development beyond the traditional honeypot sites, while reducing some of the environmental issues commonly associated with tourism (e.g. pollution, noise). Second, by recognising that the park authority is not an isolated entity and involving other stakeholders early in the process, the PDNP has overcome much of the potential resistance to its plans and has in fact been actively supported with significant resource commitments. Finally, by developing a strategy that clearly sets out the current position, the long-term objectives and the benefits that will derive from the strategy, the park has successfully attracted funding to achieve its goals.

Notes

1 Unlike its predecessors, on the safety bicycle the rider's feet were near to the ground and the pedals drove the back wheel.
2 Literally translates to 'green route'.
3 Based on an average car occupancy of two (Department for Transport, 2013).
4 Similar to public footpaths but also permits horse riding and bicycles.
5 Unsurfaced roads open to motorised traffic.
6 Footpaths are generally reserved for walkers and pedestrians and in some cases fines can be issued for improper use.
7 Now abolished, Cycling England was an independent body funded by the Department for Transport to promote cycling in England.

References

Department for Transport (2013) *National Travel Survey: 2012*, Department for Transport, London, UK.

Dickinson, J. and Lumsdon, L. (2010) *Slow Travel and Tourism*, Earthscan, London, UK.

Downward, P. and Lumsdon, L. (2001) 'The development of recreational cycle routes: an evaluation of user needs', *Managing Leisure*, vol 6, pp. 50–60.

Lee, L. (1967) *Red Sky at Sunrise*, Penguin Books, London, UK.

Lumsdon, L., Weston, R., McGrath, P., Davies, N., Peeters, P., Eijgelaar, E., and Piket, P. (2012) 'The European cycle route network EuroVelo: challenges and opportunities for sustainable tourism', European Parliament, Brussels, Belgium.

Mundet, L. and Coenders, G. (2010) 'Greenways: a sustainable leisure experience concept for both communities and tourists', *Journal of Sustainable Tourism*, vol 18, no 5, pp. 657–674.

Pooley, C., Jones, T., Tight, M., Horton, D., Scheldeman, G., Mullen, C., Jopson, A. and Strano, E. (2013) *Promoting Walking and Cycling: New Perspectives on Sustainable Travel*, Policy Press, Bristol, UK.

12 Estimating the effects of 'carrot and stick' measures on travel mode choices

Results of a survey conducted in the Dolomites (Italy)

Francesco Orsi and Davide Geneletti

This chapter is a revised version of a paper that first appeared in 2014 in the *Journal of Transport Geography* (vol 39, pp. 21–35).

Introduction

The need to preserve natural resources and recreational quality in the face of increasing visitation rates is encouraging managers of natural and protected areas around the world to consider alternatives to private motorised vehicles for transportation. In fact, alternative transportation systems (ATS) such as buses, trains and cableways may dramatically reduce the environmental and socio-economic issues associated with automobiles (e.g. noise, pollution, traffic congestion), and allow park managers and administrators to have a thorough control over visitor flows across an area (White *et al.*, 2011; Reigner *et al.*, 2012).

Yet various studies have emphasised that the actual success of an ATS depends on the quality of the service provided (e.g. cost, wait times) (Lumsdon, 2006; Lumsdon *et al.*, 2006; Shiftan *et al.*, 2006; Mace *et al.*, 2013), the kind of management measures adopted in an area (e.g. road closures) and eventually the conditions experienced by the user (e.g. traffic congestion) (Regnerus *et al.*, 2007). In particular, there is a common understanding that satisfactory modal shifts from traditional to alternative transportation can only be achieved if measures to encourage the use of ATS (e.g. multi-day passes, free tickets) are coupled with measures to discourage car use (e.g. road tolls, road closures) (Holding and Kreutner, 1998; Cullinane and Cullinane, 1999; Stradling *et al.*, 2000; Holding, 2001). This combination of incentives to ATS and disincentives on private motorised transportation generally goes under the name of 'carrot and stick' approach (Holding and Kreutner, 1998).

Shaping 'carrots' and 'sticks' that guarantee high ATS ridership and no (or minimal) losses to an area's tourist inflow is difficult because it requires an in-depth knowledge of the factors affecting visitors' choices between different transport options. In other words, managers and administrators need to understand how visitors respond to transport characteristics and management conditions in order to define measures that maximally benefit nature and society. Several authors have investigated the attitudes of visitors towards different transportation options to support the design of ATS and the definition of transportation management strategies (Steiner and Bristow, 2000; Shiftan *et al.*, 2006; Kelly *et al.*, 2007; White, 2007; Pettebone *et al.*, 2011).

However, while these studies have considered the case of multiple transport modes serving the same location, none has analysed the case of multiple modes serving multiple locations within an area. This case is in fact very interesting because mode choice may result in a profound redistribution of visitors across different access points, with notable effects on the local environment and visitor experience. In Europe, several Alpine resorts give visitors the possibility to reach hiking areas through a variety of transportation modes other than automobiles (e.g. shuttle buses, rack and pinion trains, cableways). Each of these options is unique in terms of environmental impact, transit characteristics (e.g. capacity, speed) and location served (e.g. buses reach low elevation trailheads, whereas cableways reach high elevation ones). Hence, shifting visitors from one to another may significantly modify visitor flows and bring consequences on the local environment (Lawson *et al.*, 2009).

This chapter presents a study aimed at investigating how 'carrot and stick' approaches may affect mode choice and subsequently patterns of visitor access to a hiking area served by multiple alternative transportation modes. The study is based on a stated preference survey conducted in a popular area of the Dolomites UNESCO World Heritage Site (Italy) that allowed the estimation of visitors' sensitivity to a set of management (e.g. cost, convenience, accessibility) and contingent (e.g. traffic, crowding) factors. The findings of this study can help administrators in the Dolomites shape 'carrot and stick' measures that favour shift from automobiles to ATS while ensuring that visitor flows are balanced across an area.

Study area

The study analysed the access of hikers to the southern part of the Catinaccio-Rosengarten range, in north-eastern Italy (Figure 12.1). The Catinaccio-Rosengarten range, along with the adjacent Latemar and Sciliar-Schlern ranges, covers an area of approximately 9,300 hectares and constitutes one of the nine systems of the Dolomites UNESCO World Heritage Site, which was established in 2009. The Catinaccio-Rosengarten range develops on a north-south direction and features steep rock faces and pinnacles of incomparable beauty. Its southern part is easily accessible from the villages of Vigo di Fassa (1382m) on the eastern side and Carezza-Karersee (1620m) on the western side via both road and cableways.

The national road 241 connects the two villages through the Costalunga-Karer Pass (1750m), which lies just a few hundred metres south of the range. A cable car connects Vigo to the Ciampedie Hut (2000m), while a chairlift allows a fast transfer from Carezza-Karersee to the Paolina Hut (2125m). The Costalunga-Karer Pass, the Ciampedie Hut and the Paolina Hut constitute great access points for hikes on the trail network of the range. Good accessibility, excellent tourist infrastructures and renowned natural beauty determine the success of this area, which receives thousands of visitors a day during the high season (July and August). The inscription

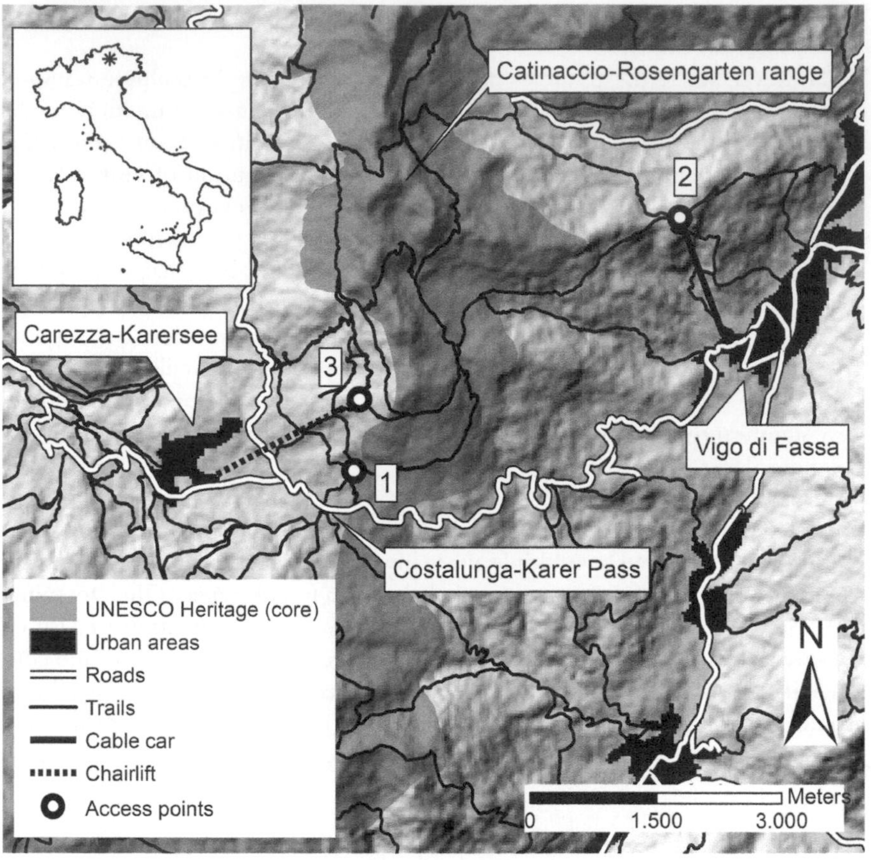

Figure 12.1 The southern part of the Catinaccio-Rosengarten range in the Dolomites (north-eastern Italy). The area has three main access points: the trailhead at the Costalunga-Karer Pass (1), the Ciampedie Hut (2) and the Paolina Hut (3) (source: adapted from Orsi and Geneletti, 2014).

onto the UNESCO World Heritage list is expected to further increase the popularity of this area and calls for transport-related measures that help maintain the outstanding value of the environment.

Method

The survey was designed considering that a visitor arriving in the gateway villages of Vigo di Fassa or Carezza-Karersee has three direct transport options for reaching the southern side of the Catinaccio-Rosengarten: driving a personal vehicle to the Costalunga-Karer Pass, riding a bus to the Costalunga-Karer Pass or using a cableway to either the Ciampedie or the Paolina Hut. The choice of one option depends on a variety of factors pertaining to the option's intrinsic characteristics (e.g. flexibility, location served), as well as to management (e.g. fares, access restrictions) and experiential (e.g. traffic, crowding) conditions.

Consistent with the findings of Shiftan *et al.* (2006) and Pettebone *et al.* (2011), five attributes were selected to describe both management and experiential conditions associated with the three transport options: cost (i.e. economic cost of an option), convenience (i.e. intrinsic ease of using an option), accessibility (i.e. daily time span over which an option is available), traffic (i.e. use level of an option) and crowding (i.e. density of visitors on the trail). Three levels were considered for each attribute (Table 12.1). Levels of attributes describing experiential conditions (i.e. traffic, crowding) were represented through digitally edited photographs, as proposed by various scholars (Manning and Freimund, 2004; Pettebone *et al.*, 2011). Considering a fractional factorial design with 72 choice sets and eight blocks, respondents were presented with nine choice sets. Each of these included the three main alternatives (i.e. drive, ride the bus, take the cableway) and a fourth option,

Table 12.1 Attributes and levels considered in the choice experiment.

	Drive	*Ride the bus*	*Take the cableway*
Cost	– No toll – Toll: 15 € – Toll: 30 €	– Free bus – Return ticket: 3 € – Return ticket: 6 €	– Return ticket: 5 € – Return ticket: 15 € – Return ticket: 25 €
Convenience	– Car parked immediately – 10 minutes to park the car – 20 minutes to park the car	– Bus passes every 20 minutes – Bus passes every 40 minutes – Bus passes every 60 minutes	– Car parked immediately – 10 minutes to park the car – 20 minutes to park the car
Accessibility	– Car access always possible – Access forbidden 10am–4pm – Access forbidden 8am–6pm	– Schedule: 6am–8pm – Schedule: 8am–6pm – Schedule: 10am–4pm	– Schedule: 6am–8pm – Schedule: 8am–6pm – Schedule: 10am–4pm
Traffic	– Photo showing no traffic – Photo showing medium traffic – Photo showing intense traffic	– Photo showing no queue – Photo showing short queue – Photo showing long queue	– Photo showing no queue – Photo showing short queue – Photo showing long queue

(*continued*)

Table 12.1 Attributes and levels considered in the choice experiment (*continued*).

	Drive	*Ride the bus*	*Take the cableway*
Crowding	– Photo showing no people on the trail – Photo showing 4 people on the trail – Photo showing 8 people on the trail	– Photo showing no people on the trail – Photo showing 4 people on the trail – Photo showing 8 people on the trail	– Photo showing no people on the trail – Photo showing 4 people on the trail – Photo showing 8 people on the trail

Source: Orsi and Geneletti, 2014

allowing respondents to skip the choice in case no travel option was satisfactory enough (Figure 12.2). The fourth alternative was presented as a 'no go' option (i.e. 'I stay at home') to reduce the bias associated with the respondent considering an imaginary travel option whose attributes would be unknown to the interviewer.

The survey was conducted over ten non-consecutive days in July and August 2012 at the three access points: a trail junction a few minutes walk from the Costalunga-Karer Pass, the Ciampedie Hut and the Paolina Hut. Here hikers willing to participate were provided with a questionnaire including a demographic part (i.e. origin, travel mode, accommodation, etc.) and the nine choice sets. Only one questionnaire was given to groups of more than one person so that responses could reflect the group's opinion (Galilea and Ortuzar, 2005).

The statistical analysis of responses was based on the consideration that the utility of the three main alternatives was given by an alternative specific constant and a linear combination of attribute parameters and levels, whereas the utility of the fourth alternative was given by an alternative specific constant only. The inclusion of a constant for the fourth alternative was aimed at estimating preference for the 'no go' option. The difference in utility formulation and intrinsic attractiveness between the three travel options and the 'no go' option suggested that the random component of utilities would not be identically and independently distributed (IID) across all alternatives, thus eliminating the conditions for the application of the multinomial logit (MNL) model. Instead, a nested logit (NL) model was applied that considers the IID assumption being valid only within the three travel options.

NL models have been repeatedly applied in both transportation (Train, 1980; Hensher and Greene, 2002) and tourism (Kelly *et al*, 2007; Thiene and Scarpa,

Option A ☐	Option B ☐	Option C ☐	Option D ☐
I drive to Passo Costalunga	**I take the bus Vigo-Passo Costalunga** or **I take the bus Carezza-Passo Costalunga**	**I take the cable car Vigo-Ciampedie** or **I take the chairlift Carezza-Paolina**	**I stay at home**
No toll	Round trip: **6 €** (per person)	Round trip: **15 €** (per person)	
Parking space **found immediately**	Bus passes **every 60 minutes**	Parking space **found immediately**	
Car access **forbidden: 8am to 6pm**	First journey: **8am**; Last journey: **6pm**	First journey: **6am**; Last journey: **8pm**	
Traffic on the road: 	Crowding at the bus stop: 	Crowding at the lift station: 	
Crowding on the trail: 	Crowding on the trail: 	Crowding on the trail: 	

Figure 12.2 Example of a choice set presented to respondents (source: Orsi and Geneletti, 2014).

2008) studies. Given a nested structure, the probability that alternative j from nest m is selected is equal to the product of the probability of selection of nest m out of all nests and the probability of selection of alternative j from within nest m, as follows:

$$P_j = P_m \, P_{j|m} \tag{1}$$

The probability that nest m is selected is

$$P_m = \frac{\exp(\lambda_m IV_m)}{\sum_{n=1}^{M} \exp(\lambda_n IV_n)} \tag{2}$$

The probability that alternative j is selected from within nest m is

$$P_{j|m} = \frac{\exp(V_j / \lambda_m)}{\sum_{i \in m}^{M} \exp(V_j / \lambda_m)} \tag{3}$$

The utility of the nest is called inclusive value (*IV*) or logsum (Hensher and Greene, 2002):

$$IV_m = \ln \sum_{i \in m} \exp(V_i / \lambda_m) \tag{4}$$

The nest parameter λ_m, which has to be estimated, provides information about how correlated error terms in the NL model are. The model was estimated first considering all observations in the sample and then considering two subsets of the entire sample: respondents interviewed at the pass and respondents interviewed at cableway stations.

The likely effects of different 'carrot and stick' measures on visitor access and travel patterns were explored through analysis of the potential choices of 10,000 virtual visitors as a consequence of eight management strategies. These were obtained by modifying the cost of the three alternatives, the frequency of the bus and the opening hours of the road, and fixing the time required to find a parking space (10 minutes), the opening hours of ATS (8 am–6 pm), traffic on the road (intermediate level), queues at bus stops and cableway stations (intermediate level) and crowding on the trail (intermediate level), as specified in Table 12.2.

The peculiar behaviour of each virtual visitor was simulated considering that its sensitivity to a specific attribute was given by the sum of the parameter estimate for that attribute and a random value comprised in the range of the attribute's standard error (i.e. ± standard error). This was done in a spreadsheet containing the parameter estimates, attribute values (to be changed in accordance to the eight strategies) and 10,000 rows (i.e. one per virtual visitor) reporting the random error terms. An additional random term X was assigned to each virtual visitor to identify its likely choice, as follows:

$$\text{Choice} = \text{if}(X < P1, 1, \text{if}(X < P1 + P2, 2, \text{if}(X < P1 + P2 + P3, 3, 4))) \tag{5}$$

where P1, P2, P3 and P4 are the probabilities of the four alternatives (drive, bus, cableway, none). The 'if' condition ensures that the alternative with the highest utility has the highest probability of being selected (not that it is certainly selected), consistent with random utility theory.

Results

A total of 223 questionnaires were collected, of which 97 (43.5 per cent) were at the Costalunga-Karer Pass, 58 (26 per cent) at the Ciampedie Hut and 68 (30.5 per cent) at the Paolina Hut. Once incomplete questionnaires were removed from the sample, the total number of completed choice sets was 1,817. Table 12.3 shows parameter estimates obtained considering all observations. The value of the nest parameter significantly different from 1 is consistent with the decision to use

Table 12.2 The eight management strategies considered for the estimation of mode choices through simulation. Each strategy is given by specific management conditions regarding road toll, road closure, bus fare, bus frequency and cableway fare.

Strategy	*Drive*	*Bus*	*Cableway*
1	No road toll Road always open	Ticket: free Frequency: 20 minutes	Ticket: 5€
2	Road toll: 15€ Road always open	Ticket: free Frequency: 20 minutes	Ticket: 5€
3	No road toll Road closed 10 am–4 pm	Ticket: free Frequency: 20 minutes	Ticket: 5€
4	Road toll: 15€ Road always open	Ticket: 3€ Frequency: 20 minutes	Ticket: 5€
5	No road toll Road closed 10 am–4 pm	Ticket: 3€ Frequency: 20 minutes	Ticket: 5€
6	Road toll: 15€ Road always open	Ticket: free Frequency: 40 minutes	Ticket: 15€
7	No road toll Road closed 10 am–4 pm	Ticket: free Frequency: 40 minutes	Ticket: 15€
8	Road toll: 15€ Road always open	Ticket: 3€ Frequency: 40 minutes	Ticket: 15€

the NL model. While most parameters were statistically significant ($p < 0.05$), standard errors were quite high in some cases, this denoting a considerable variance from the estimated mean.

Alternative specific constants (the one for the first alternative was set to zero) emphasise a general preference for the cableway over the other alternatives. Cost parameters have all negative signs (i.e. utility decreases for increasing cost) and denote a lower willingness to pay for the bus rather than for the private vehicle or the cableway. Time to find a parking space is perceived as a negative factor, and its effect on utility does not change much whether such space is sought at the pass or cableway stations. The utility of the bus option clearly decreases for increasing times between one ride and the next one. Accessibility is an important issue for visitors, as emphasised by the strong impact of restrictive conditions (i.e. road closed during most of the day, cableways open only in the central part of the day) on utility. Moderately restrictive conditions have almost no impact on the utility of the drive option, but a very positive one for the bus and cableway

Table 12.3 Coefficients estimated considering all observations in the sample.

Parameter	β	*Std. Error*
ASC bus	0.487	0.265
ASC cableway	0.857**	0.272
ASC no go	−1.9**	0.191
λ	0.677**	0.101
Cost		
Toll	−0.1**	0.016
Bus fare	−0.154**	0.034
Cableway fare	−0.135**	0.021
Convenience		
Time to find parking space at the pass	−0.05**	0.012
Bus frequency	−0.019**	0.005
Time to find parking space at the cableway	−0.043**	0.012
Accessibility		
Free car access	0.691	
No car access 10 am–4 pm	0.027	0.128
No car access 8 am–6 pm	−0.718**	0.17
Bus schedule 6 am–8 pm	0.413	
Bus schedule 8 am–6 pm	0.467**	0.117
Bus schedule 10 am–4 pm	−0.88**	0.153
Cableway schedule 6 am–8 pm	0.146	0.121
Cableway schedule 8 am–6 pm	0.242	0.132
Cableway schedule 10 am–4 pm	−0.388**	
Traffic		
No traffic	0.102	
Medium traffic	0.356*	0.138
Intense traffic	−0.458**	0.148
No queue at the bus stop	0.326	
Short queue at the bus stop	−0.085	0.102
Long queue at the bus stop	−0.241**	0.106
No queue at the cableway station	0.314	0.119
Short queue at the cableway station	−0.077	0.123
Long queue at the cableway station	−0.237	
Crowding		
0 people	0.339	
4 people	0.099	0.066
8 people	−0.438**	0.09
Number of observations	1817	
Final log-likelihood	−2011.11	
ρ^2	0.202	
Adjusted ρ^2	0.192	

*$p < 0.05$
**$p < 0.01$
Source: Orsi and Geneletti, 2014

options, whereas total freedom considerably increases the utility of the drive option, but not as much that of the bus and cableway options. Road traffic is a serious issue only when particularly intense, while queues have a great impact on the decision of using the bus or the cableway (i.e. short queues very attractive, long queues highly discouraging). The presence of other hikers on the trail is also an important factor in the selection of a travel option, with no other people constituting a strong incentive to go and many other people representing a strong disincentive. The presence of few other people is somehow accepted in this area.

The analysis of the two subsets of the overall sample based on the access point (i.e. visitors interviewed at the pass vs. visitors interviewed at cableway stations) shows that, all other conditions being equal, visitors reaching road-accessible trailheads (road-oriented visitors) are less inclined to the use of the bus than are visitors reaching cableway-accessible trailheads (cableway-oriented visitors) (Table 12.4). However, all of them prefer cableway to other travel modes. Road-oriented visitors are more sensitive to bus and cableway fares than they are to a road toll, while cableway-oriented visitors are equally sensitive to the cost of the three options. People preferring road-accessible trailheads see any limitation on road accessibility as a concern and are in favour of an extended bus schedule (i.e. 6 am–8 pm), while people preferring cableway-accessible trailheads are affected by road closures only when these are particularly binding and are fine with the current bus schedule (i.e. 8 am–6 pm). Crowding on the trail is an important issue for both groups, but road-oriented visitors are negatively affected by it even when it is just moderate.

Given the significant preference heterogeneity between road-oriented and cableway-oriented visitors, travel mode choices were separately estimated for the two groups. Figure 12.3 shows, for each management strategy and each

Table 12.4 Coefficients estimated separately for visitors preferring road-accessible trailheads (road-oriented) and visitors preferring cableway-accessible trailheads (cableway-oriented).

	Road-oriented		*Cableway-oriented*	
	β	*Std. Error*	β	*Std. Error*
ASC bus	0.005	0.445	1.09*	0.401
ASC cableway	0.428	0.546	1.72**	0.467
ASC no go	−2.26**	0.295	−1.57**	0.303
λ	0.547*	0.104	0.749*	0.162
Cost				
Toll	−0.119**	0.022	−0.095**	0.02
Bus fare	−0.252**	0.057	−0.095**	0.041
Cableway fare	−0.31**	0.069	−0.118**	0.025

(continued)

Table 12.4 Coefficients estimated separately for visitors preferring road-accessible trailheads (road-oriented) and visitors preferring cableway-accessible trailheads (cableway-oriented) (*continued*).

	Road-oriented		*Cableway-oriented*	
	β	*Std. Error*	β	*Std. Error*
Convenience				
Time to find parking space at the pass	−0.057**	0.019	−0.046**	0.017
Bus frequency	−0.008	0.008	−0.031**	0.008
Time to find parking space at the cableway	−0.076	0.029	−0.036*	0.014
Accessibility				
Free car access	0.69		0.816	
No car access 10 am–4 pm	−0.032	0.209	0.06	0.176
No car access 8 am–6 pm	−0.658*	0.251	−0.876**	0.264
Bus schedule 6 am–8 pm	0.689		0.231	
Bus schedule 8 am–6 pm	0.551**	0.19	0.442**	0.152
Bus schedule 10 am–4 pm	−1.24**	0.256	−0.673**	0.181
Cableway schedule 6 am–8 pm	0.416		0.083	
Cableway schedule 8 am–6 pm	0.426	0.315	0.187	0.136
Cableway schedule 10 am–4 pm	−0.842*	0.352	−0.27	0.146
Traffic				
No traffic	−0.054		0.286	
Medium traffic	0.395	0.212	0.361	0.196
Intense traffic	−0.341	0.215	−0.647*	0.242
No queue at the bus stop	0.316		0.358	
Short queue at the bus stop	−0.215	0.18	−0.012	0.132
Long queue at the bus stop	−0.101	0.174	−0.346*	0.154
No queue at the cableway station	−0.128		0.416	
Short queue at the cableway station	−0.17	0.31	−0.043	0.132
Long queue at the cableway station	0.298	0.315	−0.373*	0.155
Crowding				
0 people	0.459		0.324	
4 people	−0.069	0.127	0.172	0.088
8 people	−0.39**	0.133	−0.496**	0.135
Number of observations	803		1014	
Final log-likelihood	−794.666		−1084.7	
ρ^2	0.286		0.228	
Adjusted ρ^2	0.265		0.211	

*$p < 0.05$
**$p < 0.01$
Source: Orsi and Geneletti, 2014

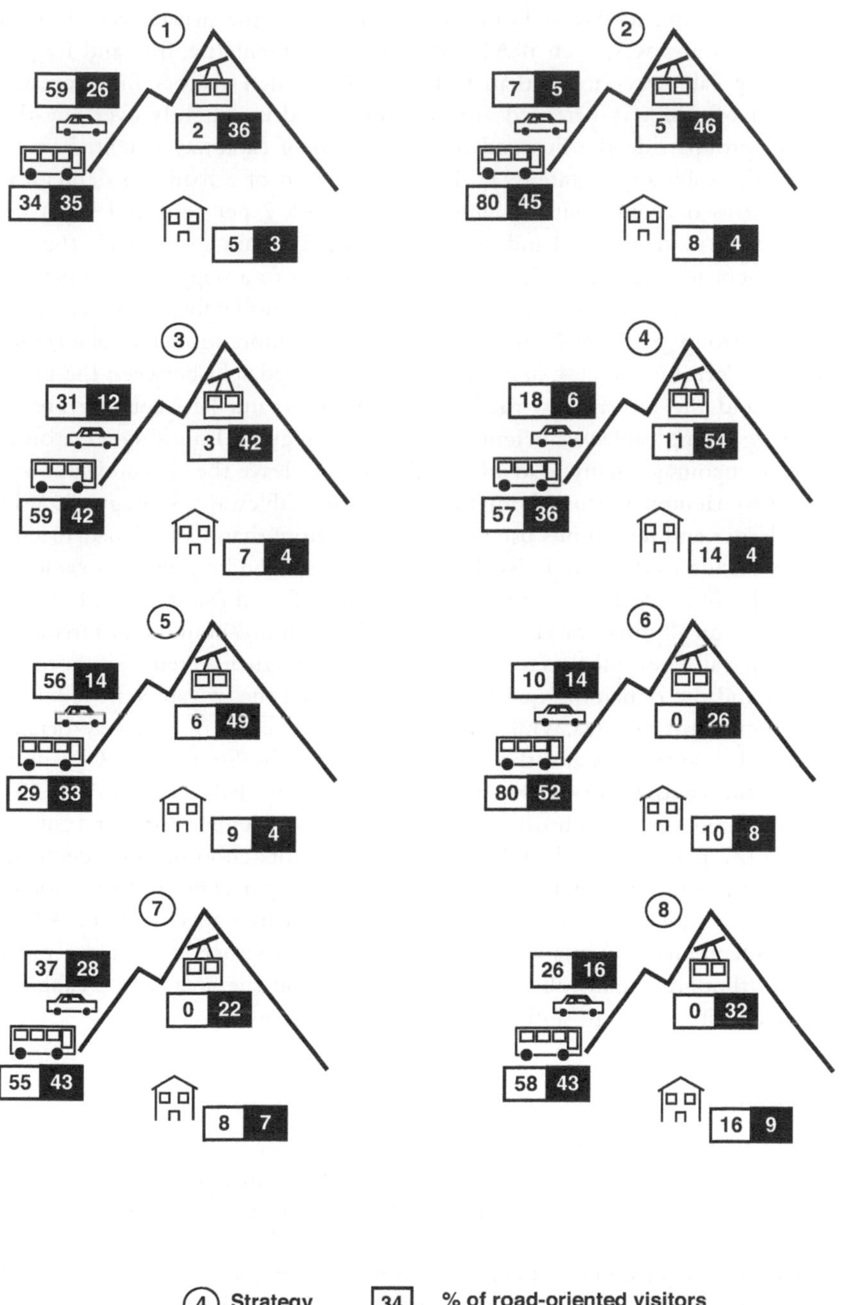

Figure 12.3 Estimated effects of the eight management strategies on modal split (expressed in percentages) for road-oriented and cableway-oriented visitors.

group, the percentages of people choosing to drive, to ride the bus, to take the cableway and to stay at home. When the use of the private vehicle is not restricted in any way, even if ATS are very convenient (i.e. free and frequent bus, cheap cableway), more than half of road-oriented visitors and more than a quarter of cableway-oriented visitors choose to drive, slightly over one-third in both groups ride the bus and over one-third of cableway-oriented visitors choose the cableway (Strategy 1). The introduction of a road toll dramatically changes this outcome, with driving dropping to a 7 per cent and 5 per cent share for the road-oriented and cableway-oriented groups, respectively, the share of bus users among road-oriented visitors growing to a staggering 80 per cent, and the share of bus and cableway users growing considerably in the cableway-oriented group (Strategy 2). Road closure proves a more moderate deterrent to the use of private vehicles: one that causes a balanced split between the private vehicle and the bus among road-oriented visitors, and between the bus and cableway among cableway-oriented visitors (Strategy 3). Imposing a fee on bus use may encourage many road-oriented visitors to leave the area and shift some cableway-oriented visitors from the bus to the cableway (Strategy 4). When road closure and fee on bus use are combined, more than half of road-oriented visitors decide to drive and less than a third to ride the bus, whereas cableway-oriented visitors do not seem to be particularly affected (Strategy 5). Enforcing limitations on all three travel options (i.e. €15 road toll, 40 minute bus frequency, €15 cableway fare) encourages three-quarters of road-oriented visitors to ride the bus and the remaining quarter to split between the private vehicle and the abandonment of the area, while one-half and one-quarter of cableway-oriented visitors choose the bus and the cableway, respectively (Strategy 6). Closing the road in the central part of the day generates a more balanced outcome, with road-oriented visitors splitting between the private vehicle (55 per cent) and the bus (37 per cent), and cableway-oriented visitors choosing in a decreasing order the bus (43 per cent) the private vehicle (28 per cent) and the cableway (22 per cent) (Strategy 7). When the bus is minimally convenient (i.e. €3 fare, 40 minute frequency), many road-oriented visitors either look at the private vehicle (though a road toll is in place) or abandon the area, while cableway-oriented visitors tend to shift to the cableway (Strategy 8).

Discussion

Results provide some relevant findings about visitor preferences towards different travel options and show a marked difference in attitude between those preferring road-accessible trailheads and those preferring cableway-accessible trailheads. In general, visitors seem to prefer the cableway over both the private vehicle and the bus. This is likely related to the extreme convenience of cableways, which enable people to quickly move from valley floors to high-elevation trailheads without the hassles of road-based transportation (e.g. time to find a parking space, wait times at a bus stop). Cost undoubtedly is a main determinant of a visitor's choice between multiple transportation alternatives, though any additional euro seems to be more acceptable when associated with the private vehicle than either the bus or the cableway. This means that the same small fee on all travel options may not prevent car use, but it may instead heavily discourage ATS use. In terms of convenience, the frequency of the bus service seems key to guaranteeing high bus ridership: administrators should be aware that buses every 60 minutes or more are not attractive at all. The daily timeframe over which a transportation option is available is also very important. A permanently open road is a fundamental condition for many visitors, just like an extended road closure is perceived as a strong limiting factor. Bus and cableway become particularly unattractive when their schedule covers only the central part of the day (i.e. 10 am–4 pm). Road traffic and queues are a true limiting factor only when massive; this suggesting that moderately intense traffic, for example, is not enough to divert visitors to other areas. Finally, visitors are particularly sensitive to crowding on the trail: while not seeing other hikers is a highly desirable condition and seeing few of them is still acceptable, the presence of many hikers considerably detracts from the utility of a travel option. This is surprising in such a popular area as that analysed in this chapter, and reminds managers and administrators that increasing visitation rates, if not well managed, could push potential visitors to other less crowded areas.

The difference in attitude between visitors preferring road-accessible trailheads (road-oriented) and visitors preferring cableway-accessible trailheads (cableway-oriented) suggests a distinction between 'traditional' hikers, who favour lower gateways, and 'modern' hikers, who choose higher gateways to easily reach high-elevation areas. Differences between the two groups are particularly evident when observing their sensitivity to cost (i.e. road-oriented visitors are more sensitive to the cost of ATS, while cableway-oriented visitors are equally sensitive to all transport alternatives), car accessibility (i.e. road-oriented visitors dislike any road closure, while cableway-oriented visitors accept moderate closures), bus schedule (i.e. road-oriented visitors invoke more extended schedules, while cableway-oriented visitors appreciate current schedules), and crowding (i.e. road-oriented visitors dislike any level of crowding, while cableway-oriented visitors accept moderate crowding).

Consistent with findings of various scholars (Cullinane, 1997; Holding and Kreutner, 1998), the simulation analysis performed for this study showed that the combination of incentives to ATS ('carrots') and disincentives on the use of private vehicles ('sticks') can achieve a considerable modal shift to more sustainable forms of mobility. This is emphasised, for example, by the estimated modal split of road-oriented visitors for Strategies 1 and 2. When no restriction on car use is enforced (Strategy 1), a seemingly convenient bus service (i.e. no fees, high frequency) can only attract 34 per cent of road-oriented visitors, while a cheap cableway (i.e. €5 fare) is chosen by no more than 36 per cent of cableway-oriented visitors. On the contrary, when a road toll is considered (Strategy 2), these figures may rise up to 80 per cent and 42 per cent, respectively.

Nonetheless, 'carrots' and 'sticks' must be carefully balanced to avoid that overly tempting 'carrots' may generate unintended consequences (e.g. crowding on the trail) or that excessively restrictive 'sticks' may encourage many visitors to leave the area. This chapter has presented the case of multiple ATS offering different ways of access into an area, therefore having the potential to deeply alter visitation patterns once visitors shift from one mode to another. In particular, cableways, given their significant hourly capacity and inherent attractiveness, may deliver unsustainable volumes of visitors to ecologically sensitive high-elevation areas. Estimates of modal split under all management strategies show that a large share of cableway-oriented visitors is willing to choose the cableway even when this is expensive and/or other alternatives are particularly attractive. In particular, the estimates for Strategy 8 emphasise that an expensive cableway can still attract twice as many people as those choosing the private vehicle. Hence, managers and administrators desirous to foster sustainable mobility should be careful in making cableways overly attractive because this would ultimately increase crowding in sensitive areas and traffic in valley floors as people drive to cableway stations from various origins.

The design of effective 'carrot and stick' strategies should also rely on a careful balance between road tolls and road closures, and an appropriate selection of bus fares and frequencies. The outcomes of Strategies 3, 5 and 7, for example, have shown that road closures, while contributing to a lower reduction in car use as compared to road tolls, ensure a more homogeneous modal split and therefore more balanced visitor flows across an area. Even though fares and frequencies of bus services are often strictly determined by budget constraints, managers and administrators should be aware that small changes in both of them might have a deep impact on ridership within different visitor groups.

While this study provided valuable insights about visitor preference towards various travel options and allowed the estimation of mode choices induced by a set of management strategies, the reliability of the results may have been impaired by some methodological shortcomings. First, the choice experiment was based on the assumption that all visitors choose their transport mode when they reach one of the two gateway villages, though this is true only for people actually based in one of those villages. In fact, combined travel options (e.g. car plus cableway) should have been considered to obtain more reliable estimates. Second, as only one questionnaire was assigned to people travelling together (e.g. couples, families, friends), groups were essentially treated like individuals, thus missing the possibility to fully capture visitors' taste heterogeneity. While this favoured the elicitation of collective attitudes, it did not allow a thorough analysis of preference variation across age, gender, origin, etc. (Marcucci *et al.*, 2011). Third, considering the number of parameters to be estimated (24), total observations (1,817) were not enough to ensure an adequate amount of statistically significant parameters ($p < 0.05$). Finally, results may have been affected by a degree of inconsistency between responses provided and actual preferences. This, which is common in stated preference surveys, is a consequence of people's attempt to comply with social norms (Kuran, 1995; Johansson-Stenman and Martinsson, 2006). During the survey, for example, many respondents showed a propensity towards the bus that is not justified by the current use level for that transportation option. However, discrepancies between statements and actual behaviour may have also been induced by unrealistic attribute levels and poorly understandable photographs used for the traffic and crowding attributes.

Conclusions

The results of the study presented in this chapter confirm the importance of combining incentives to ATS ('carrots') and disincentives on car use ('sticks') to achieve an adequate modal shift towards sustainable forms of mobility in natural areas. However, this chapter also warns about the possible risks of management strategies that, shifting too many people to a specific alternative transport mode, may result in largely unbalanced visitor flows that are eventually detrimental for both natural resources and the recreational experience. In Alpine contexts, for example, managers and administrators should try to define strategies that combine limitations on car use, accessible bus services and moderately attractive cableways if they want to reduce pollution while maintaining visitation rates in high-elevation areas under control. Yet this is not an easy task because it involves favouring environmental and recreational objectives over economic concerns. In other words, the adoption of effective 'carrot and stick' strategies ultimately depends on the political level: administrators must be willing to enforce unpopular restrictions on visitor use, must be able to raise adequate funds for the improvement of ATS and must communicate the expected benefits of a decision in front of the whole community.

Acknowledgements

This study was carried out as part of the 'AcceDo' research project, which has received funding from the Provincia Autonoma di Trento (Italy) through a Marie Curie action, European Union's 7th Framework Programme, COFUND-GA2008-226070, 'Trentino project – The Trentino programme of research, training and mobility of post-doctoral researchers'. Thijs Dekker, Stephane Hess and Riccardo Scarpa provided valuable advices on the design of the choice experiment and the statistical analysis of responses. Valentina D'Alonzo and Carlo Detassis contributed to the survey logistics and preliminary data preparation; Brigitte Scott produced the German version of the questionnaire; and Angela Renata Cordeiro Ortigara, Marika Ferrari, Romina Ferrari, Veronica Menapace, Walther Orsi and Riccardo Pasi provided fundamental assistance during fieldwork.

References

Cullinane, S. (1997) 'Traffic management in Britain's national parks', *Transport Reviews*, vol 17, no 3, pp. 267–279.

Cullinane, S. and Cullinane, K. (1999) 'Attitudes towards traffic problems and public transport in the Dartmoor and Lake District National Parks', *Journal of Transport Geography*, vol 7, pp. 79–87.

Galilea, P. and Ortúzar, J. de D. (2005) 'Valuing noise level reductions in a residential location context', *Transportation Research Part D: Transport and Environment*, vol 10, no 4, pp. 305–322.

Hensher, D. A. and Greene, W. H. (2002) 'Specification and estimation of the nested logit model: alternative normalisations', *Transportation Research Part B: Methodological*, vol 36, no 1, pp. 1–17.

Holding, D. M. (2001) 'The SanfteMobilitaet project: Achieving reduced car-dependence in European resort areas', *Tourism Management*, vol 22, pp. 411–417.

Holding, D. M. and Kreutner, M. (1998) 'Achieving a balance between "carrots" and "sticks" for traffic in national parks: The Bayerischer Wald project', *Transport Policy*, vol 5, pp. 175–183.

Johansson-Stenman, O. and Martinsson, P. (2006) 'Honestly, why are you driving a BMW?', *Journal of Economic Behavior and Organization*, vol 60, no 2, pp. 129–146.

Kelly, J., Haider, W. and Williams, P. W. (2007) 'A behavioural assessment of tourism transportation options for reducing energy consumption and greenhouse gases', *Journal of Travel Research*, vol 45, no 3, pp. 297–309.

Kuran, T. (1995) *Private Truths, Public Lies: The Social Consequences of Preference Falsification*, Harvard University Press, Boston, MA.

Lawson, S., Chamberlin, R., Choi, J., Swanson, B., Kiser, B., Newman, P., Monz, C., Pettebone, D. and Gamble, L. (2011) 'Modeling the effects of shuttle service on transportation system performance and quality of visitor experience in Rocky Mountain National Park', *Transportation Research Record*, vol 2244, pp. 97–106.

Lumsdon, L. L. (2006) 'Factors affecting the design of tourism bus services', *Annals of Tourism Research*, vol 33, no 3, pp. 748–766.

Lumsdon, L., Downward, P. and Rhoden, S. (2006) 'Transport for tourism: can public transport encourage a modal shift in the day visitor market?', *Journal of Sustainable Tourism*, vol 14, no 2, pp. 139–156.

Mace, B. L., Marquit, J. D. and Bates, S. C. (2013) 'Visitor assessment of the mandatory alternative transportation system at Zion National Park', *Environmental Management*, vol 52, pp. 1271–1285.

Manning, R. and Freimund (2004) 'Use of visual research methods to measure standards of quality for parks and outdoor recreation', *Journal of Leisure Research*, vol 36, no 4, pp. 557–579.

Marcucci, E., Stathopoulos, A., Rotaris, L. and Danielis, R. (2011) 'Comparing single and joint preferences: a choice experiment on residential location in three-member households', *Environment and Planning A*, vol 43, pp. 1209–1225.

Orsi, F. and Geneletti, D. (2014) 'Assessing the effects of access policies on travel mode choices in an Alpine tourist destination', *Journal of Transport Geography*, vol 39, pp. 21–35.

Pettebone, D., Newman, P., Lawson, S.R., Hunt, L., Monz, C. and Zwiefka, J. (2011) 'Estimating visitors' travel mode choices along the Bear Lake Road in Rocky Mountain National Park', *Journal of Transport Geography*, vol 19, no 6, pp. 1210-1221.

Regnerus, H. D., Beunen, R. and Jaarsma, C. F. (2007) 'Recreational traffic management: the relations between research and implementation', *Transport Policy*, vol 14, no 3, pp. 258–267.

Reigner, N., Kiser, B., Lawson, S. and Manning, R. (2012) 'Using transportation to manage recreation carrying capacity', *The George Wright Forum*, vol 29, no 4, pp. 322–337.

Shiftan, Y., Vary, D. and Geyer, D. (2006) 'Demand for park shuttle services-a stated preference approach', *Journal of Transport Geography*, vol 14, no 1, pp. 52–59.

Stradling, S. G., Meadows, M. L. and Beatty, S. (2000) 'Helping drivers out of their cars. Integrating transport policy and social psychology for sustainable change', *Transport Policy*, vol. 7, pp. 207–215.

Thiene, M. and Scarpa, R. (2008) 'Hiking in the Alps: exploring substitution patterns of hiking destinations', *Tourism Economics*, vol 14, no 2, pp. 263–282.

Train, K. (1980) 'A structured logit model of auto ownership and mode choice', *The Review of Economic Studies*, vol 47, no 2, pp. 357–370.

White, D. D., Aquino, J. F., Budruk, M., and Golub, A. (2011) 'Visitors' experiences of traditional and alternative transportation in Yosemite National Park', *Journal of Park and Recreation Administration*, vol 29, no 1, pp. 38–57.

13 Exploring future opportunities and challenges of alternative transportation practice

A systematic-wide transit inventory across US national parks

John J. Daigle

Introduction

Transportation management has emerged as an important issue in America's national parks (Dilsaver and Wyckoff, 1999; Daigle, 2008; Manning *et al.*, 2014). In fact, national parks are experiencing substantial increases in visitation that have led to traffic congestion, habitat degradation and air and noise pollution (Ament *et al.*, 2008; Park *et al.*, 2009). Traffic congestion and the continual development associated with the increased number of vehicles (e.g. construction of parking areas) are two of the most critical transportation issues presently challenging park managers (Dilsaver and Wyckoff, 1999; Manning *et al.* 2014).

Without transportation systems to support a growing user population, many recreation areas are experiencing not only diminished resources, but declining visitor satisfaction (Louter, 2006). Included in the mission of the National Park Service (NPS) is the directive to manage resources so as to provide enjoyment while meeting the needs of future generations of Americans. This directive is not only a key component of the agency's mission, but vital to the survival of the parks themselves, as it is the passion of Americans and their connections to these areas that create a national incentive to manage and protect them (Louter, 2006). The challenges created by increasing vehicular travel demands and associated congestion in high use parks have highlighted the need to plan and implement transit systems in national parks. These transit systems are deemed necessary to help reduce traffic congestion, demand for parking, pollution and crowding that diminish the visitor experience and threaten the environment. As Collum (2012) suggests, alternative transportation systems may both address demand and improve visitor experiences thus garnering continued support for parks.

There is a growing body of knowledge related to planning and managing alternative transportation systems in national parks. Some of the early studies coordinated within the National Park Service Alternative Transportation Program focused on the Island Explorer bus transit system in Acadia National Park (Daigle and Lee, 2000) and the use of intelligent transportation systems (ITS) to provide visitors with real-time information on parking conditions, transit arrival times, etc.

(Zimmerman *et al.*, 2003; Daigle and Zimmerman, 2004a). Other studies have been important to explore the feasibility of alternative transit (Federal Lands Alternative Transportation System Study, 2001) and evaluate perspectives of local communities (Zimmerman and Daigle, 2004; Dunning, 2005; Collum and Daigle, 2011). Knowledge also continues to build on identifying potential indicators that are important to the visitor experience (Dilworth, 2003; Turnbull, 2003; Daigle and Zimmerman, 2004b; Davenport and Borrie, 2005; White, 2007; Hallo and Manning, 2009; Holly *et al.*, 2010; Collum, 2012). Some of these studies have used multiple methods, including qualitative ones, to refine elements of the visitor experience. Finally, there is now a more comprehensive compilation of research and suggested management frameworks in planning for sustainable transportation in the national parks (Manning *et al.*, 2014).

Efforts to document transit system information and associated benefits are critical to assess operational efficiency, visitor experience, financial sustainability, and to ensure the transit system is meeting the goals and objectives of a park and the mission of the NPS. Given the costs to operate and maintain transit systems, it is especially important to share this information with managers, co-operators, policy-makers, and the general public. Notable within the current legislation on transportation management in the national parks (MAP-21) is a directive requiring the NPS to conduct a facility inventory that includes transit systems. Federal transportation programs (e.g. Paul S. Sarbanes Transit in Parks) and other federal transportation entities that provide grants to implement transportation improvements require that information be collected to ensure they are meeting their goals and objectives. This information is also vital to the broader transportation planning currently underway with the NPS to create transportation plans at the regional level.

This chapter identifies the current state of alternative transportation systems and efforts to detect the benefits they bring to the units of the NPS. A study by the Volpe Center and NPS conducting a National Transit Inventory in 2012 is examined to profile the extent and characteristics of alternative transportation systems in national parks. Also, to complement recent efforts to inventory transit systems, other efforts have been made by researchers at the Paul S. Sarbanes Transit in the Parks Center (TRIP) to measure the benefits of alternative transportation systems. Selected findings from both studies are explored to identify opportunities and challenges going into the future of alternative transportation in national parks as well as research and management implications. The next two sections provide basic information about the NPS units and briefly explore the federal legislation that has propelled efforts to study and address transportation issues in national parks, and profile the Alternative Transportation Program (ATP) to illustrate how this fits within the goals and priorities of the NPS transportation program.

Study area

The National Park Service has over 400 units and attracts nearly 300 million visitors annually (see www.nps.gov/hfc/carto/PDF/NPSmap2.pdf; Breck *et al.*, 2013). American national parks comprise over 80 million acres of public land and include extensive networks of transportation corridors – roads, trails, bike paths, waterways, public transit – that link a vast array of iconic attraction sites including viewpoints, historical and cultural sites, visitor centers, campgrounds, gateway communities (Manning *et al.*, 2014). There are currently over 5,450 miles of paved roads and 4,100 miles of unpaved roads, 1,414 bridges, 63 tunnels and extensive parking facilities (NPS, 2014). The vast majority of visitors still drive to and through the parks, experiencing the parks from the road and travelling only a short distance from their own private vehicle. However, visitors are increasingly experiencing a variety of alternative transportation systems (ATS), including shuttle buses, trams, ferries, historic vehicles, sometimes by choice, and increasingly because it is required (Manning *et al.*, 2014).

Alternative Transportation Program

Several key federal transportation bills beginning in the early 1990s enabled the NPS to work more cooperatively with federal and state transportation entities. The bills provide important guidance about the assessment of NPS transportation infrastructure such as roads and parking areas as well as the exploration of alternative transportation modes such as transit systems to accommodate visitors. The main bills are

- Intermodal Surface Transportation Efficiency Act of 1991 (ISTEA, 1991) – US $84 million per year
- Transportation Equity Act for the 21st Century of 1998 (TEA-21, 1998) – US $167 million per year
- Safe, Accountable, Flexible, Efficient Transportation Equity Act: A Legacy for Users of 2005–2009 (SAFETEA-LU, 2009) – US $1.2 billion with an additional US $500 million that was authorized by congressional resolutions for fiscal years 2010 and 2011.
- Moving Ahead for Progress in the 21st Century Act of 2012 (MAP-21, 2012) – US $240 million per year

Transportation planning responsibilities within the NPS have increased and include work on transportation enhancement, park roads and parkways, recreational trails, scenic byways, etc. This contributes to an increase in the resources allocated within the transportation bills. Under SAFETEA-LU, the Paul S. Sarbanes Transit in Parks (TRIP) program was created to assist national parks and other federal land management areas. The TRIP program provided funding for the planning, design, and implementation of alternative transportation systems, such as ferry facilities, shuttle buses, rail connections and bicycle trails. The five principal goals of the TRIP program were to (1) reduce traffic congestion; (2) enhance visitor mobility, accessibility and safety; (3) improve visitor access to education, recreation and health benefits; (4) enhance protection of sensitive natural, cultural and historic resources; and (5) reduce pollution. The national competitive TRIP program awarded approximately US $40 million to NPS out of the approximately US $120 million authorized between the fiscal years 2006 and 2010 (NPS, 2014). Demand for financial assistance from the TRIP program far exceeded the available funds. However, the TRIP program was not reauthorized by the 2012 federal transportation act (MAP-21).

The Intermodal Surface Transportation Efficiency Act of 1991 (ISTEA, 1991) and subsequent acts have encouraged the adoption of a transportation planning framework of the NPS that must integrate local, regional and statewide transportation decision-making. There are increased opportunities for national parks to work with states and local governments on transportation projects with matching grants for a number of federally-funded transportation programs. As the department of transportation in each state is responsible for setting transportation policy with regard to future projects and funding decisions, it is critically important that the national parks be a partner in the transportation planning process. Notable within the TEA-21 (1998) legislation was the directive for the Secretary of Transportation, in coordination with the Secretary of Interior, to 'undertake a comprehensive study of alternative transportation needs in National Parks and related Federal Lands' (TEA-21, 1998). This included the formation of a distinct program within the NPS called the Alternative Transportation Program (ATP). A number of studies to examine transit needs, transit strategies, and feasibility were also conducted and guided in part by this new program.

The ATP, which was launched in 1998, is responsible for coordinating policies, projects, and activities related to planning, partnering, and implementation of alternative transportation systems within and to NPS units. The program also develops strategies and recommendations for NPS-wide application on issues crossing agency and state/federal lines and jurisdictions. The mission of the program is to 'preserve and protect resources while providing safe and enjoyable access to and within the national parks by using sustainable, appropriate and integrated transportation solutions' (Krechmer *et al.*, 2001). The program provides a website (www.nps.gov/transportation/index.html) with information on transportation issues, legislation and planning documents. A key document available on the site is the National Park Service Transportation Planning Guidebook (NPS, 1999), which covers among other items NPS's planning policy and transportation planning, federal transportation legislation in relation to the NPS, success through partnerships and the basics of transportation planning. As noted by the director of the NPS upon initial publication, 'as we move forward into the next century, some of our greatest threats to national parks will come from encroaching development and activities outside the park boundaries: for that reason, our ability to understand transportation planning and laws is vital to our success as managers' (NPS, 1999).

To help cooperatively develop and integrate transportation planning into normal NPS activities, the Department of Interior (DOI) signed a Memorandum of Understanding (MOU) with the Department of Transportation (DOT) in November 1997. Several demonstration parks were identified in the MOU because of their complex transportation issues. All of the demonstration parks highlight issues that have become increasingly significant NPS-wide: to solve transportation and congestion problems, the NPS must look at these issues holistically, in a regional context, and involving all partners (see above website). Working with the various partners, especially federal transportation entities, the NPS has also been more successful at understanding and utilizing various surface transportation programs and making connections with a broader transportation research entity such as the Transportation Research Board (Daigle, 2008).

Some of the key issues identified by the ATP are the following: resource impacts must be managed; the automobile cannot always be the primary mode of transportation; visitor transit systems are not simply utilitarian in nature; baseline data generally needed to make informed decisions are often not readily available; transportation systems regularly transcend park boundaries; the park resources are the attraction, not the transportation; existing infrastructure is often at or beyond capacity; growing visitation requires complex, integrated transportation solutions; visitors expect a consistent design standard from national parks; new transportation systems are not always the solution. The need to solve these and other issues suggests that efforts be made to include: community impact studies, partnerships that involve the local community, other natural resource agencies such as the Forest Service, state and federal transportation agencies, tourism entities, friends groups, etc. as well as monitoring programs based upon management objectives (Daigle, 2008).

The National Park Service transit inventory

A Volpe Center report helped the NPS in 2012 devise a national inventory of existing transit systems and a protocol for conducting the inventory. A protocol was important to ensure consistent data collection across units and set the stage for data collection efforts in future years. In order to be considered for the inventory, a single NPS transit system had to meet the following three criteria (Breck *et al.*, 2013):

1 moving people by motorized vehicle on a regularly scheduled service;
2 operating under one of the following business models: concessions contract; service contract; partner agreement including memorandum of understanding, memorandum of agreement, or cooperative agreement (commercial use authorization is not included); or NPS owned and operated;
3 having routes and services at a given unit that are operated under the same business model by the same operator.

The NPS Alternative Transportation Program requested data from NPS units for the calendar year because most systems tend to collect information such as ridership on that basis. The protocol was designed to be a minimal burden to the unit and regional staff and therefore a modest set of reportable information was obtained to capture transit assets. Importantly, the NPS National Transit Inventory does not encompass a complete review of facilities and in the future other more detailed information such as costs and vehicle characteristics may be included.

The 2012 NPS National Transit Inventory revealed 147 discrete systems serving 72 of the 401 units of the NPS system (Breck *et al.*, 2013). As shown in Figure 13.1, the NPS alternative transportation system types are diverse with the largest type of transit being shuttle/bus/van/tram (44 per cent) followed by boat/ferry (34 per cent). The remaining transit system types include snow coach (10 per cent), plane (9 per cent), and train/trolley. Importantly, 52 systems were identified that provide sole access to an NPS system because of resource/management needs or geographic constraints. Many of the transit systems with plane service are associated with remote NPS units located in Alaska.

There is a variety of business models identified with NPS units and transit systems (Figure 13.2). The business model of 97 transit systems (66 per cent) is the concession contract where a concessioner pays the NPS a franchise fee to operate

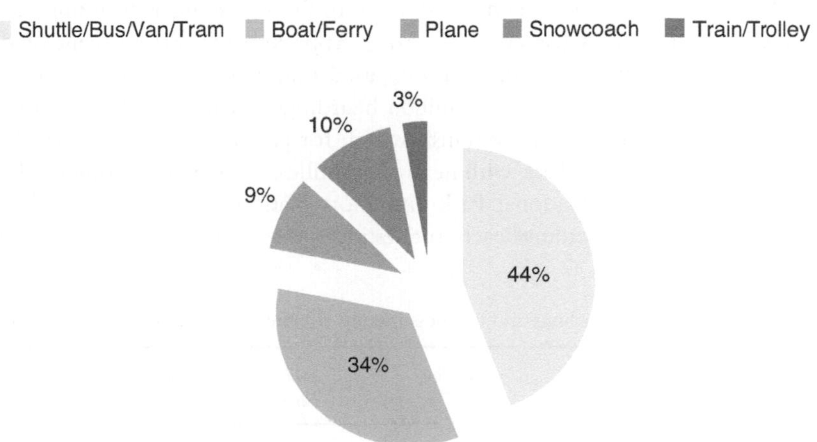

Figure 13.1 The National Park Service transit systems by mode (source: Breck *et al.*, 2013).

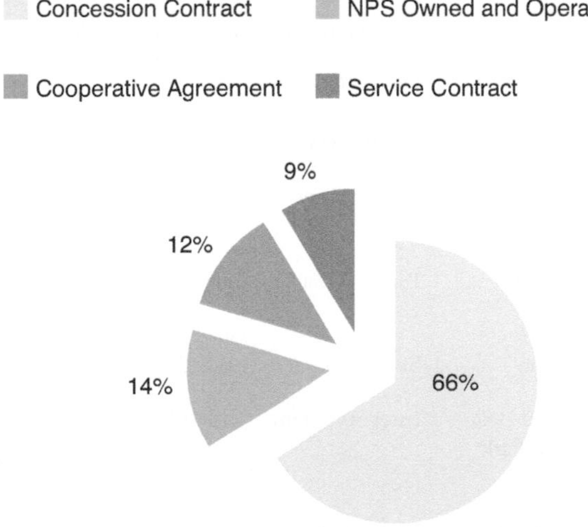

Figure 13.2 The National Park Service transit systems by business model (source: Breck *et al.*, 2013).

inside the unit. Concession contracts were included because they require resources and desire by the NPS to initiate. Also, after the bid and award process, concession contracts limit competition with other private operators and thus generally result in close working relationships with the NPS (Breck *et al.*, 2013). Twenty (13.6 per cent) of the transit systems are owned and operated exclusively by the NPS, while 17 (11.6 per cent) are operated under a cooperative agreement with another government or non-profit agency. The 'cooperative agreement' in this case includes all qualifying partner agreements (i.e. memorandum of understanding, memorandum of agreement, and cooperative agreement). Finally, 13 (8.8 per cent) of the transit systems are operated by a private firm under a service contract.

The 2012 inventory documented 36.3 million passenger boardings across all NPS transit systems (Breck *et al.*, 2013). Approximately 80 per cent of these boardings are attributable to the ten most used transit systems; with the Statue of Liberty's ferries recording over 9 million boardings during the 2012 season (see Table 13.1). The second highest transit system for passenger boardings was that at Grand Canyon National Park with nearly 3.5 million passenger boardings followed by those at Yosemite National Park and Zion National Park with a little over 3 million passenger boardings each. Interestingly, three ferry services (i.e. Alcatraz,

Table 13.1 Passenger boardings and business models for the ten highest use transit systems.

Rank	*System Name*	*Region*	*2012 Boardings*	*Business Model*
1	Statue of Liberty Ferries	Northeast	9,303,507	Concession Contract
2	Grand Canyon South Rim Shuttle Bus Service	Intermountain	6,177,000	Service Contract
3	Zion Canyon Shuttle	Intermountain	3,461,665	Service Contract
4	Yosemite Valley Shuttle	Pacific West	3,175,039	Concession Contract
5	Alcatraz Cruises Ferry	Pacific West	3,061,494	Concession Contract
6	USS Arizona Memorial Tour	Pacific West	1,460,000	Cooperative Agreement
7	Giant Forest Shuttle	Pacific West	1,439,534	Cooperative Agreement
8	Fort Sumter Ferry	Southeast	626,220	Concession Contract
9	Rocky Mountain Bear Lake & Moraine Park Shuttle	Intermountain	460,000	Service Contract
10	Acadia Island Explorer	Northeast	458,268	Cooperative Agreement

Source: Breck *et al.*, 2013

USS Arizona and Fort Sumter NPS units) fall within the ten highest use transit systems, but these generally provide sole access to the sites. The shuttle services in Rocky Mountain National Park and Acadia National Park had over 450,000 passenger boardings each in 2012. Passenger ridership in 2012 is dramatically higher as compared to initial years of operation for most of the units listed in Table 13.1. There has been a continued effort to increase the attractiveness of transit systems through marketing of services and the use of intelligent transportation systems (e.g. Acadia National Park). The degree of visitor satisfaction has been generally high as most people declared they would use these services again in a future visit to the park (Daigle and Zimmerman, 2006; Manning *et al.*, 2014).

The National Park Service National Transit Inventory (Breck *et al.*, 2013) presents a summary of transit system locations inventoried nationwide, business models, and passenger boardings for the 2012 season. Essentially the ten highest use transit systems have at least one of four categories of the business models described in Figure 13.2 with concession contracts being represented in four out of ten units. The majority of NPS units have low passenger boardings on their respective transit systems as compared to the top ten units. One hundred and seven (107) units reported between 0 and 100,000 passenger boardings during the 2012 season. Nonetheless, it is important to note that many of these units are in remote locations with low total visitation. Finally, a total of 22 transit systems nationwide had between 100,000 and 400,000 passenger boardings during the 2012 season.

The 2012 inventory documented a wide variety of funding sources used by the NPS to move people by transit. Under all concession contracts, concessioners charge visitors for service and pay a contractually required fee to the NPS (11 concessioners utilize vehicle fleets owned in full or in part by the NPS) (Breck *et al.*, 2013). In order to pay costs, 24 systems used base funds, 15 systems used transportation fees, 12 systems used TRIP grants, eight systems used Federal Lands Recreation Enhancement Funds (FLREA), and six systems used Federal Lands Highway Program (FLHP) Category III funds (Table 13.2). In the case of transportation fees, certain parks have been authorized to have a portion of visitor entrance

Table 13.2 Funding sources used to fund NPS transit systems.

Funding Source	*Number of Systems*
Base	24
Transportation Fee	15
Partner Sources	16
Federal Lands Recreation Enhancement Act	8
Transit in Parks	12
Federal Lands Transportation Program Category III	3

Source: Breck *et al.*, 2013

fee used for transit. For example, approximately US $13 of the $25 private vehicle entrance fee in Grand Canyon National Park and US $12 of the $20 private vehicle entrance fee for Acadia National Park are contributed to the respective transit systems.

The TRIP report

Between the fall of 2012 and the spring of 2013, the Paul S. Sarbanes Transit in Parks Technical Assistance Center (TRIPTAC) conducted a study to develop a process for the evaluation of the benefits of projects funded by the TRIP program. Nine study locations were chosen based on multiple criteria to reflect different federal land management agencies (including the NPS), type of unit (e.g. rural, urban), and type of ATS (e.g. bus, train, ferry). The Technical Assistance Center (TAC) worked with each unit to assess funded projects through the TRIP program including how well they achieved the measures of effectiveness (MOEs) related to TRIP program's goals and how well they achieved the management's purpose for the grant. The TAC worked with each unit to determine which data were available and which objectives and MOEs were best suited for evaluating the success of a given project.

All but one of the units included in the benefits survey were able to self-evaluate their achievements using their TRIP implementation grant; all of these units rated the status of elements as being fully achieved or, in a few cases, partially achieved by the unit (Daigle and Shapiro, 2013). No unit stated that it did not make any progress towards implementing the stated goals of its grant. Objectives and measures of effectiveness (MOEs) were also identified with the assistance of TRIPTAC, but differences were observed in availability and quality of data to quantify benefits of alternative transportation system projects.

In most cases, federal land units of TRIP grants were not organizations whose primary focus is the provision of transportation services, and the recipient staff had not been provided guidance regarding how to assess the benefits of funded projects. So on the one hand, it is noteworthy that half of the units had some data that could be used to quantify the MOEs to assess progress towards their stated goals. On the other hand, the study team was able to identify a variety of data collection activities that units could undertake to better identify and quantify benefits of transportation systems. The larger federal land management units with larger alternative transportation systems appeared to have collected more and better data than smaller units and/or units with smaller systems. This is likely due to the availability of more transportation planning expertise and financial resources.

For several of the units, data were based on current conditions only and no baseline data were available. The establishment of baseline data is critical and generally units could not quantify the benefits associated with many ATS projects (even if data were collected after ATS improvements were implemented), because information about baseline conditions did not exist. For half of the units in the study, benefits attributable to their ATS could not be quantified in any way (Daigle and

Shapiro, 2013). In several cases, only anecdotal or qualitative information existed to evaluate how conditions had improved with the implementation of the ATS or as a consequence of facility modifications.

The Daigle and Shapiro study (2013) found that individual units do not typically have the resources or expertise to systematically evaluate the benefits of ATS; therefore, it recommended that federal land management agencies create an evaluation advisory group (EAG) to support a process to evaluate alternative transportation projects. The same study suggested a framework to help select goals, objectives, and potential performance metrics so that the most efficient and effective means are identified to assess MOEs for transportation systems (Daigle and Shapiro, 2013). All data collection efforts should be driven by a need and purpose defined in an evaluation plan. The EAG could be created to help design and implement evaluation schemes for all ATS projects that are planned or implemented by a federal land management unit and this includes of course national parks. In this way, granting agency staff with more expertise on transportation planning can provide guidelines and technical assistance to units on how to quantify the benefits of projects. The federal land management agency's EAG may utilize guidelines and criteria to assist the unit in defining an evaluation plan for the submission of ATS project proposals to the granting agency.

Conclusion

There is a growing number of alternative transportation systems in protected areas and subsequently there is a need for parks to measure the benefits brought in by such systems. Transit systems are in some cases one of various ways to access a park, whereas in other cases they are simply the only means of access to a park. In high use parks, transit systems have been implemented to accommodate increased visitation and help reduce congestion and parking issues. There are diverse ATS, business models, and agreements that are used in developing and operating such ATS. In the future, it will be important to continue to inventory transit systems and coordinate these systems into a regional transportation planning effort given the amount of time, effort and budgeting required to implementing, operating and maintaining them. Recently, there have been attempts in this direction, for example in Alaska and the Northeast region, through the development of regional transportation plans.

It is prudent for park units that have, or are considering to implement ATS, to learn from other units, and to take small steps when starting an ATS (Daigle, 2010; Collum and Daigle, 2011). In many cases, collaboration with partners is necessary for getting funds and sharing technical expertise in order to increase the chance of success of the transit system. Parks should make an effort to document benefits associated with their system to not only help justify the cost of the operation and maintenance of the system, but also to improve the efficiency of the system and monitor visitor satisfaction. The performance measures typically used in transit systems are a good start, but may also need to be modified and adjusted to the special circumstances of a specific park setting.

The actual possibility to document long-term benefits of a transportation system and define the actions that may be taken to improve it relies on the ability of a park to collect quantitative data regarding the system. Even basic figures, such as the number of users each year (longitudinal data), provide a gauge to detect trends from year to year as well as a way to understand the overall impact of a system. Transit systems are a piece of the transportation infrastructure. As stated above, in NPS units there are over 5,450 miles of paved roads and 4,100 miles of unpaved roads, 1,414 bridges, 63 tunnels and extensive parking facilities. Historically, NPS's primary investment priority has been to rehabilitate structurally deficient bridges and to rehabilitate roads (and reduce the backlog of 'deferred maintenance needs' in paved roads) to preserve the useful life of these facilities (NPS, 2014). Given the cost of implementing and maintaining alternative transportation systems, documentation of benefits can be vitally important to ensuring the operational efficiency and the sustainability of this important piece of the transportation infrastructure.

In terms of enhancing alternative transportation practice in US national parks, managers need incentives if they are going to place more emphasis on systematically evaluating the benefits of ATS projects. Land managers need access to transportation expertise and guidance to learn the value of ATS data in achieving NPS's objectives, as well as information on how to develop evaluation and data collection plans (Daigle and Shapiro, 2013). A synergistic relationship should develop when land managers see the benefits transportation system data bring to managing visitors and natural resources, and to improving the efficient operation of the transportation system. The more advantages land managers perceive for collecting data to improve visitor and resource conditions, the more likely they will be to collect data.

References

Ament, R., Clevenger, A., Yu, O. and Hardy, A. (2014) 'Assessment of road impacts on wildlife populations in US national parks', in R. Manning, S. Lawson, P. Newman, J. Hallo and C. Monz, (eds) *Sustainable Transportation in the National Parks*, University Press of New England, Hanover and London.

Breck, A., Daddio, D. and Linthicum, A. (2013) 'National Transit Inventory, 2012', Alternative Transportation Program, Washington Office, National Park Service, US Department of Interior. July 2013, Report submitted by VOLPE, US Department of Transportation, Washington, DC. http://ntl.bts.gov/lib/47000/47800/47871/NPS_WASO_2013_Transit_Inventory.pdf.

Collum, K. (2012) From Automobiles to Alternatives: Applying Attitude Theory and Information Technologies to Increase Shuttle Use at Rocky Mountain National Park, MS thesis, University of Maine, USA.

Collum, K. and Daigle, J. (2011) Grand Island National Recreation Area Alternative Transportation Project. US Department of Transportation, Federal Transit Authority: Paul S. Sarbanes Transit in Parks Technical Assistance Center.

Daigle, J., Shapiro, P. and West, J. (2014) 'Developing and testing a process to evaluate the benefits of federal land management agency alternative transportation systems', *Transportation Research Board Proceedings*, 14–1766.

Daigle, J. and Shapiro, P. (2013) Developing and Testing a Process to Evaluate the Benefits of Federal Land Management Agency Alternative Transportation Systems: Technical Memorandum, US Department of Transportation, Federal Transit Authority: Paul S. Sarbanes Transit in Parks Technical Assistance Center.

Daigle, J. (2010) 'Webinar Module 4 – Alternative Transportation Systems (ATS) and the Role of Partnerships, Stakeholder Participation, and Public Involvement', www.fedlandsinstitute.org/TRIPTACArchives/TRIPTACTrainings/Default.html#mod4, accessed 12 August 2014.

Daigle, J. (2008) 'Transportation needs in national parks: A summary and exploration of future trends', *The George Wright Forum*, vol 25, no 1, pp. 57–64.

Daigle, J. and Zimmerman, C. (2004a) 'Alternative transportation and travel information technologies: Monitoring parking lot conditions over three summer seasons at Acadia National Park', *Journal of Park and Recreation Administration*, vol. 22, no 4, pp. 81–102.

Daigle, J. and Zimmerman, C. (2004b) 'The convergence of transportation, information technology and visitor experience at Acadia National Park', *Journal of Travel Research*, vol 10, pp. 151–160.

Daigle, J. and Lee, B. (2000) 'Passenger Characteristics and Experiences with the Island Explorer Bus', Technical Report 00–15. US Department of the Interior, National Park Service, New England System Support Office, Boston, MA.

Davenport, M. A. and Borrie, W.T. (2005) 'The appropriateness of snowmobiling in national parks: An investigation of the meanings of snowmobiling experiences in Yellowstone National Park', *Environmental Management*, vol 35, pp. 151–160.

Dilworth, V. A. (2003) 'Visitor perceptions of alternative transportation systems and intelligent transportation systems in national parks', Department of Parks, Recreation, and Tourism Sciences, Texas A&M University, College Station, TX.

Dunning, A. (2005) 'Impacts of transit in national parks and gateway communities', *Transportation Research Record*, vol 1931, pp. 129–136.

Federal Lands Alternative Transportation Systems Study. (2001) Cambridge Systematics, Inc., Cambridge, MA, and BRW Group, Inc., NY.

Hallo, J. C. and Manning, R. E. (2009) 'Transportation and recreation: A case study of visitors driving for pleasure at Acadia National Park', *Journal of Transport Geography*, vol 17, no 6, pp. 491–499.

Holly, M. F., Hallo, J. C., Baldwin, E. D. and Mainella, F. P. (2010) 'Incentives and disincentives for day visitors to park and ride public transportation at Acadia National Park', *Journal of Park and Recreation Administration*, vol 28, no 2, pp. 74–93.

ISTEA (1991) Intermodal Surface Transportation Efficiency Act, H.R. 2950, 102nd Congress of the United States of America.

Krechmer, D., Grimm, L., Hodge, D., Mendes, D. and Goetzke, F. (2001) 'Federal lands alternative transportation systems study – Volume 3 – Summary of national ATS need', prepared for Federal Highway Administration and Federal Transit Administration in association with National Park Service, Bureau of Land Management, and US Fish and Wildlife Service.

Manning, R., Lawson, S., Newman, P., Hallo, J. and Monz, C. (eds) (2014) *Sustainable Transportation in the National Parks*. University Press of New England, Hanover and London.

MAP-21 (2012) 'Moving Ahead for Progress in the 21st Century Act', Public Law 112–141, 112th Congress of the United States of America.

National Park Service (2014) 'Transportation Program Accomplishments', www.nps.gov/transportation_program_accomplishments.html, accessed 26 May 2014.

NPS (1999) 'National Park Service Transportation Planning Guidebook', National Park Service, Washington, DC.

SAFETEA-LU (2009) 'Safe, Accountable, Flexible, Efficient Transportation Equity Act: A legacy for Users', Public Law 109–59, 109th Congress of the United States of America.

TEA-21 (1998) 'The Transportation Equity Act for the 21st Century', Public Law 105–178, 105th Congress of the United States of America.

Turnbull, K. F. (2003) 'Transport to nature: Transportation strategies enhancing visitor experiences of national parks', *TR News*, vol 224, pp. 15–21.

White, D. D. (2007) 'An interpretative study of Yosemite National Park visitors' perspectives toward alternative transportation in Yosemite Valley', *Environmental Management*, vol 39, no 1, pp. 50–62.

Zimmerman, C., Coleman, T. and Daigle, J. (2003) Evaluation of Acadia National Park ITS Field Operational Test: Final Report. FHWA-OP-03-130. US Department of Transportation, ITS Joint Program Office, Washington D.C. p. 108.

Zimmerman, C., Daigle, J. and Pol, J. (2004) 'Tourism business and intelligent transportation systems: Acadia National Park, Maine', *Transportation Research Record*, vol 1895, pp. 182–187.

14 Participatory planning for the definition of sustainable mobility strategies in small islands

A case study in São Miguel Island (Azores, Portugal)

Artur Gil, Catarina Fonseca and Helena Calado

This chapter is a revised version of a paper that first appeared in 2011 in the *Journal of Transport Geography* (vol 19, no 6, pp. 1309–1319).

Introduction

A Management Plan (MP) is a tool to guide managers and other interested parties so that they might follow a logical decision-making process both today and in the future (Rowell, 2009). Successful management planning utilizes discussion among involved parties to systematically analyse threats and opportunities, and other difficult issues associated with a decision. The predetermined order of the steps lends logic to the target area's actions, in order to ensure the use of constantly updated information so that management may be frequently adapted to contextual changes without losing sight of its aims (Thomas and Middleton, 2003; Alexander, 2008). Stakeholders are those individuals, groups or organizations that are, in one way or another, interested, involved or affected (positively or negatively) by a particular project or action (Freeman, 1984). Involving a variety of stakeholders in planning and management brings important benefits: increased sense of 'ownership', greater public involvement in decision-making and stronger links between conservation and development. This promotes communication that allows the identification and resolution of problems (Gil *et al.*, 2011a).

While it is already a common practice to involve stakeholders and the public in many policy areas, it is a recent trend in transport planning. In the European Union (EU), the involvement of stakeholders and the public in transportation planning has been encouraged through the Directive 2001/42/EC on the 'Strategic Environmental Assessment of Plans and Programmes', and the 'International Convention on Access to Environmental Information, Public Participation in Environmental Decision-making and Access to Justice' – Aarhus Convention (2001). Among other tools for the promotion of sustainable mobility, the European Commission (EC) recommends local authorities to develop and implement Sustainable Urban Transport Plans (SUTP) to ensure long-term planning for urban transport development. Furthermore, SUTP entail a system of regular reporting and monitoring, and provide a framework toolbox to facilitate the definition of measurable objectives and quality criteria. In order to ensure greater effectiveness, the EC encourages cities to engage in a continuous dialogue with all relevant stakeholders, from citizens to private operators, in the development phase of the SUTP (EC, 2005).

According to Booth and Richardson (2001), however, transport planning is still perceived as poorly democratic because participation strategies are mostly top-down. The involvement of citizens is often limited to informing and consulting local communities, rather than encouraging active participation and partnership in the planning and decision-making process (Bickerstaff *et al.*, 2002). Although there is a growing recognition that sustainable transportation can be achieved more easily by involving all relevant stakeholders and that public acceptability is essential when radical changes are envisioned, transport planning continues to be perceived as an elitist process in many cases (Gil *et al.*, 2011b). Instead, public acceptability of sustainable mobility is obtained if the need for behavioural change is explained and citizens are convinced of their important contribution for this change to take place (Banister, 2008).

The involvement of stakeholders in mobility planning is particularly important in sensitive natural areas that are popular tourist destinations, such as small islands. Small islands are defined as land areas smaller than 10,000 square kilometres with a population under 500,000 inhabitants (Beller *et al.*, 2004). Small island environments are known to be particularly sensitive to external pressures and climate change impacts (IPCC, 2001). Their remoteness, isolation and smallness make planning and management more challenging in scientific and technical terms (Calado *et al.*, 2007, 2013). Small islands constitute a particular example of integration across space and over time of multiple natural, social and economic functions. This integration is materialized in a set of land use systems and social structures adapted to the particular natural constraints and resources framed by the available technologies (Fernandes *et al.*, 2014). Therefore, island systems constitute one of the challenges of our time: how can we balance ecological integrity with economic development and collective quality of life (Baldacchino and Niles, 2011)? The majority of European Outermost Regions (EOR) listed in Article 349 of the 'Treaty on the Functioning of the European Union' (EU) – except the French Guiana – are small islands or archipelagos constituted by small islands (Gil *et al.*, 2012). The EC communication 'The Outermost Regions: An Asset for Europe' (2008) advocates the potential contribution of these regions to overall development in Europe, taking advantage of their unique geostrategic position and turning their limits into assets.

Among Europe's outermost regions, the Azores Islands present a unique combination of remoteness, natural value and tourism importance, which make them an interesting context in which to study how stakeholders can be involved in the definition of mobility plans that benefit both humans and nature. Recently, the municipality of Delgada City, the capital of the Autonomous Region of the Azores (ARA), has been one (the only one in an insular region) of 40 Portuguese municipalities selected for the development of a Sustainable Mobility Plan (SMP), in the framework of an initiative (Sustainable Mobility National Programme) by the Portuguese

Environmental Agency (APA, 2010). The initiative, which is aimed at developing balanced and sustainable solutions to address existing mobility issues, involves 15 universities and research centres. This chapter presents the stakeholder engagement-based methodological approach that was applied and developed for the Sustainable Mobility Plan of Ponta Delgada City (SMPPD). Outcomes and assessment of the current development status of SMPPD are also described.

Study area

The Autonomous Region of the Azores, an authentic political region with legislative, executive administrative and financial powers as established by the 1976 Portuguese Constitution, consists of an archipelago of nine volcanic islands located between 37° and 40° N latitude, and 25° and 31°W longitude (Figure 14.1). Based on their geographical distribution, they are divided into three groups: the Western Group (Flores and Corvo), the Central Group (Pico, Faial, São Jorge, Graciosa and Terceira) and the Eastern Group (São Miguel and Santa Maria). They occupy an area of 2,333 square kilometres and have a population of around 240,000 inhabitants. The Azores are separated from the westernmost point of mainland Europe by 1,390 kilometres (Calado *et al.*, 2011). The municipality of Ponta Delgada is located in S. Miguel Island (Figure 14.1), the largest (around 745 square kilometres) and most populated of the archipelago, with about 134,000 inhabitants (SREA, 2009). This is also the island concentrating the highest number of tourists (ORT, 2008). During 2013, S. Miguel Island recorded 396,389 overnight stays, which represent 69 per cent of the total recorded for the whole archipelago (SREA, 2014). S. Miguel's airport is the main point of entry in the Azores region, being one of the three airports (alongside with Terceira and Faial) with flight connections to the European and American continents. The majority of passengers travel to the Azores for vacation/leisure purposes (Moniz, 2009), many attracted by nature-based tourism opportunities (Turismo de Portugal, 2007). In fact, the Regional Plan of Tourism (Regional Decree-Law 38/2008/A) places a strong focus on such opportunities (e.g. nature and landscape, volcanism, diving and whale watching) for the tourism development of S. Miguel. This is in line with the Tourism Strategy of the Regional Government of the Azores (Fonseca *et al.*, 2014), which assumes the environment as a transverse pillar for all of the Azores Region's social and economic activities and sectors (Azorean Regional Government, 2008).

S. Miguel Island has several areas of exceptional natural beauty and ecological importance, many of which with protection statuses. The Island Natural Park of S. Miguel is composed by 23 protected areas, which are more or less scattered around the island but managed in an integrated and coherent way. The philosophy underpinning the Island Natural Park ultimately requires an integrated management approach for the island as a whole, ensuring the sustainable development of the territory and the maintenance of natural values (Fonseca *et al.*, 2011).

S. Miguel is divided into six municipalities: Ponta Delgada, Ribeira Grande, Lagoa, Vila Franca do Campo, Povoação and Nordeste. Ponta Delgada City is located within the municipality of Ponta Delgada, the most populated of the Azores,

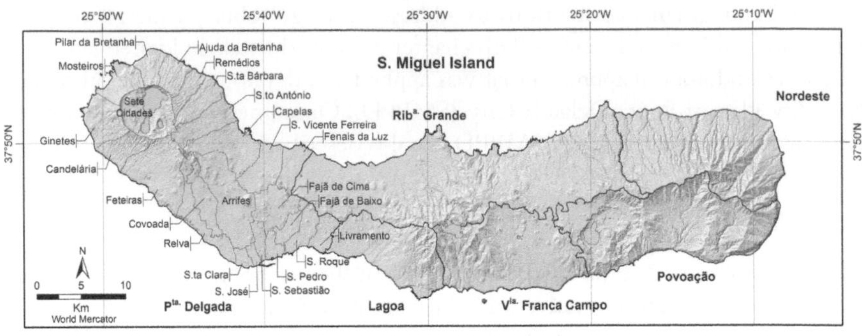

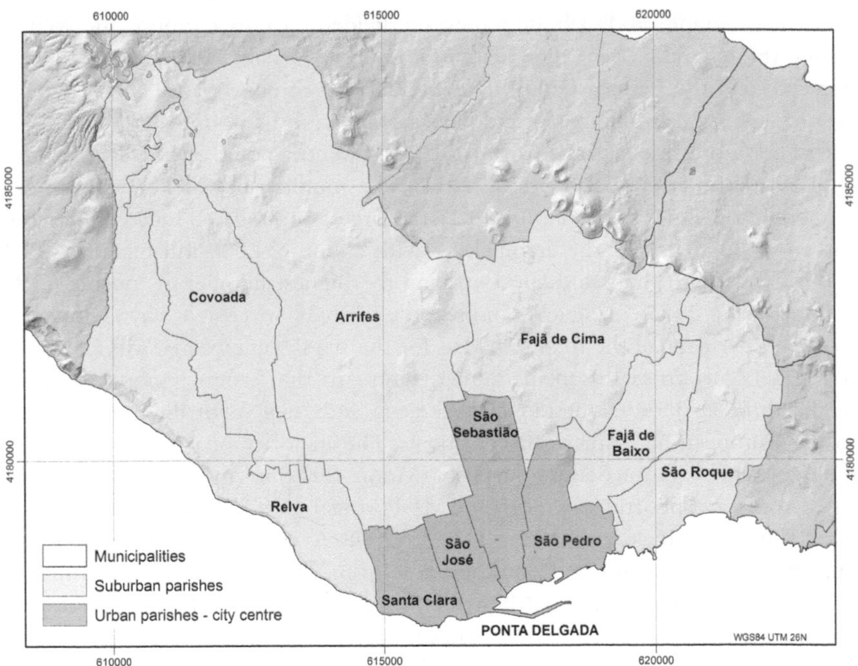

Figure 14.1 The study area.

hosting 28 per cent of the population. Most services, businesses and industries are concentrated in Ponta Delgada, making it the traffic junction of S. Miguel Island. Congestion has increased in recent years due to growing numbers of vehicles. Public transportation is limited to intercity and city buses and represents 17 per cent of commuting transport. Private car is the dominating means of transportation, being used on half of all journeys (INE, 2002). The presence of the airport in Ponta Delgada municipality also contributes to congestion. Additionally, due to the low offer in public transportation and the type of tourism (i.e. far from mass tourism), tourists visit the main attractions (including protected areas) using rented vehicles. This is evident by the high number of rent-a-car companies in S. Miguel (28), most of them concentrated in Ponta Delgada municipality (SRTT, 2014).

Understandably, the issue of mobility in Ponta Delgada extends beyond the administrative limits of the municipality, influencing the entire territory at various levels and particularly for what concerns the visitation (and sustainability) of protected areas. The study area includes the urban parishes of Santa Clara, São José, São Sebastião and São Pedro, and considers their interactions with the suburban parishes of Relva, Covoada, Arrifes, Fajã de Cima, Fajã de Baixo and São Roque (Figure 14.1).

Method

According to Calado *et al.* (2008) and to the general goals of the Sustainable Mobility National Programme (APA, 2010), the case study of Ponta Delgada City was developed in order to present an integrated and sustainable proposal based on stakeholder involvement for addressing the most relevant mobility issues in a small island city. The adopted main strategy was based on two fundamental principles:

1 Involving all the relevant public and private stakeholders directly or indirectly affected by urban mobility in an economic, social, environmental or cultural way;
2 Designing strategies that improve accessibility through the understanding of daily mobility problems and finding appropriate, effective and realistic solutions.

These fundamental principles were addressed by collecting information for improved decision-making, conducting a participatory process and developing a management tool. The SMP of Ponta Delgada (SMPPD) was developed in eight main stages. Figure 14.2 depicts the structure of the stakeholder involvement-based approach adopted for the development of the SMPPD.

Stage 1: identification of stakeholders

The potential stakeholders were identified and formally invited to participate in the development of the MP. Potential stakeholders included all actors directly or indirectly affected by mobility, such as: (1) local administration bodies; (2) regional administration bodies (environment, territorial planning, agriculture, forests, fisheries, construction, transportation and mobility, education, tourism, industry and culture); (3) private or public companies; (4) farmers, fisheries, local commerce and industrial confederations or associations; (5) non-governmental organizations (NGO); (6) landowners or their representatives/associations; (7) citizen groups; and (8) regional and local research centres and universities.

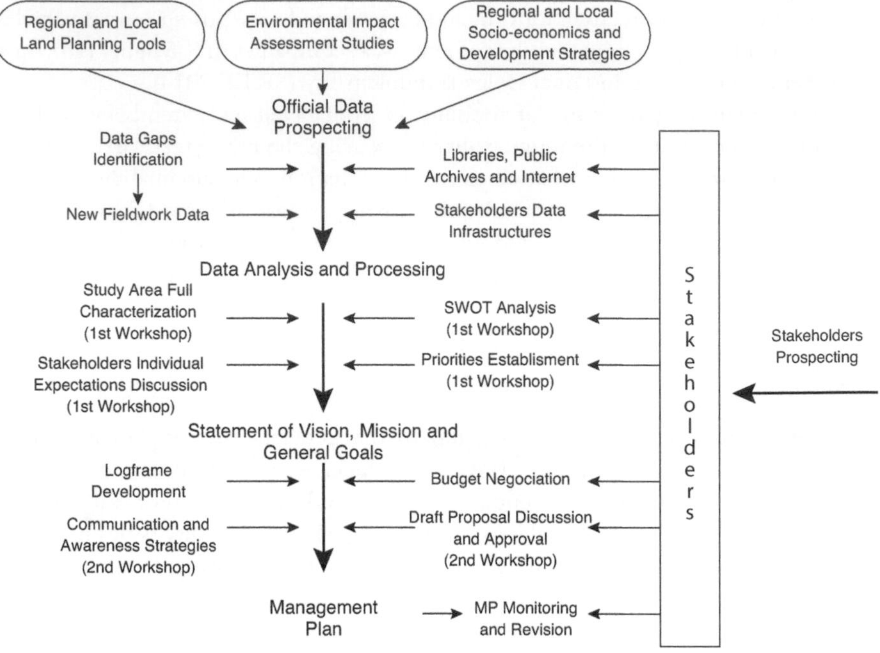

Figure 14.2 General methodological approach used for the development of the SMPPD.

Stage 2: characterization

According to the context and specificities of the MP, a detailed and systematic biophysical and socio-economic characterization of the study area was performed in order to support a robust stakeholder involvement. Biophysical and socio-economic variables were individually described and then collectively and synergistically evaluated according to the territory characteristics. To use the most updated and accurate information, all stakeholders were asked to provide access to their geographical and alphanumerical databases, as shown in Figure 14.2.

Stage 3: First Stakeholder Workshop – defining the vision, mission and general goals of the Management Plan

The Plan Manager led a one-day workshop whose ultimate objective was defining the vision, mission and general goals of the MP. At the beginning, each stakeholder or representative presented his/her expectations and objectives regarding the MP. The technical staff presented the full characterization of the study area to all stakeholders. Grouped in small working-groups (four to six persons), participants completed three tasks:

1 Analysis of the available information and identification of data/knowledge gaps regarding mobility in Ponta Delgada City, in order to ensure the robustness of the MP;
2 Development of a Strengths-Weaknesses-Opportunities-Threats (SWOT) analysis of sustainable mobility issues in Ponta Delgada City;
3 Definition of the vision, mission and general goals of the MP (proposals made by each working group were presented to all stakeholders for discussion and approval).

Stage 4: development of the Management Plan's Logical Framework draft

Based on the previous stage, the Plan Manager developed a first draft of the MP's Logical Framework ('Logframe') (Gil, 2007). This document indicated all the specific objectives necessary to accomplish each general goal; all the activities mandatory for reaching each specific goal; and all the indicators, means of verification, assumptions and responsible entities for undertaking/funding each activity. Afterwards, the Plan Manager met individually with most relevant stakeholders in order to present, discuss and pre-validate his 'Logframe' proposal. To ensure maximum support and funding for the MP, the Plan Manager gave special attention to those stakeholders able to execute or to fund the most relevant and/or expensive activities. At the end of this stage, a revised version incorporating stakeholders' suggestions was sent to all stakeholders before the second workshop.

Stage 5: Second Stakeholder Workshop – discussion, approval and ratification of the Management Plan's Logical Framework

The second workshop of SMPPD was held to discuss, validate and approve the revised 'Logframe' version. Participants were stakeholders' decision-makers or representatives with legal mandate to approve and ratify the plan on behalf of their institution. Small working groups were formed to discuss the document and the resulting suggestions were presented. After discussing and reaching a global consensus for every specific goal and activity, the final document was approved and ratified by all stakeholders.

Stage 6: budgeting and scheduling the Management Plan

After the Second Stakeholder Workshop, the Plan Manager met individually with representatives of the most important stakeholders holding political, administrative and financial decision power in order to clarify their contributions to the MP. The Plan Manager employed all possible arguments for convincing stakeholders of the urgency of carrying out the MP and its socio-economic benefits. The Plan

Manager had to search for alternative funding sources in case the stakeholder was not able to finance an activity completely.

After guaranteeing stakeholders' commitment for funding the MP implementation, the Plan Manager prepared a schedule for each activity, including the following information: (1) activity designation; (2) general goal; (3) specific goal; (4) indicator(s) and means of verification; (5) responsible stakeholder(s); (6) current status; (7) period of execution; (8) estimated costs; (9) funding resources. This allowed a straightforward monitoring of the MP implementation.

Stage 7: communication and promotion of the Management Plan

The layout of final documents was user-friendly and attractive for users and interested readers. A large number of copies were printed and sent to every stakeholder and local/regional media. To highlight the importance of the SMPPD at the municipal level, an informal public ceremony was organized with stakeholders' representatives. Furthermore, the SMPPD's full contents were disseminated online for unrestricted consultation.

Stage 8: supervision, monitoring and revision of the Management Plan

For the assessment and monitoring of SMPPD, the production of an annual report coordinated by the Plan Manager and supported by all involved stakeholders was expected. This document shall include a detailed description and assessment of concluded and ongoing activities. Although the SMPPD shall be revised every five years, it can be updated whenever necessary, according to the evolution of the mobility trends in Ponta Delgada City.

Results

Sixteen stakeholders participated in the SMPPD development, as shown in Table 14.1. These stakeholders comprehensively took full or partial responsibility to carry out the 34 operational tasks that are at the core of the SMPPD and its full achievement. The SMPPD contains a vision, a mission, five general goals and their eight associated specific goals, as follows:

- Vision: 'Solutions for a sustainable, efficient, integrated and diversified mobility, adapted to all citizens';
- Mission: 'Promoting an environmentally sustainable mobility in Ponta Delgada in order to provide a better quality of life for all citizens';
- General goals (A to E) and their respective specific goals (8 in total, from A.1 to D.8), as follows:

  A 'Promoting the efficiency of private motorized transportation in Ponta Delgada':

    – Designing and implementing a selective and strategic policy for traffic and motorized transportation parking management in the city of Ponta Delgada.

Table 14.1. List of stakeholders involved in the development of the SMPPD.

Public institutions	*Private and non-governmental organizations*
• Ponta Delgada City Council • Energy and Environment Azores Regional Agency • Urban and Suburban Parishes Management Boards • Portuguese Agency for the Environment • Regional Secretary of Environment • Regional Secretary of Public Construction and Transportation • Regional Secretary of Economy and Tourism • National Police • Ponta Delgada City Primary and Secondary School Management Boards • University of the Azores (on behalf of the Land-use Planning Research Centre) • Technical University of Lisbon (on behalf of Transport Planning Research Centre) • University of Porto (on behalf of Urban Planning Research Centre)	• Trade and Industrial Union of Ponta Delgada • Environmental NGO 'Amigos dos Açores' • Private Transportation Companies • Portuguese Association for Disabled People

B 'Promoting the utilization of public transportation in Ponta Delgada':

- Promoting the regular use of the mini-bus network within the city of Ponta Delgada.
- Promoting the creation of an integrated public transportation system aiming to ensure cost-effective connections among suburban (within Ponta Delgada municipality) and interurban (within S. Miguel Island) destinations.

C 'Promoting soft modes of mobility in Ponta Delgada':

- Promoting the rehabilitation and the improvement of sidewalks in the city of Ponta Delgada.
- Designing and implementing a 'Pedestrian Circulation Plan' in the city of Ponta Delgada.

D 'Integrating SMPPD policies and activities into land-use plans and Local Agenda 21':

- Integrating core SMPPD assumptions and measures in the Ponta Delgada Municipality's Urban Rehabilitation Program (REVIVA).
- Integrating core SMPPD assumptions and measures in the Ponta Delgada Municipality's Local Agenda 21.
- Integrating core SMPPD assumptions and measures in forthcoming Master Land Management Plans to be developed and regulated by the Ponta Delgada City Council.

E ‘Promoting mobility conditions that guarantee equal rights of all citizens in Ponta Delgada’ (transversal to all four previous general goals and therefore will be successfully reached if previous goals are synergistically attained).

Discussion

The heterogeneity of stakeholders in terms of institutional role and area of specialization, and their involvement and contribution to the SMPPD – by providing their own data/information and by sharing their experience and expectations regarding the issues at stake – allowed a coherent characterization and assessment of mobility in Ponta Delgada City. It also allowed the formulation of a more straightforward, realistic and citizen-driven SMPPD than standard Sustainable Urban Transport Plans generally are, as they prioritize a purely technical approach. Therefore, one of the most important outcomes of this project was the creation of an informal working group constituted by all SMPPD stakeholders that could support Ponta Delgada City Council decision-making on mobility and urban planning issues, on a permanent basis. Nevertheless, further steps could include the involvement in the process of the general public, in addition to relevant stakeholders.

Regarding the implementation of the SMPPD, most relevant programmed activities have already been carried out. Among these, the integration of this MP in the Municipal Agenda 21 Strategy; the reinforcement of the urban mini-bus circuit; traffic restrictions in the most traditional streets of the historical centre; the reduction of free car parking places in the city centre; the increment of pavement’s quantity and quality in order to promote walking; and incentives to encourage pedestrian and cycling mobility in Ponta Delgada City. The effects of these activities are felt by both residents and tourists visiting and staying in the city.

As public stakeholders fully funded the SMPPD financial implementation, lack of financial support from the private sector is a challenge that should be addressed as a priority in the forthcoming SMPPD revision. Furthermore, the SMPPD’s conception and development were successful because it was a consensus-based process supported by updated data provided by stakeholders or specifically produced for this plan. The experience of the SMPPD proved that involving stakeholders in sustainable transport planning is possible and can provide more satisfactory and cost-effective solutions. Owing to its success, this experience can be taken as a good practice and be replicated and adapted to other areas and cities located in small islands like S. Miguel. This would improve mobility in urban areas, guarantee the environmental sustainability of a region and contribute to the green image of the island, something that is very important for tourism.

Conclusion

Achieving sustainable development requires the involvement of multiple and diverse public and private stakeholders. Active and participatory management plans are mandatory in order to effectively address mobility issues, especially in insular territories. Such issues call for managers to proceed with transparency and rigor while sharing management responsibilities with relevant stakeholders. In the case of the SMPPD, participation and co-responsibility of all stakeholders represented the cornerstones of successful MP conception, development and implementation, as it was recognized by the Portuguese Environment Agency (APA, 2010) when designating SMPPD as an outstanding model for good practices. Ultimately, the Plan Manager and its team should be aware that, without the full involvement of stakeholders, any attempt to create and implement an MP in the area of transportation can be a wasted effort, as mobility issues demand the highest possible levels of strategy, planning and activity programming.

Acknowledgements

This book chapter was developed in the framework of a post-doctoral research project (M3.1.7/F/005/2011) led by A. Gil and supported by the FRC – Regional Fund for Science (Azorean Regional Government).

References

Alexander, M. (2008) *Management Planning for Nature Conservation: A Theoretical Basis and Practical Guide*, Springer, Dordrecht, The Netherlands.

APA (2014) 'Mobilidade Sustentável' (in Portuguese), http://sniamb.apambiente.pt/mobilidade, accessed 1 February 2014.

Azorean Regional Government (2008) 'An Assessment of Strategy for the Outermost Regions: Achievements and Future Prospects', Communication COM (2007) 507 Final, Ponta Delgada, Portugal.

Baldacchino, G. and Niles, D. (2011) *Island Futures: Conservation and Development across the Asia-Pacific Region*, Springer, Tokyo, Japan.

Banister, D. (2008) 'The sustainable mobility paradigm', *Transport Policy*, vol 15, no 2, pp. 73–80.

Beller, W., D'Ayala, P. and Hein, P. (2004) *Sustainable Development and Environmental Management of Small Islands*, UNESCO and the Parthenon Publishing Group 5, Paris, France.

Bickerstaff, K., Tolley, R. and Walker, G. (2002) 'Transport planning and participation: the rhetoric and realities of public involvement', *Journal of Transport Geography*, vol 10, no 1, pp. 61–73.

Booth, C. and Richardson, T. (2001) 'Placing the public in integrated transport planning', *Transport Policy*, vol 8, no 2, pp. 141–149.

Calado, H., Braga, A., Moniz, F., Gil, A. and Vergílio, M. (2013) 'Spatial planning and resource use in the Azores', *Mitigation and Adaptation Strategies for Global Change*, online first published: 15 November 2013, pp. 1–17.

Calado, H., Borges, P., Phillips, M., Ng, K. and Alves, F. (2011) 'The Azores archipelago, Portugal: improved understanding of small island coastal hazards and mitigation measures', *Natural Hazards*, vol 58, no 1, pp. 427–444.

Calado, H., Gil, A. and Santos, N. (2008) *Proposta de Execução do Plano de Mobilidade Sustentável da Cidade de Ponta Delgada*, University of the Azores and City Council of Ponta Delgada, Ponta Delgada, Portugal (in Portuguese).

Calado, H., Quintela, A. and Porteiro, J. (2007) 'Integrated coastal zone management strategies on small islands', *Journal of Coastal Research*, vol SI 50, pp. 125–129.

European Commission (2008) 'The outermost regions: an asset for Europe', communication from the Commission, COM(2008)/642.

European Commission (2005) 'Communication from the Commission to the Council and the European Parliament on Thematic Strategy on the Urban Environment', Communication from the Commission, COM/2005/0718.

Fernandes, J. P., Guiomar, N., Freire, M. and Gil, A. (2014) 'Applying an integrated landscape characterization and evaluation tool to small islands (Pico, Azores, Portugal)', *Revista de Gestão Costeira Integrada*, vol 14, no 2, pp. 243–266.

Fonseca, C., Pereira da Silva, C., Calado, H., Moniz, F., Bragagnolo, C., Gil, A., Phillips, M., Pereira, M. and Moreira, M. (2014) 'Coastal and marine protected areas as key elements for tourism in small islands', *Journal of Coastal Research*, vol SI 70, pp. 461–466.

Fonseca, C., Calado, H., Pereira da Silva C. and Gil, A. (2011). 'New approaches to environment conservation and sustainability in small islands: The Project SMARTPARKS', *Journal of Coastal Research*, vol SI 64, pp. 1970–1974.

Freeman, R. E. (1984) *Strategic Management – A Stakeholder Approach*. Ballinger, Cambridge, UK.

Gil, A., Fonseca, C., Lobo, A. and Calado, H. (2012) 'Linking GMES Space Component to the development of land policies in Outermost Regions – the Azores (Portugal) case-study', *European Journal of Remote Sensing*, vol 45, pp. 263–281.

Gil, A., Calado, H. and Bentz, J. (2011a) 'Public participation in municipal transport planning processes – The case of the Sustainable Mobility Plan of Ponta Delgada, Azores, Portugal', *Journal of Transport Geography*, vol 19, no 6, pp. 1309–1319.

Gil, A., Calado, H., Costa, L.T., Bentz, J., Fonseca, C., Lobo, A., Vergilio, M. and Benedicto J. (2011b) 'A methodological proposal for the development of Natura 2000 sites management plans', *Journal of Coastal Research*, vol SI 64, pp. 1326–1330.

Gil, A. (2007) Proposta Metodológica para a Elaboração de Planos de Gestão de Sítios da Rede Natura 2000, MSc Thesis, University of the Azores at Ponta Delgada, Portugal (in Portuguese).

INE (2002) *Censos 2001 – Resultados Definitivos: Região Autónoma dos Açores*, Instituto Nacional de Estatística, Lisbon, Portugal (in Portuguese).

IPCC (2001) *Climate Change 2001: Impacts, Adaptations and Vulnerability*, Intergovernmental Panel on Climate Change, Cambridge University Press, New York, USA.

Moniz, A. (2009) 'A sustentabilidade do turismo em ilhas de pequena dimensão: o caso dos Açores', Centro de Estudos de Economia Aplicada do Atlântico, Ponta Delgada, Portugal (in Portuguese).

ORT (2008) *Inquérito à Satisfação do Turista nos Açores*. Observatório Regional do Turismo, Ponta Delgada, Portugal (in Portuguese).

Rowell, T. A. (2009) 'Management planning guidance for protected sites in the UK: a comparison of decision-making processes in nine guides', *Journal for Nature Conservation*, vol 17, no 3, pp. 168–180.

SREA (2014) 'Dormidas por Ilha na Hotelaria Tradicional – Principais indicadores estatísticos Turismo' (in Portuguese), available at: http://estatistica.azores.gov.pt, accessed 23 September 2014.

SREA (2009) 'Anuário Estatístico de 2008 da Região Autónoma dos Açores', Serviço Regional de Estatística dos Açores, Angra do Heroísmo, Portugal (in Portuguese).

SRTT (2014) *Aluguer de Veículos Sem Condutor – Rent-a-car*. Secretaria Regional dos Transportes e Turismo, Direção Regional dos Transportes, Serviço Coordenador dos Transportes Terrestres, Ponta Delgada, Portugal (in Portuguese).

Thomas, L. and Middleton, J. (2003) *Guidelines for Management Planning of Protected Areas*. IUCN, Gland, Switzerland.

Turismo de Portugal (2007) *Plano Estratégico Nacional do Turismo – Para o Desenvolvimento do Turismo em Portugal*. Ministério da Economia e da Inovação, Lisboa, Portugal (in Portuguese).

15 Achieving a balance between trail conservation and road development

The case of Bhutan

Taiichi Ito

This chapter is a revised version of a paper that first appeared in 2011 in *Forests* (vol 2, no 4, pp. 1031–1048).

Introduction

The US Wilderness Act was enacted in 1964 or 40 years after Aldo Leopold first proposed wilderness status at Gila National Forest in the United States (US Wilderness Act, 1964). The act established that the absence of roads and permanent residents is essential to make an area a wilderness. Yet, in recent times, protected areas with a resident population have been recognised and called 'inhabited wilderness' (Catton, 1997). The International Union for Conservation of Nature (IUCN) has been accepting a moderate presence of humans in wilderness since 1992 (Dawson and Hendee, 2009). In fact, protected areas in IUCN's category 1b, among other objectives, aim 'to enable indigenous communities to maintain their traditional wilderness-based lifestyle and customs, living at low density and using the available resources in ways compatible with the conservation objectives' (IUCN, 2014). The recognition of indigenous communities' rights, however, may lead to a potential conflict between nature conservation and livelihoods as such communities can hardly be expected to spend their entire lives isolated from the surrounding civilisation, which relies on motorised transportation like automobiles.

While American wilderness preservationists were struggling with invasions of automobiles to protected lands, other countries have seen a very limited presence of motorised vehicles for a long time. This is the case of Bhutan, which used to be an essentially car-free country until a road from India reached Paro, in the west, in 1962. Since then, the road system has gradually expanded to formerly isolated communities in the north, and further up to the border with China in a matter of few years. In contexts like this, the improvement of the road network is clearly a double-edged sword for residents as easier access means a loss of traditions, culture and, in some cases, health. The effects of road development may also have an impact on visitors, by removing some of the wilderness characteristics they in fact came to experience. Finding a reasonable trade-off between road development and the maintenance of wilderness conditions therefore seems key to preserving the quality of life of the local population and the recreational experience of visitors. This chapter explores the history of road development in Bhutan as well as the relationship between traditional trails and roads, and provides guidance for the introduction of sustainable transport modes in the country. First, the influence of road expansion on trails and on their users such as residents and trekkers is analysed based on historical documents written in English and Japanese. Second, the current conditions of the trails and trailheads in Jigme Dorji National Park are examined through the results of GPS-based field surveys conducted during the monsoon seasons of 2008, 2009, 2010, 2011 and in March 2014. Finally, ideas about how to establish sustainable transportation in Bhutan are discussed in the light of alternative transportation systems (e.g. monorails, gondolas, cableways) developed in mountain protected areas worldwide.

Study area

Bhutan is a small mountainous country located in the Himalayas between China and India, with a size of 38,394 square kilometres and a population of 740,000 inhabitants. The whole country of Bhutan is often seen as a big national park for at least two reasons. One is cultural, as the country maintains a beautiful landscape and its inhabitants still wear traditional clothes in their daily lives. The other is administrative, as the tariff system that has been required of visitors since 1974 (when Bhutan started hosting foreign tourists) includes a royalty resembling the entrance fee to a protected area. In fact, the royalty is spent to support free medical care and free education in isolated communities, which are often located in national parks.

The first protected area in Bhutan, Manas Wildlife Sanctuary, was established in 1964. Today biological corridors connect five parks, four wildlife sanctuaries and a nature reserve (Figure 15.1), which occupy 51 per cent of the total country's surface (Wangchhuk, 2010; Thachen, 2013) and do not charge entrance fees, partly because they host communities. In the case of Jigme Dorji National Park, which has a total area of 4,316 square kilometres, most of the 6,000 inhabitants live in Gasa district or surrounding areas (Tshering, 2009). Unlike many national parks in developed countries, most of residents in the park still live in roadless areas.

Nonetheless, expansion of the road system is rapidly changing the landscape as well as the lifestyle of the Bhutanese. While Bhutan has been a roadless country or inhabited wilderness until 1962 when the first road was built between the Indian border town of Phuentsholing and Paro, now the road network has been expanded to cover all districts including Gasa, which is located in the heart of Jigme Dorji National Park.

Historical analysis of trail-to-road transitions in Bhutan

Historical information on the trails and roads in Bhutan is limited because visitation was restricted to anyone but royal guests until 1974. Among such guests, J. C. White in 1905–1907, B. K. Todd in 1952, Nakao in 1958, D. Doig in 1962 and M. Matsuo in 1969 published documents that help us have a better idea of what the condition of trails and roads was like. Table 15.1 shows the routes these visitors followed when visiting the country.

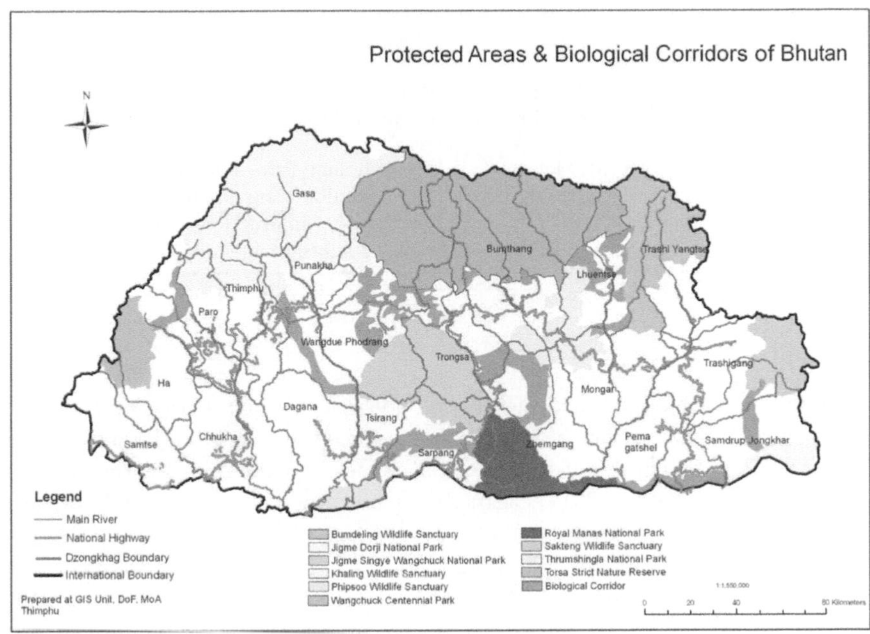

Figure 15.1 Protected areas of Bhutan (courtesy of Phuntsho Thinley).

Table 15.1 Early visitors to Bhutan and the routes they followed.

Year	*Months*	*Explorers*	*Routes*[1]
1905	March	J. C. White Political Officer	Gangtok (Sikkim)–Natu La–Ha–Paro–Thimphu–Punakha–Tongsa–Thimphu–Lingzhi La–Pangri Dzong (Tibet)–Gangtok
1906	May		Gangtok–Gauhati (Assam)–Dewangiri–Trashigong–Tashi Yangtse–Bodo La–Sela (Tibet)–Gangtok
1907	November		Gangtok–Pangri Dzong–Tremo La–Drygel Dzong–Phuntsholing
1952	June–November	B. K. Todd Businessman	Kalingpong (Sikkim)–Samti–Sele La–Ha–Paro–Thimphu–Chubuka Dzong–Phuntsholing–Kalingpong
1958	June–November	S. Nakao Ethnobotanist	Kalingpong–Phuntsholing–Chubuka Dzong–Paro–Ha (Sele La, Hara Chu La)–Paro–Tremo La–Paro–Thimphu–Lingzhi Dzong–Laya–Gasa–Punakha–Trongza–Thimphu–Buxa

(*continued*)

Table 15.1 Early visitors to Bhutan and the routes they followed (*continued*).

Year	*Months*	*Explorers*	*Routes*[1]
1961	June–August	D. Doig Journalist	Rajabhat Khawa (India)–Phuntsholing–Chhukha Dzong–Paro–Thimphu
1969	October–December	M. Matsuo Civil Engineer	Phuntsholing–Paro–Thimphu–Paro–Wangdue Phodrang–Phunakha–Trongsa–Trashigang–Samdrup Jongkhar–(via India)–Phuntsholing–Paro (Air) Kolkata

1 Place names vary with documents.

White's logbook, published in *National Geographic* (White, 1914), let the world know about Bhutan. In fact, he visited Bhutan three times, in 1905, 1906 and 1907, and each time, he chose different passes by which to enter and exit the country. Hence, his report (White, 2011) clarified the passes and trails used by the Bhutanese and Tibetan traders.

More than 40 years later, in 1952, Todd was invited as the first American visitor to the royal wedding of the third king, Jigme Dorji Wangchuck, and published an article in the *National Geographic* magazine where he mentioned that passes from Tibet to Bhutan were already restricted. In 1958, S. Nakao, an ethnobotanist from Japan who was invited by Queen Kesang Choedon Wangchuck, could document the route conditions in more detail than White or Todd could. He wrote that there were bungalows to stay in each night, and houses were scattered along the way. After arriving in Paro, Nakao spent five months intensively exploring Tremo La, near the Jomolhari mountain, Lingzhi, Laya, and Sela La. He mentioned that at Tremo La, more than 100 mules passed every day. He had a keen eye and noted that Bhutanese passers-by were farmers selling extra rice in Tibet, whereas Tibetans were professional merchants. He was in fact the last visitor who could experience this bustling trade because the border was eventually closed in 1959.

When Nakao returned to Jigme Dorji's house in Ha from his trip in early August, he was informed that the Indian Prime Minister Pandit Jawaharlal Nehru and his daughter Indira Gandhi had planned to visit Bhutan and to follow Nakao's route. In fact, they did their journey in September and proposed the construction of a road between Phuentsholing and Paro, an idea that the third king, Jigme Dorji Wangchuck, did not like (Kuwabara *et al.*, 1978).

Three years later, Doig (1961) was the first British journalist to visit Bhutan from Phuentsholing. He was travelling with an Indian engineer commissioned to select a corridor for building a road. In his writings, he reported that in the same year 'the King and Prime Minister of Bhutan visited New Delhi for talks with India's Prime Minister, Mr. Jawaharlal Nehru, and Indian government officials' and that 'Bhutan agreed to accept aid from India for a more extensive development program'. He also specified that a new road would link Paro Dzong to Jainti (India), therefore allowing 'the first vehicles to penetrate a land where the wheel has been virtually unknown for transportation' (Doig, 1961). The road, which was eventually constructed in 14 months and opened in 1962 (Department of Road, 2008), measured '112 miles northward from the Indian border to span the 45-mile crow's-flight distance, reducing the journey to six hours by jeep' (Scofield, 1974).

When Matsuo and his group visited Bhutan during October–December 1969, they left a detailed report on the east–west road that was under construction (Kuwabara *et al.*, 1978). As a professor of civil engineering, Matsuo's observations of the road conditions were critical and precise. He wrote that the road was paved from Paro to Thimphu, but not to Wangdue Phodrang. While Matsuo was concerned about road maintenance as the road was being constructed over fragile granite, he appreciated the gently laid out old mule tracks that were designed in harmony with nature. He also commented on wages and the other working conditions of the workers who came from India and Nepal.

In short, by 1969, two north–south routes had been built, and another route in the middle and the Wangdue Phodrang–Ngatshang part of the east–west route were under construction (Figure 15.2). The north–south highways overlapped traditional tracks, whereas the east–west roads often crisscrossed trails, thus rapidly reducing roadless areas from the Indian side. Scofield (1976), who made a round trip of the east–west highway by jeep, explained that construction of an east–west road in Bhutan had been triggered by the takeover of neighbouring Tibet by China in the 1950s. He also mentioned that road construction was possible because each family provided four-week labour and 25,000 people were imported from Nepal and paid under an Indian aid program. The result was not only the creation of the east–west national highway as a backbone of the country, but also an increase of Indian influence from the south, as symbolized by the construction of three south–north roads

Figure 15.2 Road network in Bhutan in 1969.

(Figure 15.1). All of the northern border passes previously used by Western missions in the 1900s were officially closed: the Bhutanese were forced to turn south to exchange their rice for textiles or salt, and to use roads instead of trails.

After the coronation of the fourth king, Jigme Singe Wangchuck, in 1972, Bhutan officially opened its doors to non-Indian foreign tourists, though the access of visitors was limited to ground transportation from neighbouring Indian towns until the Paro airport was officially opened in 1983 (Brunet *et al.*, 2001). Since October 3, 1974, the first tourists reached the country for a one-week stay and in that same year, the network of protected areas, including the Jigme Dorji Wildlife Sanctuary, started being developed (Tshering, 2009).

When Nakao revisited Bhutan in 1981, again arriving from Phuentsholing, he was delighted to see that what required an eight-day walk in 1958 was now an easy six-hour drive to Paro (Nakao, 2004). According to Ueda (2006), Bhutan had 700 automobiles by the time of Nakao's second visit. Today, the proportion of foreign visitors that enter through Phuentsholing or Samdrup Jongkhar is limited as 76 per cent of tourists reach Paro by airplane (International Tourism Monitor, 2007). Hiking from border towns is no longer possible as no trails or bungalows remain. Yet, trails that are used for trading with Tibet, located on the north side of the east–west highway, remain and have been extended to allow travel within the border.

Trail-to-road transitions and trailhead conditions in Jigme Dorji National Park

Jigme Dorji National Park, originally established as a wildlife sanctuary in 1974, was designated as a national park in 1993 and its boundaries were modified (Tharchen, 2013). The park has three south–north trails connected to the Snowman Trek, which runs along the northern border (Figures 15.1 and 15.3). Until the official border closure in 1959, these trails were used by traders with Tibet as well as by yak herders making transhumance or seasonal migrations. The western part of the Jomolhari Trek from Drukgyel Dzong includes a stone-paved branch to Tremo La, whereas the eastern part, from Dodina, leads traders to Lingzhi La. The Laya-Gasa Trek is connected with Tibet through Tomo La and Wagya La near Masa Gang peak (Table 15.2).

The western part of the Jomolhari Trek is the most popular route in Jigme Dorji National Park. Its official trailhead is Drukgyel Dzong, but an unpaved feeder road goes five kilometres beyond it and reaches a suspension bridge over the Paro River at Mishi Zampa, therefore allowing local taxis to get to this point. In fact, many trekkers prefer to use cars to reduce the number of hours walked on their first day (Mayhew, 2011), confirming a common behaviour of destination-oriented trekkers (Shiratori and Ito, 2001). Paving this road would move the trailhead to the bridge and, if a bridge for cars were built, the trailhead would move further north as it has already occurred in the Laya–Gasa Trek. Such a development would dramatically decrease the roadless area, allowing trekkers to easily bypass the road-side communities, and eventually favouring the so-called 'windshield wilderness' (Louter, 2006).

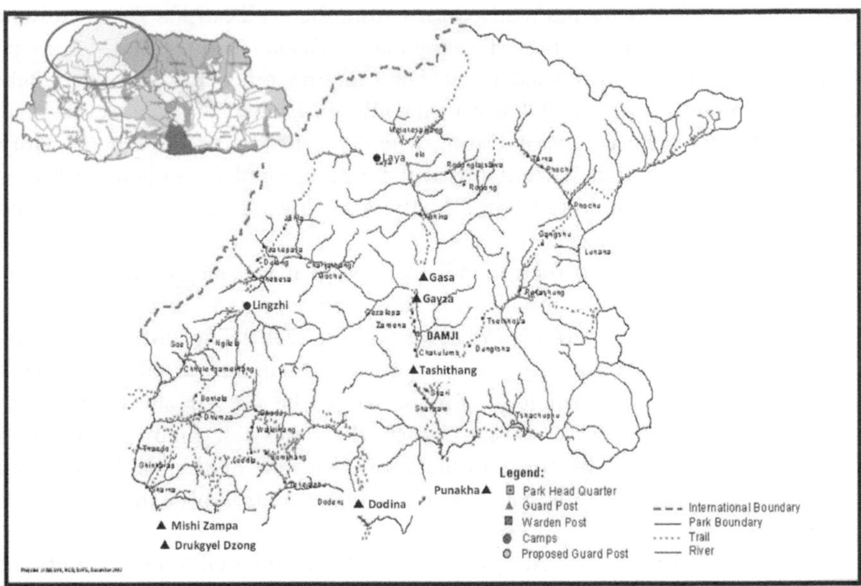

Figure 15.3 Map of Jigme Dorji National Park (based on Tshering, 2009).

Table 15.2 The three trails within and near the Jigme Dorji National Park.

Trails	*Trailheads in 1998 (Distance from Gateway Towns)*	*Trailheads in 2011*	*Trailheads-Park Boundary Relations*	*Passes to Tibet*
Jomolhari Trek (West)	Drukgyel Dzong (14 km from Paro)	Drukgyel Dzong/Mishi Zampa with 4WD	Outside the Park	Tremo La
Jomolhari Trek (East)	Dodina (13 km from Thimphu)	Dodina	On the Boundary	Lingzhi La
Laya-Gasa Trek	Tashithang (25 km from Punakha)	Gayza/Gasa	Inside the Park	Ya La

The official trailhead of the eastern part of the Jomolhari Trek is at Dodina, 15 kilometres from Thimphu. Western travellers used this trail to leave Bhutan in 1905, whereas Nakao used it in 1958 to get to Lingzhi and then reach Punakha through Laya and Gasa. As the local users of this trail are affiliated with the Chari Gonpa or monastery, a further extension of the road is unlikely, ensuring that the trailhead and the park boundary will coincide.

The third trek, the Laya–Gasa Trek, has changed drastically since Nakao walked it in 1958, and following the road construction, the trailhead has shifted toward Gasa every year. The first edition of *Lonely Planet Bhutan* (Armington, 1998) mentioned that the road from Punakha ended at a forestry office at Tashithang, where the district boundary is located. The fourth edition (Mayhew, 2011), however, mentioned that the trailhead was at Damji, where the headquarters of Jigme Dorji National Park are located, but noted that the road reached the suspension bridge beyond the Gayza community. As of 2011, the road almost reached Gasa and would be extended to Laya.

Nonetheless, road construction beyond Damji may cause occasional landslide problems during the monsoon season (Figure 15.4e), as emphasised by damages following the severe rainfall of August 2010. The district administrator of Gasa, Dasho Sonam Jigme, proposed to maintain the old mule track to let people bypass the roadblock (Kuensel 2010), because an alternative route would require a one-week walk to reach the east–west highway. His proposal was also inspired by the consideration that yak herders and horsemen prefer to use trails, and walking on roads is not exactly a pleasant experience for trekkers.

Comparison of road and trail routes to Gasa obtained by GPS shows significant differences. The trail via suspension bridge (Figure 15.4b) is much shorter than the road, though its use has declined once the road to Gasa Dzong was completed (Figure 15.4d). Moreover, after road completion, the kiosk near the suspension bridge was taken down and the trail cut off (Figure 15.4a).

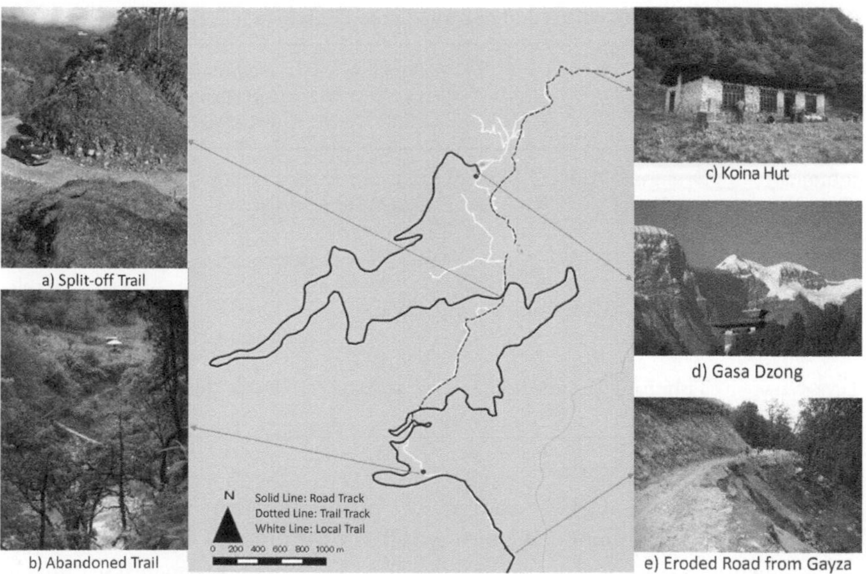

Figure 15.4 Comparison of road and trail routes to Gasa obtained by GPS. Locations are depicted where the trail has been cut off (a) or abandoned (b), where important tourism sites are found (c, d) and where road construction may cause occasional landslide problems (e).

Sustainable transportation in Bhutan

Bhutan has 3,636 kilometres of road network and further road extension is proposed (The Planning Commission Secretariat, 2000). The Bhutan 2020 strategy (Planning Commission, 1999) raises concerns about the social impacts of road construction specifying that some villages, located along mule tracks, could be isolated by the construction of nearby roads. Yet, the strategy's emphasis is on transportation development, recognising that '75% of rural population lives within half-day walk from nearest road' and that an expanded road network would transform 'the lives of many who have traditionally lived in isolation' (Planning Commission, 1999).

Recent government transportation policy (Royal Government of Bhutan, 2010) mentions words such as 'electric tram/train, ropeways or cable car network, and helicopter operations'. With the exception of electric tram/train, the other modes of transportation are already used in Bhutan, though helicopters are only operated in case of emergency. Regarding ropeways, there is one (Tashira Ropeway) on the Gangte Trek that climbs 1,340 metres to carry goods as well as people (twice a day) up to the village (Mayhew *et al.*, 2011). In addition, a cable railway was installed to carry building materials at Tango monastery, which can be reached by a one-hour walk from the trailhead (Figure 15.5). While these transportation facilities are suitable on steep slopes, an electric tram/train system would fit relatively well on flat areas, though it could serve steeper terrains if provided with a pinion and rack system. The environmental impact of each of these transportation systems is definitely less than that of road-based transportation, but they require a certain amount of investment and proper maintenance for use as public transit systems.

While monorail systems have not been considered in the above mentioned transportation policy, small monorails of the kind used for agriculture and forestry purposes seem promising as an alternative mode of transportation in mountain

Figure 15.5 The cable railway installed at Tango Monastery for carrying building materials.

protected areas with residents and trekkers (Figure 15.6). These small monorail systems, often used to carry fruits over terraced orchards, were developed in Japan in 1966, but have recently evolved to carry passengers as well as heavy construction materials in mountainous areas. According to the Japan Monorail Technology Association (2014), ten small companies located in citrus producing regions in Japan are supplying diverse monorail systems. Yamada *et al.* (1996a) showed that the use of monorails can dramatically reduce risks for forest workers operating in very steep terrains. The advantages brought in by monorails to communities are shown by the experience of the Hinohara Village (Tokyo Prefecture), which developed five lines of monorail to carry old people who live in houses located two kilometres from the road (Tokyo Shimbun, 2010). The technical and economic advantages of monorails are many. The corridor required by a monorail is as narrow as a trail and the train can climb 45-degree slopes with just a small engine (Figure 15.6). The construction and maintenance costs of a monorail are overwhelmingly lower than those of road-based transport modes (Yamada *et al.* 1996b) and three workers can set up 60 to 90 metres of rail per day and remove 90 to 120 metres per day (Furukawa, 2006). Finally, a monorail system can easily be built and maintained by the locals, this making it a 'technology with human face' or 'intermediate technology' (Schumacher, 1973).

Beyond the mere technological issues, however, the protection of shrinking roadless areas, the design of new roads and the establishment of alternative transportation systems should all be based on a thorough understanding of local people's preferences and expectations. Automobile access to Chimi Lhakhang, for example, was denied once the area has been identified as a holy place (Imaeda, 2001). Cultural and religious heritage

Figure 15.6 Eight-passenger monorail crossing a trail at Palau Eco Park (Courtesy of Masanori Take, Sep. 2014).

may in fact restrain access as shown by limitations to mountain climbing in nearby Nepal. However, imposing a strict protection status in areas with a significant resident population is difficult and can rarely be justified (Neumann, 1999). The residents of Gasa, for example, invoked the extension of the road to Gasa Dzong (Kato, 2009), as stated in the management plan (Tshering, 2009): 'an additional extension of about 22 kilometres from Damji to Gasa is expected to be completed in the next three years'. Unfortunately, when the road was eventually completed near Gasa Dzong in 2011, the old trail was cut off (Figure 15.4). In fact, park managers need persuasive frameworks to find compromises between road expansion and trail preservation that satisfy both residents and trekkers, and help preserve not only biodiversity but also cultural diversity as stipulated in the Gross National Happiness policy (Gurung and Seeland, 2008).

In this respect, many decades ago, Benton MacKaye, a forester and regional planner in the eastern United States, had suggested using trails as a tool to control mechanical civilisation (MacKaye, 1921). Using a water-based metaphor, he proposed controlling metropolitan flow with 'dams' and 'levees' (MacKaye, 1990). The 'dams' are interpreted as filters for traffic control such as trailheads or gates, while the 'levees' can be zones consisting of steep slopes or green belts to confine road influences. It is important to mention that MacKaye's idea developed from concerns about the rapid expansion of roads and automobiles in the United States after World War I: a situation that somehow resembles that of Bhutan after 1962. MacKaye's 'dams and levees' concept considers two approaches encompassing six access options to control development in the protected areas of Bhutan (Figure 15.7). These approaches, which need to be employed together according to circumstances, are explained below considering the case of the Laya-Gasa trek.

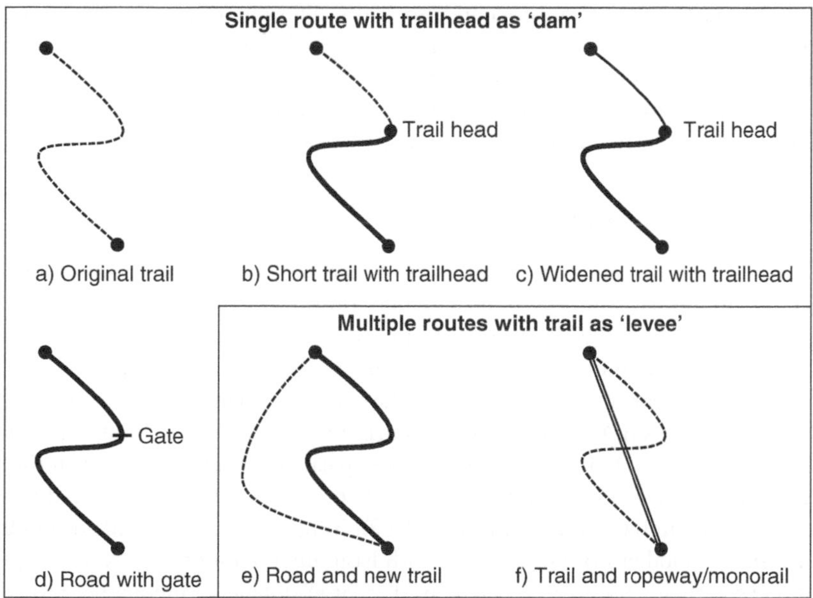

Figure 15.7 The two approaches and six access options that can be used to guarantee trail sustainability (source: adapted from Ito, 2011).

Single route with trailhead as 'dam'

When residents in isolated roadless areas such as Laya prefer traditional life without auto access, impacts on the natural environment as well as on cultural heritage can be kept to a minimum by simply leaving the traditional trail and not pursuing any other development (Figure 15.7a). When residents accept a one-day walk instead of a longer one (e.g. two days to Gasa), the road is extended to a certain point such as the Koina hut (Figure 15.4c), which is located halfway between Gasa and Laya (Figure 15.7b). In case a development is desired by the local community, an intermediate option adopted in Bavaria (Germany) could be applied to Bhutan. This is a consensus building process used in roadless 'Almwirtshaft' (summer livestock farms in the mountainous zone of Bavaria), which are supported by the local government not only for ecological or economic reasons, but also for the protection of the cultural landscape and the traditional environmental knowledge (Ichida-Iwata, 1997). According to this model, a subsidy is used to maintain huts, which are used by both farmers and hikers, and to transport heavy materials such as farming tools. Some 'Alms' are located in rather inaccessible areas, and farmers take their livestock on foot and carry heavy equipment by helicopter. The 'Almwirtshaft' model considers that only small trail improvements are acceptable for allowing the passage of small tractors (Figure 15.7c). In Bhutan, this kind of moderate development could be pursued to preserve distinct cultural landscapes consisting of small temples ('chorten') and traditional cantilever bridges. Small monorail systems as those described above, could be used for transportation of heavy materials as they have minimal widths (i.e. one metre).

If the road is extended to isolated communities, traffic could be regulated using gates as it is done in various protected areas around the world to preserve particularly sensitive sites (Daishima and Ito, 2006). The gate can be opened at a specific time of the day to specific vehicles such as public buses or cars with permits (Figure 15.7d). This system requires gatekeepers, and their wages should be covered by bus fares or parking fees. Such revenues could also be used to subsidise horsemen for sustaining their traditional environmental knowledge.

Multiple routes with trail as 'levee'

Both trails and roads are recognised in landscape ecology, but the former represent an environmental resource, while the latter constitute a disturbance (Forman and Godron, 1986). MacKaye's 'levees', interpreted as environmental resources, go side by side with roads, but are sufficiently separated from these to be protected from the sound and sight of automobiles (MacKaye, 1990). When a trail is overtaken by a road, as it happens in the area near the Indian border, new trails should be considered (Figure 15.7e). For instance, the 1938 Appalachian Trail Agreement in the US stipulated that no road be built within a mile of a trail in federal land, and within a quarter-mile in state land (Anderson, 2002). In fact, while roads and automobiles can easily be hidden from one's view with basic interventions (e.g. design, vegetation cover), their sound can be perceived at large distances. Adequately protected trails could be used by horsemen or yak herders in Bhutan and, if a certain number of trekkers use them, bungalows or rest houses, like those observed by Nakao and Doig in the 1950s, could be built and managed by local communities (Ito and Kato, 2010). Such trails may also work as emergency access routes in areas particularly susceptible to natural disasters (e.g. landslide) where infrastructures may remain out of service (e.g. road blocked by debris) for long periods.

Before starting any road development project, the use of alternative transportation modes such as monorails or ropeways should be considered (Figure 15.7f). These do not always rely on fossil fuels and, when powered by electricity (which may not be available anyway), generate minimal noise emissions, therefore responding to the government's call for environmentally friendly transportation systems (Royal Government of Bhutan, 2010). In fact, while the strict definition of 'wilderness' excludes any motorised access, the environmental impact of alternative transportation systems like monorails seems largely acceptable.

Conclusion

Until 1961, the southern border towns of Bhutan were trailheads for de facto protected areas or inhabited wilderness. Now roads are expanding farther north and in many cases have entered protected areas. If the northern borders were opened, roads from Tibet would be built as it happened in Mustang (Nepal), to create the Trans-Himalayan Highway (Mayhew and Bindloss, 2009). This would result in further reduction and fragmentation of wilderness areas. Secure establishment of trailheads and the management or creation of trails is indispensable to maintain biodiversity, cultural diversity and heath for residents as well as trekkers. In order to keep roadless conditions, other more sustainable modes of transportation, such as monorails, should be considered.

References

Anderson, L. (2002) *Benton MacKaye*, The Johns Hopkins University Press, Baltimore, MD.

Armington, S. (1998) *Bhutan, 1st ed.*, Lonely Planet, Victoria, Australia.

Brunet, S., Bauer, J., De Lacy, T. and Tshering, K. (2001) 'Tourism development in Bhutan: Tensions between tradition and modernity', *Journal of Sustainable Tourism*, vol 9, no 3, pp. 243–263.

Catton, T. (1997) *Inhabited Wilderness*, University of New Mexico Press, Albuquerque, NM.

Daishima, K. and Ito, T. (2006) 'Automobile restriction in protected areas as a management tool' in Jim, C.Y., Corlett, R.T. (eds) *Sustainable Management of Protected Areas for Future Generations*, IUCN: Gland, Swizerland and Friends of Contry Parks: Hong Kong, China, pp. 211–218.

Dawson, C. P. and Hendee, J. C. (2009) *Wilderness Management, fourth edition*, Fulcrum Publishing, New York, NY.

Department of Roads (2008) 'About Us, Ministry of Works & Human Settlement,' www.dor.gov.bt/Others/aboutus.htm, accessed 5 July 2011.

Department of Tourism (2007) International Tourism Monitor Annual Report, Department of Tourism, Thimphu, Bhutan.

Doig, D. (1961) 'Bhutan, the mountain kingdom', *National Geographic*, vol 120, pp. 384–414.

Forman, R.T.T. and Godron, M. (1986) *Landscape Ecology*, John Wiley & Sons, New York, NY.

Furukawa, K. (2006) Monorail for Thinning Operation. Forest Report of Gifu Prefectural Research Institute for Forests, vol 649 [in Japanese].

Gurung, D. G. and Seeland, K. (2008) 'Ecotourism in Bhutan: extending its benefits to rural communities', *Annals of Tourism Research*, vol 35, no 2, pp. 489–508.

Ichida-Iwata, T. (1997) 'Agriculture, nature conservation and leisure in mountainous region: a case study in Bavaria', in *Mountain Village Development and Environmental Conservation*, Nansosha, Tokyo, pp. 272–290 [in Japanese].

Imaeda, Y. (2001) *Enchanted by Bhutan*, Iwanami Shoten, Tokyo [in Japanese].

Ito, T. (2011) 'Road expansion and its influence on trail sustainability in Bhutan', *Forests*, vol 2, no 4, pp. 1031–1048.

Ito, T. and Kato, G. (2010) 'The role of hut system in promoting community-based tourism in Bhutan', *Global Partnership*, vol 5, pp. 1–12.

IUCN (2014) International Union for Conservation of Nature, www.iucn.org, accessed 21 October 2014.

Japan Monorail Technology Association (2014) 'About Us', www.monorail-kk.jp, accessed 30 August 2014.

Kato, G. (2010) Development of Trekking in Bhutan as a Form of Sustainable Tourism, MS thesis, University of Tsukuba, Japan [in Japanese].

Kuensel (2010) *Remote Gets Remoter: Road Slip Brings a Host of Problems in the Northernmost District*, Kuensel Corporation, Thimphu, Bhutan.

Kuwabara, T., Matsuo, M., Kurita, Y., Yoshino, H. and Tani, Y. (1978) *Crossing Journey of Bhutan*, Kodansha, Tokyo [in Japanese].

Louter, D. (2006) *Windshield Wilderness: Cars, Roads, and Nature in Washington's National Parks*, University of Washington Press, Seattle, WA.

MacKaye, B. (1921) 'An Appalachian trail: a project in regional planning', *Journal of the American Institute of Architects*, vol 9, pp. 325–330.

MacKaye, B. (1990) *The New Exploration: A Philosophy of Regional Planning*, Appalachian Trail Conference and the University of Illinois Press, Harpers Ferry, WV.

Mayhew, B. (2011) *Bhutan, 4th ed.*, Lonely Planet, Victoria, Australia.

Mayhew, B. and Bindloss, J. (2009) *Trekking in the Nepal Himalaya*, Lonely Planet, Victoria, Australia.

Nadik, T. D. (2008) 'Tariff Change and Trends in Tourists Arrivals', Tourism Council of Bhutan, Thimphu, Bhutan.

Nakao, S. (1959) *Unexplored Bhutan*, Mainichi Shimbun, Tokyo [in Japanese].

Nakao, S. (2004) *Exploring Natural History, Works of Nakao Sasuke, vol 3*, Hokkaido University Press, Sapporo, Japan [in Japanese].

Neumann, R. P. (1999) *Imposing Wilderness*, University of California Press, Berkeley, CA.

Ogata, Z. (1969) *A Rough Sketch of Bhutan*, Fuyo Publishing, Tokyo, Japan [in Japanese].

Planning Commission Secretariat (2000) Bhutan National Human Development Plan Report 2000, Royal Government of Bhutan, Thimphu.

Planning Commission (1999) Bhutan 2020: A Vision for Peace, Prosperity and Happiness, Part II, Royal Government of Bhutan, Thimphu, Bhutan.

Rinzin, C., Vermeulen, W. J.V., Glasbergen, P. (2007) 'Ecotourism as a mechanism for sustainable development: the case of Bhutan', *Environmental Sciences*, vol 4, no 2, pp. 109–125.

Royal Government of Bhutan (2010) 'Economic Development Policy of the Kingdom of Bhutan', http://rtm.gnhc.gov.bt/wp-content/uploads/2013/10/EDP.pdf, accessed 10 Oct. 2011.

Schumacher, E. F. (1973) *Small is Beautiful*, Harper & Row Publishers, New York.

Scofield, J. (1974) 'Bhutan crowns a new dragon king', *National Geographic*, no 133, pp. 546–571.

Scofield, J. (1976) 'Life slowly changes in a remote Himalayan Kingdom', *National Geographic*, vol 135, pp. 658–683.

Shiratori, K. and Ito, T. (2001) 'Motorized Access Control as a Wildland Recreation Management Tool: Access Changes and Visitor Behavior at Daisetsuzan National Park', in T. Sievanen, C., Konijnendijk and L. Langner (eds) *Forest and Social Services—the Role of Research*, Finnish Forest Research Institute, Helsinki, Finland. vol 815, pp. 29–41.

Tharchen, L. (2013) *Protected Areas and Biodiveristy of Bhutan*, CDC Printers, Kolkota, India.

Todd, B. K. (1952) 'Bhutan, land of the thunder dragon', *National Geographic*, vol 102, pp.713–754.

Tokyo Shimbun (2010) 'Monorail to Okutama-Hinohara', newspaper article dated May 18 [in Japanese].

Tshering, K. (ed) (2009) Integrated Conservation Management Plan; Jigme Dorji National Park 2009–2012 Draft, Jigme Dorji National Park: Damji, Bhutan.

Tshering, K. and Christ C. (2012) 'Ecotourism Development in the Protected Areas Network of Bhutan', Royal Government of Bhutan, Thimphu, Bhutan.

Ueda, A. (2006) *Concept of Development in Bhutan*, Akashi Publisher, Tokyo [in Japanese].

US Wilderness Act (1964) Public Law 88–577 (16 U.S. C. 1131–1136) 88th Congress, Second Session.

Wangchhuk, L. (2010) *Facts about Bhutan, 2nd ed.*, Absolute Bhutan Books, Thimphu, Bhutan.

White, J. C. (1914) 'Castles in the air', *National Geographic*, vol 25, pp. 365–453.

White, J. C. (2011) *Sikkim and Bhutan, 1909*, Reprinted by Kessinger Publishing, Whitefish, MT.

Yamada, Y., Aoi, T., Minato, K., Yoshimura, T. and Owari, T. (1996a) 'Effects of using a monorail for riding in forestry operations – effect of reducing the working burden', *Journal of Japanese Forestry Society*, vol 78, no 3, pp 314–318 [in Japanese with English summary].

Yamada, Y., Yoshimura, T., Aoi, T., Minato, K. and Owari, T. (1996b) 'Economic effects of using a monorail for riding in forestry operations', *Journal of Japanese Forestry Society*, vol 78, no 4, pp 419–426 [in Japanese with English summary].

16 Case studies

Lessons learned

Francesco Orsi

Chapters 8 through 15 have provided a wide set of case studies targeting a variety of themes pertaining to the book's core topic: from transportation modes to visitor preferences, from access policies to mobility plans. The selected case studies offer an overview of different contexts, including designated (e.g. national parks) and non-designated natural areas, in Europe, the United States and Asia. This chapter aims to briefly review each of the case studies, highlighting peculiar elements that can inform future research and drive concrete actions to foster sustainable transportation in natural settings.

Chapter 8 explores the challenges associated with analysing tourism traffic and estimating the effects of traffic management measures on visitor flows in tourism regions. The latter are territories characterised by an outstanding environmental and cultural value that offer a wide array of spatially distributed tourist attractions. The fact that these regions may be rather densely populated makes them a perfect context for establishing sustainable forms of mobility because the existing public transit system can be improved to adequately serve both residents' and visitors' needs. As opposed to remote designated areas (e.g. national parks of the American West), tourism regions are often working lands, namely areas where people live and work. Here daily life and tourism coexist, thus meaning that any management strategy should account for the needs and expectations of local inhabitants.

The research approach presented in the chapter unveils the importance of analysing transportation-related data and policies at a scale that is large enough to capture the inherent complexity of transportation dynamics. This is certainly not easy as it involves building large databases on traffic and visitor behaviour, and simulating the overall effects of a given policy. The chapter also shows that a combination of incentives ('carrots') to alternative transportation and disincentives ('sticks') on car use did not reduce tourist demand, but did reduce the environmental impact of tourism. While this contradicts the doubts of many operators about the detrimental effects of sustainable transportation measures on tourism, the actual possibility to preserve an adequate demand ultimately depends on the quality of the offer. In fact, strong investments in sustainable forms of mobility can create a unique tourist product, which may attract the vast tourist segment that looks for car-free vacation opportunities.

Chapter 9 presents two successful mobility initiatives – Bus Alpin and AlpenTaxi – promoted in Switzerland to guarantee car-free access to remote tourist destinations through on-demand public transport services. While 'last mile' transfers are often an insurmountable obstacle for public transit providers, the two Swiss experiences demonstrate that a combination of local-scale transport services, fares and sponsorships can sustain alternative transportation even beyond settlements.

Nonetheless, besides the guidelines provided by the authors, three preconditions seem very important for the actual success of such experiences in other countries. The first one is the presence of a widespread and reliable public transportation network on which to build these 'last mile' services. Such a network would encourage many to choose alternative transportation, therefore guaranteeing adequate ridership to these services. The second one is the concurrent adoption of measures aimed at limiting private motorised vehicles. In fact, 'last mile' services get particularly competitive when they offer access to areas that are not reachable via private transportation (e.g. road forbidden to cars). The third one is the ability of a tourism region (given physical characteristics and tourism approach) to attract tourist segments practicing recreational activities that can take a greater advantage of 'last mile' services. The possibility to easily walk one-way itineraries (e.g. hikes starting at point A and ending at point B), for example, is very important for many visitors and can take great advantage of 'last mile' public services.

Given the concepts outlined in Chapter 2, on-demand services like those adopted in Switzerland greatly contribute to the sustainability of a transportation system as they enable people who cannot use a private vehicle to reach remote locations. Hence, administrators and park managers should not just look at these services as fair transportation modes, but rather as opportunities to attract all those visitors who do not want to or cannot use a private vehicle.

Chapter 10 presents a socio-psychological method to test the attitudes of national park visitors towards changing their transportation behaviour, and in particular reducing their reliance on the private vehicle. The study provides helpful suggestions about how to interpret visitors' behaviour and addresses transportation management measures. The assessment of visitor attitudes showed that, depending on the visitor segment, the awareness of a behaviour's consequences (i.e. negative impacts of car use) may be higher or lower than the actual willingness to modify such behaviour and that, no matter the segment, the awareness of the difficulty of modifying behaviour largely prevails over the willingness to modify it.

These findings raise concerns about the accuracy of stated preference surveys in estimating modal shift in protected areas, suggesting that the introduction of a measure to favour alternative transportation may have unexpectedly positive (i.e. more people than expected choose alternative transportation) or negative (i.e. less people than expected choose alternative transportation) consequences. Nonetheless, administrators and managers should be aware that the characteristics (e.g. group size, age, preferred recreational activity) of a segment are a key determinant of the actual possibility to choose alternative transportation (e.g. wilderness lovers may need to reach trailheads that cannot be served by shuttle bus services) and they should plan policies accordingly. The idea of planning ad hoc measures for each segment based on its ability to spend and its willingness to change transportation behaviour seems strategic to building comprehensively sustainable transportation systems. Despite all considerations on visitor preferences, however, we should not forget that transportation policies in natural settings might have a very positive impact on cultural change, showing people that alternative transportation is not just acceptable but also convenient.

Chapter 11 explores the initiatives undertaken by the Peak District National Park (UK) to encourage cycling to and within the park. The strategy adopted in this context, encompassing actions on the cycling path network, the cyclist-friendly infrastructure, sustainable transport packages and promotion, addresses all of the key elements contributing to a marked increase in the share of bicycle transportation in the medium to long term.

However, while cycle-tourism is a rapidly growing tourism sector, many natural areas around the world still lack the basic facilities that allow this activity to be performed by all visitor groups in a safe manner. The decision to fill this gap will in many cases be driven by the recognition of the huge economic revenues that cycle tourism may bring to an area. Yet economic interests cannot lead to unreasonable projects (e.g. cycling paths in pristine environments), which generally determine unintended consequences (e.g. unregulated development around the cycling path causing damages to flora and fauna). In particular, administrators must be aware that the promotion of cycling in the absence of adequate transport facilities to/from the natural area (e.g. cycling path between the city and the natural area, trains with bike racks) may lead to the paradox of a sustainable transportation (i.e. the bicycle) favouring unsustainable transportation (i.e. cars of cyclists reaching the area).

The Peak District National Park, in line with other protected areas surrounded by cities, has considered the importance of allowing city dwellers to bike to the park: in fact, this seems an effective way to both reduce motorised transport and sustain the economy of the region (i.e. cyclists rely on a series of businesses on their way to the park). Finally, it is important to observe that the bicycle is a private vehicle and as such one that prevents managers from having a thorough control over visitor movements across an area. Hence, as natural areas become popular cycling destinations, ad hoc management strategies are needed to minimise impacts on the environment and recreational quality.

Chapter 12 investigates visitor preferences towards transport modes and access points to a hiking area in the Dolomites (Italy), and analyses mode choices under a series of management strategies. The chapter conveys two main messages. The first one, which reflects the findings of previous studies, is that modal shift from the private vehicle to alternative transportation can be significantly increased only if the use of the former is discouraged through some disincentive. The second one is that any incentive to alternative transportation and disincentive on car use must be carefully designed to avoid both a significant economic loss and an unsustainable reconfiguration of visitor flows.

In particular, the chapter focuses on the Alpine context, where the coexistence of multiple transport modes (e.g. buses, cableways) serving multiple locations may generate unexpected consequences as visitors shift from one mode to another (e.g. excessively high use of cableways may increase crowding in high-elevation areas). Hence, this chapter provides further evidence of how big the power of transportation as a management tool can be. Any decision regarding transport characteristics (e.g. capacity, frequency, fares) may have a deep impact on the movement of people, eventually determining use levels across an entire area. As already emphasised in Chapter 3, the fact that a transport mode has a low environmental impact (e.g. minimal or no carbon emissions) does not mean per se that its adoption will be sustainable. Indeed, managers and administrators have to define transport management strategies ensuring that standards of quality are met throughout the natural area.

Chapter 13 presents the current state of alternative transportation in US national parks by considering the types of transit systems in units of the National Park Service (NPS), the types of business models and funding sources, and the amount of passenger boardings per transit system and unit. Given the visitation rates to and the diversity (i.e. from small moderately wild units to huge wildernesses) of the US national park system's units, the review provides administrators and park managers worldwide with some tips about how alternative transportation could/should be managed.

In particular, various business models, ranging from full ownership to concession contract to cooperative agreement, are shown as valid alternatives to provide effective service depending on the context. The significant fee (i.e. up to US $20–25) required to drivers for entering many national parks and the fact that a considerable share (i.e. around 50 per cent) of it is used in some cases to run transit systems suggest that in highly popular protected areas fees can be increased to sustain highly efficient alternative transportation systems. In this respect, it is crucial to find an acceptable trade-off between the entrance fee and visitors' willingness to pay so that both visitation rates and revenues can be sustained. The idea of charging most of a transit system's cost to private vehicle users, though unpopular, might be an option to acknowledge the responsible behaviour of alternative transportation users and eventually favour modal shift.

Chapter 14 presents a multi-stage process for the involvement of stakeholders in the definition of sustainable mobility strategies. The experience described, which may seem out of the scope of this book as it focuses on an urban area, is in fact perfectly relevant because it raises the important issue of sustainable transportation in gateway communities. The latter, which lie just outside pristine territories, have a stake in the management of a natural area, provide a number of tourist-related services, and serve as the official entry point for tourists, therefore constituting the first impression a visitor can get of an area. In small islands, as the distance between the city and pristine lands may be significantly limited, the sustainable management of transportation in and around the city is key to better preserving the environment and the recreational experience of incoming visitors.

The case study of Ponta Delgada shows that involving a wide range of stakeholders in the development of mobility plans is highly effective because updated data can be shared and a consensus between conflicting interests can be reached more easily. However, as stated in the chapter, the participatory process should not be limited to institutional stakeholders (e.g. city council, tourism agency), but involve the public and, in particular, residents and tourists. In fact, the peculiar role of gateway cities, which host a permanent population and welcome temporary visitors, calls for transportation management strategies that meet the expectations of both categories.

Chapter 15 explores the possibilities of coexistence of roads and trails in Bhutan, a country of immense natural value, growing tourist appeal and minimal transport infrastructure. In particular, two main approaches (i.e. single route with trailhead as 'dam', multiple routes with trail as 'levee'), encompassing six specific options, are proposed to meet the transport demand of different communities, while ensuring the survival of traditional trails. Such trails would allow the existence of small communities and road-less areas, and would therefore favour nature-based tourism in the country.

By proposing strategies, the chapter also raises crucial questions regarding the possibility to conciliate economic development and traditions, and the actual meaning of the term sustainable transportation in remote and largely pristine territories. In these contexts, a massive tourist presence, though relying on alternative forms of mobility, may have a significant impact on both natural resources and the local culture. The design and management of transportation infrastructures in this case should be informed by a comprehensive understanding of the potential large-scale consequences associated with moving visitors from one area to another. To this extent, the strategies proposed in the chapter seem very interesting because they depict the possibility of a two-level development, where a 'high-speed' road network providing easy long-distance mobility is not in conflict with a 'low-speed' trail network ensuring local and tourism-oriented mobility. Nonetheless, the success of similar strategies inevitably depends on national-scale policies that envision standards for visitor use across the country and govern visitor flows so as to meet such standards (including quotas on visas for entering the country).

Part IV

Policies – Strategies and policies for sustainable transportation

17 From conventional to sustainable transportation management in national parks

Robert Manning, Christopher Monz, Jeffrey Hallo, Steven Lawson and Peter Newman

This chapter is based on the following book:
Manning, R., Lawson, S., Newman, P., Hallo, J. and Monz, C. (eds) (2014) *Sustainable Transportation in the National Parks: From Acadia to Zion*, University Press of New England, Hanover, NH.

Transportation and national parks

Transportation and nature-based tourism attractions such as national parks are intimately linked. In the United States, for example, nearly 300 million visitors per year travel to, from and within the national parks. American national parks, which comprise over 80 million acres of public land, include extensive networks of transportation corridors – roads, trails, bike paths, waterways, public transit – that link a vast array of iconic attraction sites – viewpoints, historical and cultural sites, visitor centres, campgrounds and gateway communities. The inherent complexities of the intersection between transportation and national parks demand explicit management attention, encompassing an informed and sustainable approach.

However, transportation is more than a means of access to national parks and outdoor recreation: it can be a form of recreation itself, offering most visitors their primary opportunities to experience and enjoy the natural and cultural landscapes embodied by national parks. For example, the iconic roads of many of the US national parks (e.g. Going-to-the-Sun Road in Glacier National Park, Tioga Road in Yosemite National Park, and the Park Loop Road in Acadia National Park, among many others) were specifically designed for visitors to experience the parks in their vehicles and are important manifestations of the historic and contemporary linkages between transportation, national parks, and outdoor recreation. All of these roads were a response to demand for 'driving for pleasure', what is historically one of America's most popular recreation activities (Manning, 2011).

Transportation can be even more than this: it is also a potentially powerful tool for managing national parks and protected areas. As suggested in Chapter 2, transportation ultimately determines where park visitors travel (and where they do not) and can be used by park managers to help deliver the 'right' number of visitors to the 'right' places at the 'right' times (Manning, 2007; Lawson *et al.*, 2009; Manning, 2009; Lawson *et al*, 2011; White *et al.*, 2011; Whittaker *et al.*, 2011; Meldrum and DeGroot, 2012; Reigner *et al.*, 2012). In this way, transportation can be used to manage parks and outdoor recreation in a sustainable way by protecting park resources and the quality of the visitor experience (Manning, 2007, 2011).

The intersection between transportation and national parks in the US is as old as the national parks themselves (Runte, 1990, 2010, 2011; Carr, 1998, 2007; Nash, 2001; Havlick, 2002; Sutter, 2002; Louter, 2006; Johnson, 2012). While railroads provided the first access to many of the newly established national parks in the latter part of the nineteenth century, it was ultimately the automobile that opened the parks to the public at large. More recently, there has been growing sensitivity about potential impacts of cars in national parks, including air and noise pollution, disturbance of wildlife, and road and parking lot congestion. Serious proposals to eliminate cars from some of the most popular areas of the national parks (e.g. Yosemite Valley) have been vigorously debated in recent decades, with mixed results. For example, cars can still enter Yosemite Valley, but the National Park Service (NPS) closed the eastern end of the valley to private vehicles and implemented an alternative transportation system (ATS) in the form of a propane-powered shuttle bus system.

From conventional to sustainable transportation management

Conventional transportation management in protected areas is essentially 'demand-driven' as illustrated in Figure 17.1a. In this approach, transportation management responds to current and projected visitation with facilities and services designed and operated according to demand. This means that if demand exceeds supply, then facilities and services are expanded (e.g. roads are widened, parking lots are made bigger, shuttle bus service is improved). The demand-driven approach is designed to maximize efficiency and enhance the convenience of travel to and within national parks. However, a demand-driven approach to transportation management in national parks can facilitate unsustainable levels of visitation that impact park resources and cause visitor crowding and congestion (Burson *et al.*, 2000; Roof *et al.*, 2002; Ament *et al.*, 2008; Park *et al.*, 2009/10; Lawson *et al.*, 2011; Pettengill *et al.*, 2012; D'Antonio *et al.*, 2013; Taff *et al.*, 2013). In other words, conventional transportation planning in protected areas may lead to enhanced transportation facilities and services but may also unintentionally lead to degraded areas and a diminished recreational quality.

A more sustainable approach to transportation management is depicted in Figure 17.1b. According to this model, management objectives and associated indicators and standards for park resources and the visitor experience are formulated. Management objectives describe the level of resource protection that is desired in a park (or a site within a park) and the type and quality of visitor experience that is desired. Indicators are measurable, manageable variables that can be used as proxies for management objectives, and standards define the minimum acceptable condition of indicator variables (Manning *et al.*, 1996; Manning, 2007; Whittaker *et al.* 2012). Once indicators and standards have been formulated, then transportation planning and management is conducted within this park management framework. This leads to intentional improvements in both transportation and park

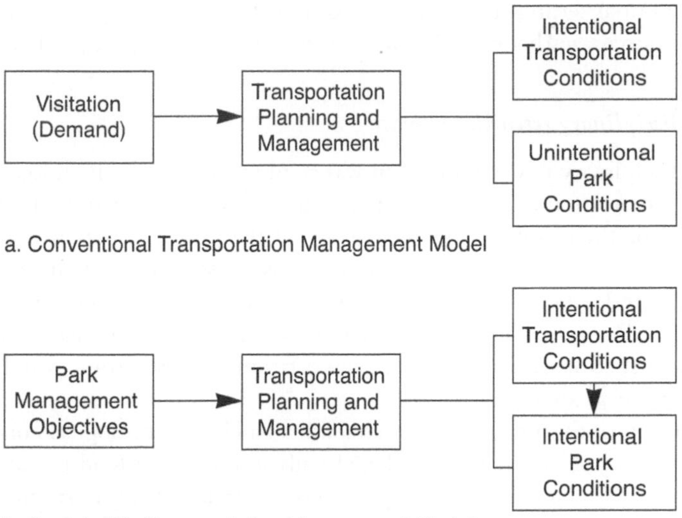

Figure 17.1 Conventional (demand-driven) versus sustainable transportation management in protected areas (source: Manning *et al.*, 2014).

resource and experiential conditions, and allows transportation to be used as a tool to help manage parks in a more sustainable manner (i.e. to meet resource protection and visitor experience objectives). The remainder of this chapter describes two programs of research conducted in the US national parks to illustrate the shift from conventional, demand-driven transportation management to a more sustainable approach to transportation management.

Conventional transportation management: the downstream consequences of demand-driven transportation

A recent study at Rocky Mountain National Park in Colorado (USA) illustrates the potential resource and experiential problems of conventional demand-driven transportation management (Lawson *et al.*, 2011; D'Antonio *et al.*, 2013). Rocky Mountain National Park protects over 400 square miles of the iconic Continental Divide landscape and includes the popular Trail Ridge Road, the highest continuous road in the US, and over 300 miles of hiking trails. In 1978, due to high and growing visitor demand in the Bear Lake Road Corridor of Rocky Mountain, a shuttle bus system was developed by the NPS to service the area (Gamble *et al.*, 2007). Ridership and the capacity of the shuttle bus system to the Bear Lake Road Corridor have increased substantially since its implementation, resulting in increased visitation to scenic and hiking destinations in this corridor. The transportation system in the Bear Lake Road Corridor now delivers high numbers of recreationists to some low-capacity destinations, such as lakeshores and viewpoints.

This situation has the potential to lead to unintended consequences for the resource and experiential settings in the Bear Lake Road Corridor: degradation of fragile natural resources, aesthetic implications of these resource impacts, and crowding.

An interdisciplinary research approach

An interdisciplinary research program was conducted to assess the resource-related impacts of recreation at sites within the Bear Lake Road Corridor and determine the effects of these impacts on the quality of the visitor experience. Data were collected that address ecological conditions, visitor standards for these ecological conditions, and a GPS-based assessment of visitor locations and densities. These three sets of data were combined in a variety of ways to integrate natural and social science information on the effects of transportation on park resources and the quality of the visitor experience.

A first step in this process was to establish a baseline of ecological conditions in visitor use areas off designated, hardened trails and sites that tend to exhibit more rapid ecological change. To accomplish this, off-trail locations within the study area were assessed and mapped for recreation disturbances. The second step was to determine areas where the intensity of resource damage has become unacceptable based on visitor standards for resource conditions. This step was accomplished by determining visitor standards for two important metrics of off-trail disturbance – loss of vegetation cover and the proliferation of informal trails – via a visitor survey that incorporated visual research methods (Manning *et al.*, 2004; Manning 2011). Subsequent GIS analysis allowed for a determination of locations where standards are currently violated. The third step involved the use of GPS-based tracking data to determine the percentage of time visitors experienced various levels of resource disturbance.

The study conducted at Rocky Mountain National Park showed that many recreation disturbances occur around lakes and this may result in impacts to lake ecosystems, such as increased turbidity and nutrient content (Hammitt and Cole, 1998). Moreover, the level of informal trail proliferation, and the associated ecological impacts to soil and vegetation, appear to be problematic throughout the trail system (D'Antonio *et al.*, 2013).

Findings from the visitor assessment of visual simulations of resource impacts were integrated with the assessments of current conditions to determine where resource change violated visitor thresholds of acceptability. Visitor contact with areas where informal trail density violates acceptability thresholds was estimated by determining the spatial overlap between visitor use and locations where high informal trail densities occurred. Overall, counts of the frequency of occurrence (i.e. the number of visitor use points that fall in areas where standards are violated) indicate that, on average, visitors spend 13 per cent of their trip to the Bear Lake Road Corridor (one of the most popular destinations in the park) in areas where the standard for informal trails is violated. In fact, the density of informal trails is not uniform throughout the trail system: changes in percentage reflect experiences in specific destinations. For example, when considering only visitors hiking to

Emerald Lake (one of the most popular destinations in the park), the results showed that visitors spend a much higher portion (23 per cent) of their hike in locations that violate standards for ecological conditions.

Recreation sites and areas of dispersed use were also analysed to identify where areas of visitor use overlap with areas that violate visitor standards for vegetation loss. Findings indicate that visitors who experienced areas of dispersed use were exposed to areas that violate the acceptability standard 48 per cent of the time, whereas visitors who were exposed to visitor-created trails experienced sites violating the standard 97 per cent of the time (D'Antonio *et al.*, 2013).

Impacts of demand-driven transportation management

The shuttle bus system in the Bear Lake Road Corridor of Rocky Mountain National Park has continued to expand since its inception in 1978, thus representing an example of conventional demand-driven transportation management. While this is a convenience to visitors, the shuttle bus system now delivers so many visitors to this area that its environmental integrity (e.g. soil compaction, vegetation loss, water pollution, disturbance of wildlife) has been threatened and the quality of the visitor experience (e.g. aesthetic degradation, loss of wildlife viewing opportunities) has been undermined. The 'downstream' effects of transportation management must be taken into account by shifting to a more sustainable approach to transportation management in which the environmental and social capacities of park sites are determined and an appropriate transportation system is designed and managed accordingly.

Sustainable transportation management: an integrated study of park road capacity

Denali National Park and Preserve (Denali) in Alaska (USA) is a large park comprised of approximately six million acres. However, the park has only one road: a narrow, low-speed route that takes a sinuous path over the park's dramatic and diverse terrain. Extending 91 miles from the park entrance to the old mining community of Kantishna where it dead-ends, the Denali Park Road traverses boreal forests and subarctic tundra, crosses rolling mountains and sheer cliffs, and meanders through scenic vistas and prime wildlife viewing areas. The first 15 miles of the road are paved and open to private vehicles, but after this, the road transitions to gravel and visitors are required to travel by bus. The Denali Park Road gives visitors of all abilities the opportunity to travel through a vast, rugged wilderness. As they travel the road, visitors observe wildlife in their natural habitat and enjoy outstanding scenery. While the current transportation system allows access to the park using the Denali Park Road, the number of bus trips is restricted.

Since the inception of the mandatory transportation system at Denali in 1972, some stakeholders have expressed concerns about the regulatory policy. Some have suggested that the policy did not provide for growth in park visitation or flexibility to meet changing needs of visitors, bus operators, and park resources.

Other stakeholders feel that the limit on vehicle trips did not adequately protect park resources or provide opportunities for visitors to choose park experiences that provide more opportunities for solitude. The transportation system for the Denali Park Road has never been comprehensively evaluated. Visitation at Denali is projected to increase and, along with it, the demand to travel the Denali Park Road.

An interdisciplinary program of research

The issues related to the management of the Denali Park Road have biological, sociological, and physical elements that require better understanding. Thus, managers decided to comprehensively re-evaluate road use limits in relation to concerns for wildlife and preservation of the high-quality experience associated with touring the park road. The goal of the research was to assess the effects of changes in traffic volume and patterns on important indicators of resource and social values by combining the results into a predictive model of detailed road traffic scenarios.

One of these studies was aimed at defining indicators and standards for the visitor experience (Manning and Hallo, 2010). Investigators employed qualitative interviews and quantitative surveys of park road users to identify and measure experiential indicators and standards. The standards for selected indicators could then be applied to predictive modelling to assess impacts on visitor experience of alternative management scenarios.

The visitor survey used normative theory and methods and visual simulations of a range of park conditions (Manning, 2004; 2011). The survey measured standards for five potential indicator variables: (1) number of buses seen at one time on the Denali Park Road; (2) number of buses stopped at the same place to observe wildlife; (3) number of buses and people stopped at a rest area; (4) wait time at wildlife stops to see wildlife (as all buses/visitors 'take their turn'); and (5) percentage chance of seeing a grizzly bear. The first three of these variables were addressed through a series of photographic simulations to depict a range of use levels and associated impacts. For each series of photographs, respondents were asked to evaluate the acceptability of each of the study photographs. For the variables 'wait time at wildlife stops to see wildlife' and 'percentage chance of seeing a grizzly bear,' a range of conditions was described numerically. Respondents were asked to evaluate the acceptability of the visual and numerical options using a 9-point Likert-type scale ranging from –4 ('very unacceptable') to +4 ('very acceptable'). As an example, standards for the number of buses stopped to observe wildlife on the Denali Park Road were measured using a series of eight study photographs, as shown in the examples in Figure 17.2.

Figure 17.3 shows the resulting social norm curves derived from the average acceptability ratings. These findings reveal that increasing numbers of buses are generally found to be increasingly unacceptable. For all respondents, mean acceptability ratings fall out of the acceptable range and into the unacceptable range between four and five buses.

Figure 17.2 Examples of photo simulations showing a range of buses stopped on the Denali Park Road to observe wildlife. Photographs like these were used by researchers in Denali National Park to assess standards for the visitor experience (source: Park Studies Laboratory, University of Vermont).

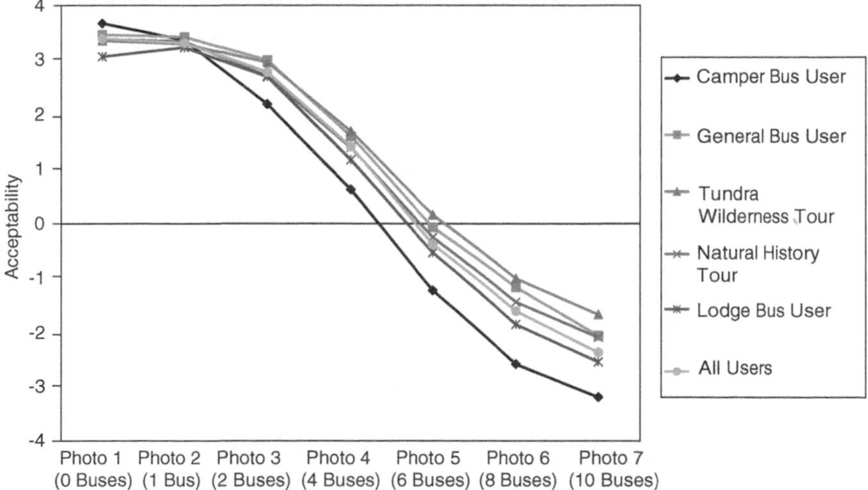

Figure 17.3 Social norm curves for the acceptable number of buses at wildlife stops on the Denali Park Road.

Potential impacts of buses on park wildlife were the subject of the second study (Phillips *et al.*, 2010). Roads and vehicles may affect wildlife in many ways, including degrading the quality of adjacent habitat, restricting movements, and altering behaviour (Trombulak and Frissell, 2000; Forman *et al.*, 2003). The objective of the study was to examine potential relationships between traffic volume and patterns and wildlife behaviour. To do this, GPS technology was used to study the fine-scale movement patterns of Dall's sheep and grizzly bears, as well as the distribution and abundance of other large mammals along the park road.

Grizzly bears were captured from a helicopter using standard aerial darting techniques and Dall's sheep from a helicopter using net gunning techniques. Twenty bears and 20 sheep were fitted with GPS collars that collected one location per hour during the summer/fall visitor use season. Location data from bears and sheep were used to examine movements and road crossing behaviour in relation to vehicle numbers and traffic patterns.

Hourly summaries of vehicle numbers by road section were obtained using traffic counters placed at six locations along the road. Information about the number and distribution of large mammals (i.e. grizzly bears, caribou, Dall's sheep, moose, and wolves) was collected along the road from touch-panel interfaces installed in 20 buses. Bus drivers entered the species observed when they stopped to view wildlife along the road. Data entered into the panels were geo-coded automatically by GPS automatic vehicle locator (AVL) units installed on each bus. Managers implemented a 'quiet night' of minimal or no traffic as an experimental control during one summer season. Traffic was limited to urgent or emergency travel from 10 pm on Sundays until 6 am on Mondays to examine potential impacts on the number of wildlife viewing opportunities for visitors on morning trips into the park.

Important findings from the wildlife component of the study include: (1) duration of time when bears were inactive was shortest nearest the road and increased as distance from the road increased; (2) Dall's sheep responded negatively to increased traffic volumes by increasing their movement rates when approaching the road and shifting away from the road at higher traffic levels, (3) the potential restriction of movement by sheep because of traffic impediments may be of greater concern to park managers than loss of habitat; and (4) large mammals were observed more frequently on mornings after night-time traffic levels were reduced suggesting that vehicles on the park road may be impacting the possibility to view wildlife.

The third study in the program of research developed a computer-based simulation model of buses on the road (Morris *et al.*, 2010) to estimate the effects of alternative levels and patterns of road traffic on wildlife and the quality of the visitor experience. The model was developed using traffic microsimulation, which is similar to other simulation approaches, but includes aspects that allow for a better understanding of traffic events (Gimblett *et al.*, 2001; Lawson *et al.*, 2003; Cole, 2005; Itami, 2005; Morris *et al.*, 2005).

First, the geometry of the park road was constructed in the simulation model by referencing a GIS layer of road information. Second, records from approximately 4,000 trips made by 87 buses equipped with AVLs were used to examine driving rules, speed behaviour, rest stops and designated stops (e.g. stops at scenic vistas and campgrounds) dwell times, and wildlife-encounter stop dwell times of vehicles on the park road.

To integrate the model with important indicators for the visitor experience and resource protection, standards were incorporated, based on findings from the experiential and wildlife studies described above. To ensure protection of crossing opportunities for Dall's sheep, park managers determined that a gap between vehicles that is longer than ten minutes each hour should be maintained as a standard at three traditional migration corridors along the road. The simulation model incorporated these three sheep crossing locations. Park managers chose to analyse three potential standards for crowding, as indicated by the number of buses for a specified level of road use at a particular location. These crowding levels correspond to data derived from the visitor survey.

The resulting simulation model was then used to evaluate the effects of increasing use and alternative bus-scheduling patterns on wildlife and the quality of the visitor experience. By examining change in violation rates for the crowding standards for viewscapes, wildlife, and rest stops, the sensitivity of the standards to different levels of bus use modelled by the simulation experiments can be assessed. The carrying capacity of the park road is defined by each of the crowding standards and the sheep-crossing gap time described previously.

Development of a sustainable transportation plan

The interdisciplinary program of research described above was used to help the NPS develop the Denali Park Road Vehicle Management Plan (NPS, 2012). The plan divided the road into three sections or zones and formulated a series of wildlife and experiential indicators and standards for each zone. For example, for Zone 1, an indicator is the number of vehicles at a wildlife stop, and the associated standards are that (1) at least 75 per cent of wildlife stops will have three or fewer vehicles, averaged over five years and (2) no one year will have less than 70 per cent of wildlife stops with three or fewer vehicles.

The NPS developed four alternative plan options for public review. The option chosen allows the number of vehicles to rise to 160 per 24-hour period. Park staff are committed to monitoring resource and experiential indicators and implementing management actions to ensure that standards for wildlife and the visitor experience are maintained. In this way, transportation is being used to help manage the sustainability of the park by specifying the levels and patterns of vehicle use on the road that will maintain vital resources and experiential standards.

Conclusions

The long and important relationship between transportation and national parks is primarily demand driven: transportation facilities and services are generally developed and expanded to facilitate public use. In particular, roads, parking lots, and even more contemporary public transit systems are often developed and 'improved' to expand public access to the parks. However, transportation can have substantive negative impacts on park resources and the quality of the visitor experience. The shuttle bus system in the Bear Lake Road Corridor at Rocky Mountain National Park (USA) is an example of this demand-driven approach to transportation management. The recent program of research at Rocky Mountain National Park outlined above found that the shuttle bus system delivers more visitors than many sites within this area can accommodate, and this has led to unacceptable resource degradation and undermined the quality of the visitor experience.

A more sustainable approach to transportation management should formulate management objectives and associated indicators and standards for resource conditions and the quality of the visitor experience. Then, transportation facilities and services are designed and managed to help ensure that standards are maintained and management objectives are achieved. This approach was illustrated in the interdisciplinary program of research that was conducted on the Denali Park Road in Denali National Park (USA). This program of research helped provide an empirical foundation for the new Denali Park Road Vehicle Management Plan that protects wildlife and provides a high quality visitor experience.

The research described in this chapter, along with studies at a number of other national parks, suggests related principles or best management practices for sustainable transportation in national parks (Manning *et al.*, 2014). For example, transportation management in national parks and related areas must address both resource and experiential issues, and this often requires an interdisciplinary research approach. Moreover, transportation can be an important component of the recreation experience in national parks, offering many visitors their primary means of experiencing the parks. This means that transportation must be managed to help ensure a high quality visitor experience and that transportation management in national parks and related areas is likely to be quite different from conventional utility transportation. The primary objective of conventional utility transportation is efficient point-to-point travel, while the primary objective of transportation management in a national park is to protect park resources and provide a high quality visitor experience.

Transportation can also be used as a powerful park management tool, delivering the 'right' number of visitors to the 'right' places at the 'right' times: this potential power of transportation should be designed into park management more deliberately and explicitly as a means to manage park carrying capacity and sustainability. Alternative Transportation Systems (ATS) can be an attractive alternative to automobile use in national parks and related areas as they can reduce the potential impacts of transportation on air quality, noise, and wildlife, and enhance the quality of the visitor experience by reducing road and parking lot congestion and associated stress among visitors. However, ATS must be designed and managed carefully to ensure they contribute to protection of park resources and the quality of the visitor experience. In these and other ways, transportation can contribute to – rather than detract from – the sustainability of national parks and related areas.

References

Ament, R., Clevenger, A. P., Yu, O. and Hardy, A. (2008) 'An assessment of road impacts on wildlife populations in US national parks', *Environmental Management*, vol 42, no 3, pp. 480–496.

Burson, S. L. III, Belant, J. L., Fortier, K. A. and Tomkiewicz, W. C III. (2000) 'The effect of vehicle traffic on wildlife in Denali National Park', *Arctic*, vol 53, pp. 146–151.

Carr, E. (1998) *Wilderness by Design: Landscape Architecture and the National Park Service*, University of Nebraska Press, Lincoln, NE.

Carr, E. (2007) *Mission Sixty-six: Modernism and the National Park Dilemma*, University of Massachusetts Press, Amherst, MA.

Cole, D. N. (2005). 'Computer Simulation Modeling of Recreation Use: Current Status, Case Studies, and Future Directions', Technical Report RMRS-GRT-143. USDA Forest Service, Fort Collins, Colorado, USA.

D'Antonio, A., Monz, C., Newman, P., Lawson, S. and Taff, D. (2013) 'Enhancing the utility of visitor impact assessment in park and protected areas: A combined social-ecological approach', *Journal of Environmental Management*, vol 124, pp. 72–81.

Forman, R.T.T., Sperling, D., Bissonette, J., Clevenger, A., Cutshall, C., Dale, V., Fahrig, L., France, R., Goldman, C., Heanue, K., Jones, J., Swanson, F., Turrentine, T. and Winter, T. (2003) *Road Ecology: Science and Solutions*, Island Press, Washington, D.C.

Gamble, L., Lawson, S., Monz, C. A. and Newman, P. L. (2007) 'Modeling the Effects of Alternative Transportation on Resource Protection and Visitor Experience in Rocky Mountain National Park', Alternative Transportation in the Parks and Public Lands Program, Project Proposal, Rocky Mountain National Park, Estes Park, CO.

Gimblett, R., Daniel, T., Cherry, T. D. S. and Meitner, M. J. (2001) 'The simulation and visualization of complex human-environment interactions', *Landscape and Urban Planning*, vol 54, no 1–4, pp. 63–78.

Hammitt W. E. and Cole D. N. (1998) *Wildland Recreation: Ecology and Management* (2nd ed.), John Wiley, New York, NY.

Havlick, D. G. (2002). *No Place Distant: Roads and Motorized Recreation on America's Public Lands*, Island Press, Washington, D.C.

Itami, R. M. (2005) 'Port Campbell National Park, Australia: Predicting the effects of changes in park infrastructure and increasing use' in D. Cole (Comp.), Computer simulation modeling of recreation use: Current status, case studies, and future directions (pp. 57–59). General Technical Report RMRS-GTR0143. USDA Forest Service, Rocky Mountain Research Station, Fort Collins, Colorado, USA.

Johnson, C. (2012) 'Getting there: Yosemite and the politics of transportation planning in the national parks', *The George Wright Forum*, vol 29, no 3, pp. 351–361.

Lawson, S., Chamberlin, R., Choi, J., Swanson, B., Kiser, B., Newman, P., Monz, C., Pettebone, D. and Gamble, L. (2011) 'Modeling the effects of shuttle service on transportation system performance and quality of visitor experience in Rocky Mountain National Park', *Transportation Research Record: Journal of the Transportation Research Board*, vol 2244, pp. 97–106.

Lawson, S. R., Manning, R. E., Valliere, W. A. and Wang, B. (2003) 'Proactive monitoring and adaptive management of social carrying capacity in Arches National Park: An application of computer simulation modeling', *Journal of Environmental Management*, 68, no 3, pp. 305–313.

Lawson, S., Newman, P., Choi, J., Pettebone, D. and Meldrum, B. (2009) 'The numbers game: Integrated transportation and user capacity research in Yosemite National Park', *Transportation Research Record*, vol 2119, pp. 83–91.

Louter, D. (2006) *Windshield Wilderness: Cars, Roads, and Nature in Washington's National Parks*, University of Washington Press, Seattle, WA.

Manning, R. E. (2004) 'Recreation planning frameworks' in M. J. Manfredo and J. J. Vaske (eds), *Society and Natural Resources: A Summary of Knowledge* (pp. 83–96), Modern Litho, Jefferson, MO.

Manning, R. E. (2007). *Parks and Carrying Capacity: Commons Without Tragedy*, Island Press, Washington, DC.

Manning, R. E. (2009) *Parks and People: Managing Outdoor Recreation at Acadia National Park*, Island Press, Washington, DC.

Manning, R. E. (2011). *Studies in Outdoor Recreation: Search and Research for Satisfaction* (3rd ed.), Oregon State University Press, Corvallis, OR.

Manning, R. and Hallo, J. (2010) 'The Denali Park Road Experience: Indicators and Standards of Quality', *Park Science*, vol 27, no 2, pp. 33–41.

Manning, R., Lime, D., Freimund, W. and Pitt, D. (1996) 'Crowding norms at frontcountry sites: A visual approach to setting standards of quality', *Leisure Sciences*, vol 18, no 1, pp. 39–59.

Manning, R. E., Lawson, S., Newman, P., Budruk, M., Vallerie, W., Laven, D. and Bacon, J. (2004) 'Visitor perceptions of recreation-related resource impacts', in R. Buckley (ed) *The Environmental Impacts of Ecotourism* (pp. 259–271), CABI, London.

Manning, R., Lawson, S., Newman, P., Hallo, J. and Monz, C. (2014) *Sustainable Transportation in the National Parks: From Acadia to Zion*, University Press of New England, Hanover, NH.

Meldrum, B. and DeGroot, H. (2012) 'Integrating transportation and recreation in Yosemite National Park', *Society News, Notes & Mail*, vol 29, no 3, pp. 302–307.

Morris, S., Gimblett, R. and Barnard, K. (2005) 'Probabilistic travel modeling using GPS', in A. Zerger and R. M. Argent (eds) MODSIM 2005 International Congress on Modelling and Simulation (pp. 149–155), Modelling and Simulation Society of Australia and New Zealand, University of Arizona Press, Tucson, AZ, USA.

Morris, T., Hourdos, J., Donath, M. and Phillips, L. (2010) 'Modeling traffic patterns in Denali National Park and Preserve to evaluate effects on visitor experience and wildlife', *Park Science*, vol 27, no 2, pp. 48–57.

Nash, R. (2001) *Wilderness and the American Mind*, Yale University Press, New Haven, CT.

NPS (2012) 'Busses and Shuttles in National Parks', National Park Service, Washington, DC, www.nps.gov/transportation/busses_shuttles.html.

Park, L., Lawson, S., Kaliski, K., Newman, P. and Gibson, A. (2009–2010) 'Modeling and mapping hikers' exposure to transportation noise in Rocky Mountain National Park', *Park Science*, vol 26, no 3, pp. 59–64.

Pettengill, P., Manning, R., Anderson, L., Valliere, W. and Reigner, N. (2012) 'Measuring and managing the quality of transportation at Acadia National Park', *Journal of Park and Recreation Administration*, vol 30, pp. 68–84.

Phillips, L., Mace, R. and Meier, T. (2010) 'Assessing impacts of traffic on large mammals in Denali National Park and Preserve', *Park Science*, vol 27, no 2, pp. 42–47.

Reigner, N., Kiser, B., Lawson, S. and Manning, R. (2012) 'Using transportation to manage recreation carrying capacity', *The George Wright Forum*, vol 29, no 4, pp. 322–337.

Roof, C., Kim, B., Fleming, G., Burstein, J. and Lee, C. (2002) 'Noise and Air Quality Implications of Alternative Transportation Systems: Zion and Acadia National Park Case Studies', (Publication No. DTS-34-HW-21M-LR1-B), US Department of Transportation, John A. Volpe National Transportation Systems Center, Cambridge, MA.

Runte, A. (1990) *Yosemite: The Embattled Wilderness*, University of Nebraska Press, Lincoln, NE.

Runte, A. (2010). *National Parks: The American Experience (4th ed)*, Taylor Trade Publishing, Boulder, CO.

Runte, A. (2011). *Trains of Discovery: Railroads and the Legacy of Our National Parks (5th ed)*, Roberts Rinehart Publishers, Boulder, CO.

Sutter, P. (2002) *Driven Wild: How the Fight Against Automobiles Launched the Modern Wilderness Movement*, University of Washington Press, Seattle, WA.

Taff, B. D., Newman, P., Pettebone, D, White, D., Lawson, S.W., Monz, C. and Vagais, W. (2013) 'Dimensions of alternative transportation experience in Yosemite and Rocky Mountain National Parks', *Journal of Transport Geography*, vol. 30, pp. 37–46.

Trombulak, S. C. and Frissell, C. A. (2000) 'Review of ecological effects of roads on terrestrial and aquatic communities', *Conservation Biology*, vol 14, pp. 18–30.

White, D. D., Aquino, J. F., Budruk, M., and Golub, A. (2011) 'Visitors' experiences of traditional and alternative transportation in Yosemite National Park', *Journal of Park and Recreation Administration*, vol 29, no 1, pp. 38–57.

Whittaker, D., Shelby, B., Manning, R., Cole, D. and Hass, G. (2011) 'Capacity reconsidered: Finding consensus and clarifying differences', *Journal of Park and Recreation Administration*, vol 29, no 1, pp. 1–20.

18 Identifying key factors for the successful provision of public transport for tourism

Andreas Kagermeier and Werner Gronau

As some of the case studies in Part 3 of this book have shown, establishing suitable mobility services in rural and protected areas is anything but simple owing to a combination of social (e.g. individual preferences) and technical (e.g. frequency of service) aspects that make alternative transportation either unattractive or inefficient. This chapter attempts to bring together the experience gained around the world and to identify which key factors are required to ensure the successful provision of public transport in protected areas. The focus is on general aspects on the demand side and their implication for shaping services, and on aspects concerning internal cooperation between stakeholders on the offer side, including the crucial issue of economic sustainability, which will be addressed in detail in the next chapter.

Basic conditions for the provision of public transport in rural areas

Leisure traffic undoubtedly has a high affinity towards private motorised modes of transport (Gather and Kagermeier, 2002, p. 9; Guiver 2007, p. 276). In Germany, for example, buses and trains account for only 10 per cent of motorised leisure day trips, and only one out of seven holidays involve the use of terrestrial public transport (Kagermeier, 2003, p. 264). At the same time, the share of non-captive visitors (i.e. visitors who choose to use their own cars) is higher than it is in everyday transport. Hence, one of the reasons for such massive reliance on private modes of transport is that tourist groups with a high degree of car ownership are overrepresented in leisure traffic. Additionally, trips undertaken by whole families and groups give members of the family or group who do not usually have access to a car the opportunity to go on car journeys, too. In other words, the tendency to use private motorised vehicles appears to be a structural precondition, even before other aspects on the demand or offer side are taken into account.

Furthermore, users of leisure transport facilities are often passengers by choice, i.e. they are free to choose between different modes of transport, and are given a wide range of destinations to choose from. Hence, the issue of providing public transport for leisure is quite different from that of providing public transport services for everyday needs. While in the latter case, potential passengers must be certain of reaching their destination, this being the condition for maintaining a job for example, in the former case access to a tourist destination does not have to be guaranteed because visitors are free to choose which place they want to visit. Similar to retailers concerned about customers' ability to reach their store, operators of leisure facilities and stakeholders in tourist regions want to attract customers/visitors, therefore it is up to them – the offer side – to ensure accessibility. Unlike retailers and many other service providers, who can choose their location based on good accessibility, however, tourist destinations are geographically constrained. Nature-based tourist destinations such as protected areas, for example, are usually located a long way from metropolitan areas and high-capacity public transport hubs. If potential visitors who can choose from a variety of modal alternatives and destinations are to be addressed, then high-quality transport services must be offered. The level of quality of the transport service must be reflected in all of the basic 'hard' conditions, i.e. the frequency and quality of the service, and the price charged.

Another crucial basic condition requiring consideration is that the demand for leisure-related transport is highly volatile, featuring high, short peaks at the weekends and during the holiday season, and levels close to zero in the low season. Moreover, nature-based tourism is heavily influenced by meteorological parameters. A cold, rainy weekend may cause demand to fall significantly, whereas a hot, sunny weekend may attract visitors in droves.

The issue of varying demand in everyday transport can be resolved by offering flexible demand-responsive solutions. These measures include using different sized vehicles, providing optional schedules and only serving routes when passengers have actively requested use of the public transport service. In order to reduce costs, the service is only given when the need for it has been voiced (Enoch, 2004, p. 32; Farrington *et al.*, 2008). Demand response transport can reduce the cost of a transport service, but still meet people's everyday mobility needs. However, demand-responsive transport usually fails to attract passengers of choice. In other words, services that are adequate for captive passengers in a demand-driven everyday setting usually fail to attract visitors to a leisure or tourist destination.

As a result, providers of public transport in rural protected areas are faced with a dilemma. In order to encourage visitors of protected areas to use public transport, a high-quality service must be provided, but this is very expensive and the temporally volatile and often spatially dispersed demand fails to cover its full costs.

Seeking to resolve this fundamental dilemma, there are often suggestions to use 'carrot and stick' strategies (Dickinson and Dickinson, 2006, p. 199; Lumsdon, 2006, p. 754; Gronau and Kagermeier, 2007, p. 132; Guiver 2007: 277; Scuttari *et al.*, 2013, p. 617). These involve actions at two levels: on the one hand, incentives are used to make public transport more attractive (and visitors more willing to pay); on the other hand, the volume of demand required to make a public transport service profitable is increased by introducing disincentives for car use (e.g. ban on cars, reduction of parking spaces).

Conditions on the demand side

In addition to addressing the fundamental issues described above, the planning of leisure-based transportation calls for a careful analysis of the demand-side mobility offers. A number of demand side-related aspects requiring consideration in the attempt to successfully provide public transport in protected areas are discussed below.

Identifying the target groups

First of all, it is important to consider the fact that people's choice of travel mode for leisure and tourism purposes is often less rational than it is for everyday mobility purposes (Gronau and Kagermeier, 2004, p. 129). Enjoyment of the actual journey and the ability to drive at will may be strong motives for using a car or a motorbike for leisure purposes. Hence, the motives and subjective dispositions of target groups need to be analysed. One of the tools for identifying subjective aspects on the demand side that has been increasingly applied over the past decade is based on the concept of lifestyle groups (see Chapter 19 for more details about this concept). Visitors who tend to be emotional about their private motorised vehicle and enjoy the fun factor of driving usually exhibit less elasticity to respond to public transport offers. For example, it is clearly harder to convince racing fans to use a bus to get to the racetrack than it is to achieve the desired modal split for visitors to a festival run by an eco-friendly non-governmental organisation. It is therefore more meaningful to concentrate on target groups that are likely to respond to the stimulus represented by a public transport service (Stanford, 2014).Visitors to protected areas tend to have a nature-oriented mindset, but often demonstrate a matter-of-fact attitude towards their choice of transport (Gronau and Kagermeier 2007, p. 129). As Lumsdon *et al.* (2006, p. 150) showed by undertaking a factor analysis of stated preferences by visitors to the Peak District National Park, the convenience aspect is quite important to visitors. However, their feelings about the use of private cars are quite rational, merely judging cars as a mode of transport, without demonstrating positive emotional feelings.As the findings of Schmied and Götz (2006, p. 56) show, therefore, nature-oriented tourists, whether of a traditional type or belonging to the LOHAS (Lifestyle Of Health And Sustainability, Emerich, 2011) group, already tend to use environmentally friendly means of transport.

The attractiveness of public transport systems clearly depends on the activities that different visitor groups engage in. Some visitors may like nature-oriented activities (e.g. surfing, rafting, diving or skiing) that involve transporting bulky equipment, which is not easily accommodated on a bus or a train.Yet these activities are usually practiced by a limited share of protected area visitors. Most visitors in fact, apart from observing nature and visiting nature-oriented exhibitions or nature trails, favour walking, hiking or cycling. As far as these latter activities are involved, the use of public transport may even be convenient as it prevents the need to return to the place where private vehicles were parked, thus enabling visitors to pursue linear walks and courses (Guiver *et al.*, 2007, p. 281).

Catchment area

In many natural areas around the world, the provision of alternative transportation options is limited to a shuttle bus service covering the 'last mile'. This means that the core area is completely free of cars and that more park and ride facilities have to be installed at the edge of the core area. Hence, even if the core area is freed from the negative impacts of the use of private cars, this solution is unsatisfactory from a more comprehensive sustainability perspective (Lund-Durlacher and Dimanche, 2013, p. 505). In terms of a tourist's ecological (and especially carbon) footprint, the use of an environmentally friendly means of transport for the whole journey is evidently of much greater effect.

When planning a public transport system for the whole journey, merely providing an adequate connection to the next public transport hub (e.g. railway station) is not enough, instead it is important to account for the origin of visitors and the quality of the public transport services provided in the relevant catchment area. Empirical evidence from Germany suggests that the same quality of service (especially the frequency of the service) may result in quite different modal-split figures. A leisure destination having a metropolitan catchment area served by an efficient suburban railway system may receive twice as many visitors using public transport as a destination whose catchment area is served by a significantly poorer rail system (Gronau and Kagermeier, 2007, p. 131).

Thus, when car use is restricted, whether by charging high parking fees or forbidding access, the actual possibility of attracting volumes of visitors that may sustain the tourism industry living off a protected area ultimately depends on the availability of a high quality public transport network serving the whole catchment area.

Frequency of visit and marketing issues

In everyday mobility, a location is reached on a daily basis, several times a week or weekly. Hence, the choice of locations and means of transport is influenced and steered to a great extent by routine. Leisure activities such as visiting a protected area usually occur much less frequently. Sometimes a visit is made only once in a lifetime, sometimes only once a year, or sporadically anyway. Reaching potential customers via marketing channels is difficult even for everyday traffic in rural areas and raising awareness of existing services requires comprehensive and innovative marketing approaches (Gronau and Kagermeier, 2004). Reaching one-time or sporadic visitors is an even tougher challenge and calls for ad hoc marketing approaches that consider a clear market segmentation based on the type of visitor. Marketing approaches for protected areas in the surroundings of large cities or metropolitan areas have to focus on day trippers who may decide to visit the protected area spontaneously. Marketing approaches for protected areas located in more remote regions, where the majority of visitors are holidaymakers, have to rely on a rather different mix of activities.

When addressing visitors from neighbouring cities or metropolitan areas, it is essential that residents from the catchment area are already aware of the public

Figure 18.1 Innovative marketing with adhesive foils covering the whole body of tramways to raise awareness among Cologne residents about the National Park Eifel (Germany) (source: photo by Nationalpark Eifel/M.Wetzel).

transport service before they decide to visit the protected area. The fact that day trips are often decided spontaneously, without an intensive search for information about sites to visit (including transport opportunities), and that the degree of private car availability is high when starting from one's own domicile means that there are fewer opportunities for marketing communication during the decision-making process. For example, an intensive marketing campaign initiated by the Eifel National Park (Germany) together with the public transport company of the neighbouring Cologne area (Regionalverkehr Köln GmbH) relied not only on traditional marketing methods (e.g. leaflets, information on the Internet and posters in the city of Cologne), but also on adhesive foils covering the whole body of tramways to raise awareness among Cologne residents (Figure 18.1) (Wetzel, 2008, p. 11). This awareness campaign contributed to double the proportion of visitors who reached the national park by public transport between 2005 and 2007 (Erdmann and Stolberg-Schloemer, 2007, p. 14).

In protected areas far from metropolitan areas, the lion's share of visitors are holidaymakers staying in neighbouring villages and resorts. In this case, as the activity programme is usually planned before or during the holiday, the marketing communication can be largely based on the potential visitor's proactive information-seeking process. The main challenge is ensuring that visitors have the relevant information when deciding to undertake the visit. This, of course, means that detailed and appealing information about the public transport option has to be available together with all printed or electronically disseminated information about the protected area. At the same time, information centres – whether run by a regional or local tourism organisation or an organisation that manages the protected area – have to combine information about sightseeing opportunities with the offer of public transport provision. These requirements place high emphasis on an effective cooperation between transportation companies, tourism boards and nature conservation organisations. All service providers along the value chain in addition to the direct stakeholders must be included in the process. Employees of accommodation facilities play a vital role in communicating public transport opportunities, not just by disseminating published material (e.g. flyers), but especially by providing verbal information. As the study undertaken by Froehlich (1998) in the Bayerischer Wald National Park (Bavarian Forest, Germany) shows, only very few hotel owners were aware of the existing offer of public transport and were able to inform visitors about it, which meant that they provided only minimal support to visitors using public transport to reach the core area of the national park (Gronau *et al.*, 1998). Instead, all stakeholders should be seen as, and view themselves as, ambassadors of a public transport system.

The fact that the target group of marketing initiatives uses the public transport option provided very occasionally despite a significant proportion of one-time visitors is reached by such initiatives means that the response time required to react may be much longer than in the case of everyday transport. Empirical findings show that it may take up to three years/seasons for demand to peak (Kagermeier and Gronau, 2007, p. 228).

Organisational and structural aspects on the offer side

In addition to the challenges on the demand side, the internal organisational and structural aspects of the stakeholders involved in public transport provision, with their different and sometimes conflicting objectives, make it even harder to satisfy the needs of all parties involved and therefore achieve a successful public transport scheme. Below is a list of relevant stakeholders with an explanation of their potential role towards the establishment of successful public transport systems in natural areas.

1 Organisations and institutions in charge of protected areas: these organisations and institutions usually tend to place great emphasis on protecting the natural resources of a site. To them, the environmental aspect plays the greatest role, and the number of visitors to the site (and the economic impact of the valorisation of the natural heritage) is often marginal. These organisations commonly regard a public transport service as an instrument to reduce the negative impact of visitation on the local environment (e.g. noise, air pollution, soil sealing). They usually only have a limited budget that does not allow them to provide considerable support for implementing and operating a public transport service.

2 Destination management organisations: for tourism-oriented organisations (often organised as public institutions or public-private organisations or enterprises), the primary focus is on economic income and the effect tourism has on jobs. They are mainly interested in promoting the destination and assuring an adequate quality of service as well as developing new products and attracting investors to increase hotel bed capacity. Regardless of whether they are publicly or privately financed, they usually only have a limited budget, which does not enable large amounts to be invested in public transportation.
3 Private tourism service providers: as the tourism industry is characterised to a great extent by small- and medium-sized enterprises (SME), the financial capacity of private businesses to contribute to the tourism service chain is usually quite limited. Fierce competition and customers' limited willingness to pay resulting in low fares and thus a limited yield, along with an often quite low employment rate in rural regions around protected areas make it difficult for them to finance the necessary reinvestment in their own business. They are mainly interested in augmenting their yield by increasing the number of tourists or the earnings per tourist.
4 Transport and mobility service providers: given the difficult situation of the public transport sector in rural areas (e.g. low and dispersed demand) (Gronau and Kagermeier, 2004; see also Chapter 19), the tourist demand is a welcome supplement to the core business of both public and private transportation companies. In fact, as the largest share of public transport in rural areas is the daily transport of pupils, transport service providers are often keen to run tourist-oriented services during weekends and school holidays to increase their earnings.
5 Politicians and society: official political discourse and public opinion usually tend to concentrate on the global aspects of protecting the environment. Against the backdrop of the sustainability discussion and the current climate change debate, politicians are keen to reduce the negative impacts of leisure-related traffic. Whether the focus is on the global carbon dioxide aspect or the local effects of noise and air pollution emissions (which are often a nuisance to the local population and thus their potential voters), they usually wish to encourage tourists (i.e. non-voters) to shift to public transport. However, their position often appears like 'Sunday speeches': when it comes to substantial financial commitments, politicians often refer to the private sector or user-financed approaches for public transport.
6 Tourists: the final stakeholder in the series of groups involved is the tourist or visitor to a protected area. Visitors mainly seek convenience (i.e. possibility to easily access a destination) almost regardless of whether it is a leisure park, an exhibition, a beach or a protected area (Hergesell and Dickinger, 2013). Visitors often choose to drive to their destination and therefore tend to see any encouragement to modal shift towards alternative forms of mobility as a nuisance.

Tourists often perceive themselves as 'cash cows' that are already bringing money into the region and are therefore usually rather unwilling to pay more for transportation. Considering their hedonistic leisure motives, they generally want to enjoy their holidays as comfortably and pleasantly as possible. If a destination fails to promise carefree, comfortable accessibility, potential visitors are usually free to choose another competing location or destination for their leisure activities.

Ultimately, however, the crucial issue when establishing public transport schemes in natural and protected areas is not just ensuring that all local and regional stakeholders cooperate in the conceptual design, running and marketing of the product, but securing financial sustainability for such schemes.

One partial solution for increasing revenues of transport services in protected areas could be that of avoiding mono-functional routes serving only the protected area, but combining these with a broader service covering the surrounding smaller villages. An example of this approach is provided by the Naturpark line in the Lippe region (North Rhine-Westphalia, Germany), which will be explored in detail in Chapter 19 (Freitag, 2004). In that case, cooperation between different public and private nature organisations, and tourism and transport-oriented organisations, institutions and enterprises led to a successful public transport service that not only caters to both visitors and inhabitants, but also generates additional funding for covering the cost of operating the transport service.

Another promising approach for resolving the dead-end problem of insufficient funding was developed in the Black Forest (Germany). The concept of the KONUS[1] card enables tourists to the Black Forest to gain access to all public transport services without extra charge. The idea behind KONUS is the result of cooperative activities between public and private tourism and transport organisations, institutions and businesses as well as a joint decision taken by local and regional politicians. These actors agreed to give a certain proportion of the local tourist tax (€0.30 per overnight stay; Hotz, 2008, p. 3) to the regional transport organisation. All local and regional stakeholders were involved in developing improved schedules and routes to optimise the service for tourists, enabling them to easily reach all key places throughout the Black Forest region including the Black Forest National Park. In 2013, almost 10,000 hotels and residences in almost 140 communities were able to offer their overnight guests this public transport mobility option and the transport enterprises received almost €4 million to run the services (Schwarzwald Tourismus GmbH, 2013). The result of efficient internal marketing and the intensity of cooperative activities between all of the local and regional stakeholders involved is that private service providers truly act as ambassadors for the service. They promote KONUS as an added value on their websites (Hotz, 2008, p. 15) 'so that potential visitors are aware of this public transport mobility option from the beginning. KONUS even managed to persuade leading national tour operators such as TUI and Thomas Cook to include the KONUS mobility option in their advertisements as an added value' (Hotz, 2008, p. 16).

Conclusion

This chapter has provided a broad overview of the factors determining the ability to provide viable public transport services in natural and protected areas. What has clearly emerged in these pages is that it is much more challenging to develop and operate such services in a leisure context than it is for everyday transport purpose. This is partly due to the basic conditions governing leisure transport, namely:

- a high degree of choice riders;
- a high degree of freedom in the choice of destination;
- a volatile demand with high peaks and long off-season periods.

Solutions from everyday transport options cannot easily be applied to balance these basic conditions. In everyday travel, people who use public transport are usually captive riders with less freedom to choose which location to visit, and demand is continuous on workdays. In leisure mobility, neither the (cost-reducing) demand-responsive transport solution attracts the necessary number of tourists, nor can 'stick' measures (i.e. restrictions on car use) be applied extensively without having a negative impact on economic performance. While a high level of quality (functioning as a stimulus or 'carrot') is required to address a target group, this same group usually has a limited willingness to pay for the offer, increasing the gap between the service level required to address customers and the insufficient funding available to provide such a level of service.

Apart from these basic conditions, additional demand-side characteristics make the challenge even more difficult. The target group's lifestyle characteristics and the usually inadequate connections in rural catchment areas have to be considered. The one-time or occasional nature of such visits presents an additional challenge to marketing, which has to include all local and regional stakeholders from different spheres. It is essential to integrate all of the relevant local and regional stakeholders with regard to their different rationalities and goals: their comprehensive cooperation seems to be key to achieving viable solutions for the provision of public transport in natural and protected areas.

Note

1 KONUS is the acronym of: 'KOstenlose NUtzung des öffentlichen Nahverkehrs für Schwarzwaldurlauber', i.e. use of public transport without extra charge for vacationers in the Black Forest.

References

Dickinson, J. E. and Dickinson, J. A. (2006) 'Local transport and social representations: Challenging the assumptions for sustainable tourism', *Journal of Sustainable Tourism*, vol 14, no 2, pp. 192–208.

Emerich, M. (2011) *The Gospel of Sustainability: Media, Market, and LOHAS*, University of Illinois Press, Champaign, IL.

Enoch, M. (2004) Innovations in Demand Responsive Transport. INTERMODE Final Report, Manchester, UK.

Erdmann, C. and Stolberg-Schloemer, B. (2007) 'Besucherbefragung im Nationalpark Eifel und in seiner angrenzenden Region 2007 – Analyse und Vergleich mit der Besucherbefragung 2005', Rheinisch-Westfälische Technische Hochschule (RWTH), Aachen.

Farrington, J., Gray, D. and Kagermeier, A. (2008) 'Geographies of rural transport', in R. D. Knowles, J. Shaw and I. Docherty (eds.) *Transport Geographies: Mobilities, Flows and Spaces*, Blackwell, Oxford, UK, pp. 102–119.

Freitag, E. (2004) 'Die Touristiklinie im Kreis Lippe. Evaluierung eines Freizeitverkehrsangebotes im ländlichen Raum', in A. Kagermeier (ed.), *Verkehrssystem- und Mobilitätsmanagement im ländlichen Raum*, MetaGis, Mannheim, Germany, pp. 193–204.

Froehlich, S. (1998) Mit Kind und Kegel in den Nationalpark, unpublished Diplom Thesis Munich University of Technology, Munich.

Gather, M. and Kagermeier, A. (2002) 'Freizeitverkehr als Gegenstand der Mobilitätsforschung', in M. Gather and A. Kagermeier (eds.), *Freizeitverkehr, Hintergründe, Probleme, Perspektiven*, MetaGis, Mannheim, pp. 9–12.

Gronau, W. *et al.* (1998) 'Möglichkeiten verkehrsgestaltender Maßnahmen im Nationalpark Bayerischer Wald', in H. Popp, and A. Kagermeier (eds.), *Akzeptanz der Erweiterung des Nationalparks Bayerischer Wald*, unpublished results of a student research project at the Institute of Geography at Munich University of Technology, Munich, pp. 158–226.

Gronau, W. and Kagermeier, A. (2004) 'Mobility management outside metropolitan areas: case study evidence from North Rhine-Westphalia', *Journal of Transport Geography*, vol 12, pp. 315–322.

Gronau, W. and Kagermeier, A. (2007) 'Key factors for successful leisure and tourism public transport provision', *Journal of Transport Geography*, vol 15, pp. 127–135.

Guiver, Jo, Lumsdon, L., Weston, R. and Ferguson, M. (2007) 'Do buses help meet tourism objectives? The contribution and potential of scheduled buses in rural destination areas', *Transport Policy*, vol 14, no 4, pp. 275–282.

Hergesell, A. and Dickinger, A. (2013) 'Environmentally friendly holiday transport mode choices among students: the role of price, time and convenience', *Journal of Sustainable Tourism*, vol 21, no 4, pp. 596–613.

Hotz, S. (2008) 'KONUS: Busse & Bahnen gratis für Schwarzwald-Urlauber', presentation given at the symposium Chancen des nachhaltigen Tourismus in Regionen: Klimafreundlich Reisen durch umweltgerechte Mobilität und ÖPNV, 5 September 2008, Düsseldorf, www.zukunft-reisen.de/klimafreundlich_reisen.html, accessed on 1 May 2014.

Kagermeier, A. (2003) 'Freizeit- und Urlaubsverkehr: Strukturen – Probleme – Lösungsansätze', in Ch. Becker, H. Hopfinger and A. Steinecke (eds.), *Geographie der Freizeit und des Tourismus. Bilanz und Ausblick*, Oldenbourg, München/Wien, pp. 259–272.

Kagermeier, A. and Gronau, W. (2007) 'Erfolgsfaktoren intermodaler Mobilitätsangebote für Freizeit und Tourismus', in F. Walter, M. Naumann and A. Schuler (eds.), *Standortfaktor Tourismus und Wissenschaft*, Erich Schmidt Verlag, Berlin, pp. 219–232.

Lumsdon, L. M. (2006) 'Factors affecting the design of tourism bus services', *Annals of Tourism Research*, vol 33, no 3, pp. 748–766.

Lumsdon, L. M., Downward, P. and Rhoden, St. (2006) 'Transport for tourism: Can public transport encourage a modal shift in the day visitor market?' *Journal of Sustainable Tourism*, vol 14, no 2, pp. 139–156.

Lund-Durlacher, D. and Dimanche, F. (2013) 'Mobilities and sustainable tourism: an introduction', *Journal of Sustainable Tourism*, vol 21, no 4, pp. 505–510.

Schmied, M. and Götz, K. (2006) 'Soft Mobility Offers in Tourism – Demand, Supply and the Consumer's Role', in Federal Ministry of Agriculture, Forestry, Environment and Water Management (BMLFUW) (ed.) *Environmentally Friendly Travelling in Europe. Challenges and Innovations Facing Environment, Transport and Tourism*, BMLFUW, Vienna, pp. 54–64.

Schwarzwald-Tourismus GmbH (2013), *Black Forest export: Sustainability*, www.schwarzwald-tourismus.info/Presse/Pressemeldungen-nach-Themen/Schwarzwald-mobil/Schwarzwaelder-Exportschlager-Nachhaltigkeit, accessed on 1 May 2014.

Scuttari, A., Della Lucia, M. and Martini, U. (2013) 'Integrated planning for sustainable tourism and mobility. A tourism traffic analysis in Italy's South Tyrol region', *Journal of Sustainable Tourism*, vol 21, no 4, pp. 614–637.

Stanford, D. J. (2014) 'Reducing visitor car use in a protected area: a market segmentation approach to achieving behaviour change', *Journal of Sustainable Tourism*, vol 22, no 4, pp. 666–683.

Wetzel, M. (2008) 'Mobil im Nationalpark: Natur erfahren mit Bus und Bahn'. Presentation given at the symposium Chancen des nachhaltigen Tourismus in Regionen: Klimafreundlich Reisen durch umweltgerechte Mobilität und ÖPNV, 5 September 2008, Düsseldorf, www.zukunft-reisen.de/klimafreundlich_reisen.html, accessed on 1 May 2014.

19 Increasing the economic feasibility of public transport supply in natural areas

Werner Gronau and Andreas Kagermeier

Introduction

There is today a strong need to preserve the world's remaining natural areas for the well being of future generations. Yet these areas are often the setting for intense tourist activity relying on travel modes that imply negative externalities, including atmospheric pollution and noise (Countryside Agency, 2003). This is clearly not just a threat to the ecological functioning of many environmentally fragile areas, but also a detrimental factor for the qualities (e.g. tranquillity, unspoiltness) that attract visitors (Guiver *et al.*, 2007). The reduction of the impacts of visitation in natural settings is largely connected to the actual possibility of shifting a significant share of visitors from private to public modes of transport. This is often achieved through so-called 'carrot and stick' measures, which combine incentives to public transit (e.g. passes, increased frequency of service) and disincentives on car use (e.g. road restrictions, tolls) (Steiner and Bristow, 2000; Dickinson *et al.*, 2004). Unfortunately, however, the maintenance of ad hoc public transport services in rural areas is very expensive and many local authorities and national parks are reducing their financial support to rural public transport (Reeve, 2006, p. 3). This means both a reduced service for people living in rural areas and the impossibility for visitors to reach (and move within) these areas without a private vehicle. A solution to this dilemma could come from the improvement of scheduled services whose financial sustainability is ensured by tourist ridership. This chapter explores whether and how tourists can actually increase the economic feasibility of rural public transport supply and therefore contribute to the sustainability of transportation in rural regions. Empirical evidence about the opportunities of fruitful cooperation between public transport agencies and the tourism sector towards better transport provision in natural settings is provided.

The vicious circle of public transport demand and supply in rural areas

When discussing the situation of public transport in rural areas several framework conditions have to be considered. First of all, there is a problem of perception, which is evident when comparing public transport in urban and natural settings. While in the former case public transport is perceived as an alternative even by non-users (Gronau *et al.*, 2004, p. 316), in the latter case it is scarcely perceived as an alternative at all. In fact, rural regions present a dispersed settlement structure, which results in an ad hoc demand for public transport supply, and a low overall population density, which generates low demand and raises doubts about the real necessity of public transport services (Gronau *et al.*, 2007, p. 128). Beside these already quite unfavourable conditions, the implementation of an efficient public transport system in natural areas is faced with the strong competition of private vehicles. In fact, rural areas show a significantly higher degree of car-ownership than metropolitan areas do and traffic jams, which could contribute to a larger use of public transport, are relatively rare. All these conditions create a high competitive environment for any public transport supply in rural regions.

Unfortunately, the aspects mentioned above usually result in poor user intensity as well as a low productivity of rural transport services. As Dickinson (2006, p. 194) summed it up: 'it is not surprising in a rural context that public transport proves a poor competitor to the car. In areas with low population density, economic and use level criteria are unlikely to be met and dispersed destinations make it hard to offer a transport alternative that will appeal to the majority of people'. The common response by the responsible institutions is a rather poor supply, especially outside the peak hours. Additionally, low productivity, by negatively affecting the service quality and expenditures for maintenance, results in unattractive services and subsequently a further decrease in the already low demand.

This interdependency of demand and supply via the user intensity leads to the regularly discussed vicious circle of public transportation in rural areas (Figure 19.1a). The overall low economic feasibility due to the low and dispersed demand results in a low quality public transport supply in terms of service, maintenance and number of trips. This problem is evident in many countries around the world. In England, for example, the Department for Transport admits 'that public transport is dirty, unreliable and slow' (Department of Environment, Transport and Regions, 1998). This low- and poor-supply level leads to a less attractive public transport supply, which cannot properly satisfy the demand and eventually determines a low number of users. The small number of users cannot create a sufficient profit and the supply level is reduced once more.

This process results in a steady rise in car-dependency through the decrease of public transport supply and public transport attractiveness at the same time. The only way to end this seemingly relentless decline in public transport quality in rural regions is to identify (and attract) additional user groups for the service. The increased number of users in fact may contribute to a virtuous circle where higher revenues allow the product to be upgraded and even more demand to be created (Figure 19.1b).

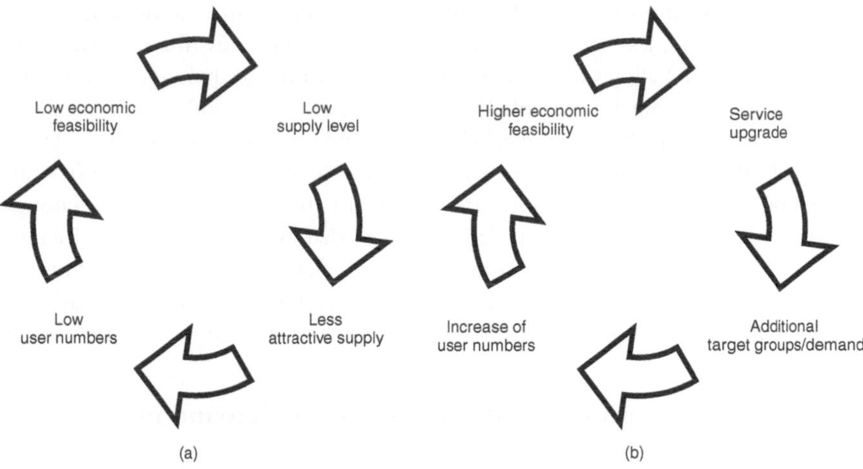

Figure 19.1 The vicious (a) and virtuous (b) circle of public transport supply.

Tourists and excursionists as a chance for an increased public transport demand in natural areas

When considering the possibility of increasing the economic feasibility of public transport through the identification of additional users, a question arises as to which new group to target. This chapter, which originates from findings of the German Ministry of Research funded research project IMAGO, focuses, among several possible options, on the group of tourists and excursionists. The advantages offered by public transportation to this group are evident. For example, public transport enables hikers and cyclists to follow linear itineraries (i.e. start and end do not coincide) without the need for them to return to the point where they left their vehicle (Guiver *et al.*, 2007, p. 279). Additionally, many tourists enjoy using more than one transport mode during a journey and public transport reduces the risk of getting lost while travelling in a region that is not well known (Downward *et al.*, 2004, p. 418). Further, Downward *et al.* (2006, p. 145) observed that the opportunity of 'both looking at the passing scenery whilst travelling, and touring the region', instead of concentrating on the traffic, is considered by a significant number of people as a clear advantage. The use of a public transport is obviously favoured by people who do not like parking restrictions or other kinds of fee (Guiver *et al.*, 2007, p. 277). Besides the opportunities offered by public transport to tourists, there is a wide range of positive impacts that the tourist group may bring to the public transport sector. In fact, this target group usually uses public transport within hours of low demand, during weekends, in the late morning or the early afternoon, therefore supplementing the already existing demand, without the need for additional capacity. Further, as leisure and tourism are strongly subsidised by the public sector, a public transport that is also targeting tourists and excursionists can open up new funding opportunities. Moreover, less time constraints during the weekends (there is usually no commuting traffic) allows more panoramic routes (with slightly longer travel times) to be followed, favouring the cohabitation of 'regular' users and tourists aiming at particular off-the-track sites. Nonetheless, it is important to emphasise that expectations of both target groups (i.e. regular users and tourists) have to be met efficiently: relabeling existing public transport offers to tourist and excursionists is not enough. In this respect, Lumsdon (2006, p. 757) observed that 'what emerges is a substantial mismatch between the needs of utilitarian and the leisure user. Thus, the common practice of designing networks primarily for utility purposes and then marketing them to the tourists is not likely to be attractive unless adapted to meet their need'.

Framework conditions for attracting leisure demand in public transport

While the section before outlined various arguments to support the opportunities provided by the combination of public transport for utility purpose on the one hand and tourism and leisure related use on the other, the linkages between transport and tourism are in fact weak. That is because transport for local passengers is managed by local transport planning systems, which inevitably place a greater emphasis on trips for utility purpose (Lumsdon, 2006, p. 751). Lumsdon (2006, p. 754), in his research on the design of tourism bus services, showed how stakeholders in the field agree that the design of a multifaceted conventional bus network serving any purpose would be more appropriate. However, precisely such an integrated approach towards providing a multifaceted bus network is still an exception. In order to establish public transport systems that can attract leisure demand, various factors should be considered, as follows:

- Appropriate design of the network: the chronic lack of funds often results in a limited number of buses, which have to be used for various purposes and are not user-friendly (Lumsdon, 2006, p. 757). Instead, the network should be designed in a way that guarantees a regular and reliable service to the potential users.
- All year service: regularity is definitely important for attracting new users. The availability of an all year service, though less frequent during the low season, may ensure high ridership (Lumsdon, 2006, p. 757).
- Marketing plan: marketing communication is fundamental to guaranteeing the success of a tourism bus network. For example, high quality booklets providing a thorough timetable information may play as catalogues of places to visit (Lumsdon, 2006, p. 757).
- Creation of a specific brand: the adoption of names (e.g. Hadrian's Wall Bus, Moorbus) and logos that connect visitor minds with specific places may be an option to build customer confidence over time (Lumsdon, 2006, p. 757).

- Priority access: giving priority to visitors arriving by public transport (e.g. bus stop in front of the visitor centre, whereas the parking lot is located on a peripheral area) can raise people's interest in the public transport option (Lumsdon, 2006, p. 761).
- Service delivery: the overall quality of the service provided (e.g. reliability, personal security, cleanliness) is key to attracting tourists and favouring modal shift from the private vehicle (Lumsdon, 2006, p. 759).

The aspects outlined above set the starting point for the development of a public transport product as described in the next section.

Empirical findings from a German case study

This section presents a case study from the region of Lippe in the north-eastern part of North Rhine Westphalia (Germany), where the existing public transport service was restructured to satisfy the needs of tourists. Such redesign was started in the framework of the research and demonstration project IMAGO, which was funded by the German Ministry of Research in 2003. The project focused on the introduction of high quality public transport supply within rural regions, by utilising existing high quality local urban bus networks and spreading them over the surrounding areas, while at the same time connecting the existing high quality lines. The so-called 'Naturparkbus' was developed on the previously existing 792 regional bus line. This line used to connect two major towns at an hourly interval and two major regional railroad lines with each other. Unfortunately, demand during weekends and public holidays, when no commuting was involved, was particularly low and local authorities were about to shut down the connection. The project and related funding gave the project team the opportunity to work on possible solutions to ensure adequate public transport supply on this line even on weekends and public holidays. The peculiar characteristics of the region where the bus line runs, within the natural reserve of the 'Teuteburger Wald', and the various tourism sites and attractions in the surroundings (i.e. an open-air museum, a memorial for a Roman battle, a medieval castle, a bird park, and a natural monument) (Figure 19.1) greatly supported the idea of converting the existing regional bus into a tourism product aiming at tourists and excursionists during the weekends.

The project team, in cooperation with the local public transport authority KVG-Lippe, aimed to create a multifaceted bus network facilitating local as well as leisure and tourism-related demand. In the first phase, an analysis of attractions found within walking distance of the existing bus route and a detailed mapping of bus stops were performed. These were aimed at both identifying possible additional attractions that were not yet accessible due to lack of a nearby bus stop, and evaluating the accessibility of the various sites by the existing bus network, especially in relation to the private vehicle. The analysis showed how in many cases bus stops had been located based upon the operator's benefit rather than the customers' point of view. For example, many of the small and narrow roads providing access to specific attraction sites were not conveniently serviced by public transport, forcing

Figure 19.2 The 'Naturparkbus' in the region of Lippe in the north-eastern part of North Rhine Westphalia (Germany) is an example of how a schedule public transport service can be redesigned to meet tourist demand (source: KVG-Lippe).

public transport users to walk rather long distances to reach those sites. Additionally, many bus stops were lacking any facility beyond the bus stop sign, therefore providing minimal support to the users. In order to fix these problems, the location of bus stops was optimised and bus stops themselves were upgraded, whereas the bus schedule was redesigned to ensure a better connectivity with the existing train network. Nowadays, at sites such as 'Herrmansdenkmal' or 'Externsteine', bus users directly reach the ticket office at the site's entrance, while visitors using private vehicles have to cover a certain distance to get to the site from the large-scale car park. By ensuring access to regional trains, especially in the morning and the afternoon, the local bus network has eventually turned into an interesting day trip opportunity for urban dwellers from Bielefeld and Paderborn.

The second phase of the project, following on the optimisation of the supply-side, focused on the demand side. As already mentioned in this chapter, high quality marketing plans, including the use of ad hoc brands, have proven to be a very efficient instrument to increase awareness about a transport option and subsequently boost the demand. That is why the 'regional bus 792' was renamed 'Naturparkbus' (Figure 19.2), a new graphic was applied to it and a flyer outlining the leisure and tourism-related opportunities disclosed by the bus line was published. The ad hoc designed bus replaces the standard one on weekends and public holidays turning the 'regional bus line 792' into the 'tourism line 792' served by the 'Naturparkbus'.

Hence, daily users are still familiar with the weekday line 792, but at the same time tourists and excursionists can easily identify the 'Naturparkbus' as a dedicated leisure and tourism bus option. Furthermore, the attractive new look of the buses ensured an increased attention among visitors in the region, especially at the designated sights. The bus itself, with its new and easily accessible bus stops, raised visitors' awareness and therefore supported the promotion of the product.

The third and final phase of the project was about the evaluation of the newly established transport option. Besides basic data on occupancy rates, ticket sales and user characteristics, special attention was paid to the economic feasibility and the environmental sustainability of the product. The evaluation took place two years after the introduction of the renovated bus line on four different days during the week and the weekends within the summer holiday season, using a quantitative survey among users. The analysis of data and the survey showed that already in the third year the new product was very well accepted by the market, the demand in terms of average occupancy rate had doubled as compared to what it used to be before, though starting from only around ten passengers per trip. However, occupancy rates were heavily influenced by the weather conditions: while on rainy days demand dropped to just a few passengers per trip, on sunny days it often outnumbered the supply. Further, after only three years of service, tourists and excursionists represent the major share of users, whereas the number of everyday users has remained constant.

In other words, the new marketing approach, by slightly changing the original routing scheme, was able to double the number of users in only two years. What consequences this increase of passengers had on the economic feasibility of the bus service can best be shown by looking at the tickets used by the tourists and excursionists. Tourists usually purchase the relatively expensive single or day ticket,

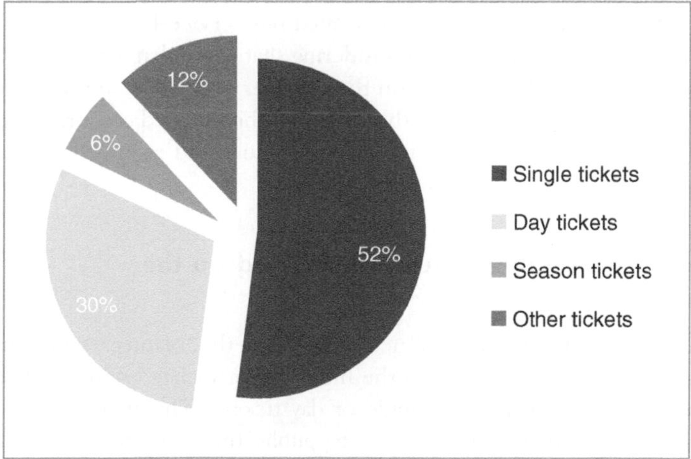

Figure 19.3 Shares of different ticket types used on the bus line 792.

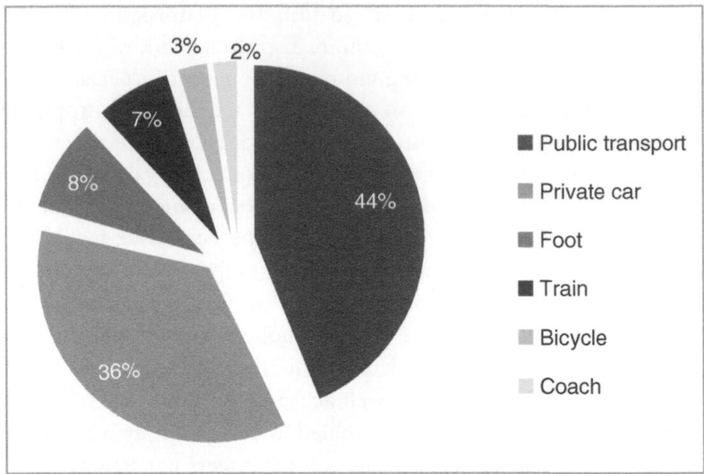

Figure 19.4 Transport modes used by 'Naturparkbus' passengers on their last visit to the area.

whereas season and other tickets (e.g. 'Fliegender Hermann' ticket and special tickets for mobility impaired and retired people) represent a negligible share of total sales (Figure 19.3). This means that the additional user group made up of tourists and excursionists creates a significantly higher income than regular users (who mainly choose season tickets). In fact, the economic feasibility of the product almost quadrupled while the degree of capacity utilisation doubled. Therefore, the new target group, by dramatically increasing the total economic revenues of the bus service, safeguarded the long-term existence of the service and ensured a mobility opportunity for the locals. However, the renovated bus service had also a very positive impact on the environment when considering that more than one third of its users (36 per cent) had relied on the car on their former visit to the area (Figure 19.4).

Hence, the strategy adopted for the 'Naturparkbus' proved comprehensively successful as it attracted more users, increased revenues and eventually encouraged a significant modal shift among tourists.

Discussion on economic feasibility based on the case study

While acknowledging the economic feasibility of the 'Naturparkbus', one has to recognise that this is mostly due to the high price associated with the tickets generally purchased by tourists (i.e. single or day tickets). This inevitably raises questions about the need of tariff systems for public transport in the leisure context that are both remunerative for the transport company and fair for the user. To this purpose, there may be a need to considerably discount day and weekly tickets so that public transport becomes more attractive for car users and low-income people (Lumsdon, 2006).

Financial support in the early phases of planning and setting up of the network, as well as for marketing issues, is undoubtedly needed to kick-start a service that would otherwise encounter considerable difficulties in getting popular. Further, as a service may require years to achieve decent market shares (Gronau and Kagermeier, 2007, p. 133), public transport authorities must be ready to sustain operations for a given period before a service gets well-known among its potential users. This simply means that an adequate budget must be dedicated to the service, the best possible product must be designed and attractive marketing activities must be conceived, as it happens in any other sector of the economy. In other words, instead of introducing a low-quality public transport supply with high amounts of financial incentives for a short period of time and shutting it down after one or two years of operation due to poor economic feasibility, it is preferable to use financial incentives to set up a quality public transport product and support it through effective marketing. The 'Naturparkbus' experience showed that the quality of the service provided is ultimately the main determinant of its success and ability to positively contribute to a natural area's preservation through enhanced modal shift. Whether and how low fares foster accessibility and an increase in ridership can be debated, but devoting the available financial resources to the development of public transport for tourist use is a way to obtain an overall improvement of the service that benefits both people and the environment.

Conclusion

Public transport supply in rural regions is generally poor and this often leads to a situation where even basic transportation needs of the local population cannot be met. At the same time, many rural regions around the world are becoming attractive for tourism as they offer an unspoilt environment to practice various recreation activities (e.g. cycling, hiking). The availability of efficient public transportation systems in such areas would then have a dual function: it would guarantee daily transportation for residents and would allow tourists to reach popular spots and undertake their preferred recreation activities. This chapter, originating from the empirical results of the German research and demonstration project IMAGO, has shown that redesigning a public transportation system to meet the tourism demand can increase the economic feasibility of the service and contribute to the protection of the local environment. The great advantage of this kind of intervention is that minor investments in marketing activities and limited modifications to the service's original routing scheme may have very positive outcomes, including increased ridership and revenues. The fact that tourists generally purchase more expensive tickets (e.g. single or day tickets) as compared to daily users results in revenues growing more strongly than the number of users. While each natural area presents specific characteristics and therefore calls for ad hoc strategies to ensure adequate public transportation, targeting the tourist segment seems an effective way to breaking the vicious circle of poor demand and poor supply that plague public transport services in rural regions and often determines their disruption.

References

Countryside Agency (2003) 'The State of the Countryside 2003', The Countryside Agency, Cheltenham, UK.

Department of Environment, Transport and Regions (1998) 'A New Deal For Better Transport For Everyone: The Government's White Paper on Transport', Stationary Office, London, UK.

Downward. P. and Lumsdon, L. (2004) 'Tourism transport and visitor spending: a study in the North York Moors National Park, UK', *Journal of Travel Research*, vol 42, no 4, pp. 415–420.

Downward, P., Rhoden, S. and Lumsdon, L. (2006) 'Transport for tourism: can public transport encourage a modal shift in the day tourist market?', *Journal of Sustainable Tourism*, vol 14, no 2, pp. 139–156.

Dickinson, J. E. and Robbins, D. (2007) 'Using the car in a fragile rural tourist destination: A social representations perspective', *Journal of Transport Geography*, vol 15, no 2, pp. 116–126.

Dickinson, J. E. and Dickinson, J. A. (2006) 'Local transport and social representatives: challenging the assumptions for sustainable tourism', *Journal of Sustainable Tourism*, vol 14, no 2, pp. 192–208.

Dickinson, J., Calver, S., Watters, K. and Wilkes, K. (2004) 'Journeys to heritage attractions in the UK: a case study of National Trust property visitors in the South West', *Journal of Transport Geography*, vol 12, no 2, pp.103–113.

Freitag, E. (2004), 'Die Touristiklinie im Kreis Lippe. Evaluierung eines Freizeitverkehrsangebotes im ländlichen Raum', in A. Kagermeier (ed), *Verkehrssystem- und Mobilitätsmanagement im ländlichen Raum*, MetaGis, Mannheim, pp. 193–204.

Gronau, W. and Kagermeier, A. (2004) 'Mobility management outside the metropolitan areas: Case study evidence from North Rhine-Westphalia', *Journal of Transport Geography*, vol 12, no 4, pp. 315–322.

Gronau, W. and Kagermeier, A. (2007) 'Key factors for successful leisure and tourism public transport provision', *Journal of Transport Geography*, vol 15, no 2, pp. 127–135.

Guiver, J., Lumsdon, L., Weston, R. and Ferguson, M. (2007) 'Do buses help meet tourism objectives? The contribution and potential of scheduled buses in rural destination areas', *Transport Policy*, vol 14, no 4, pp. 275–282.

Lumsdon, L. (2006) 'Factors affecting the design of tourism bus services', *Annals of Tourism Research*, vol 33, no 3, pp. 748–766.

Reeves, R. (2006) *Tackling Traffic*, Council for National Parks, London, UK.

Steiner, T. J., and Bristow, A. L. (2000) 'Road pricing in national parks: A case study of the Yorkshire Dales National Park', *Transport Policy*, vol 7, no 2, pp. 93–103.

20 Elements that encourage bicycling and walking to and within natural areas

Natalie Villwock-Witte

Introduction

The United States of America (USA) possesses natural areas of unique beauty and iconic landscapes, which attract large numbers of visitors from around the world. While there are several federal agencies that manage public lands in the USA, some of the most well-known public land units in the country, including the iconic Glacier National Park (NP), Yosemite NP, Grand Canyon NP, and many others, are all under the jurisdiction of the National Park Service (NPS). The NPS's mission is to: 'preserve unimpaired the natural and cultural resources and values of the national park system for the enjoyment, education, and inspiration of this and future generations' (NPS, 2014).

Today it is rather evident that to sustain the allure of natural areas and achieve the above-mentioned mission, the USA must work towards providing more sustainable modes to access and travel within these areas. Historically, access to many parks, including Glacier and Yosemite NP, was provided by rail. However, with the introduction of the automobile, visitor facilities at these parks were redesigned to accommodate car travel. Unfortunately, studies have found that one of the largest contributors to greenhouse gases emitted within these national treasures is vehicle emissions from the very mode of travel that brings most visitors to the sites (NPS, 2014). This is a significant concern considering the potential impact of climate change on national parks. For example, some predictions indicate that by 2020, there will no longer be glaciers in Glacier NP (Minard, 2009)! Further, the many issues associated with car traffic (e.g. noise, congestion, land consumption) are major threats to wildlife (Ramp *et al.*, 2006) and the recreational experience of visitors (Daigle and Zimmerman, 2004).

An ongoing challenge tied to the NPS's mission is to minimize the impact of transportation on resources, without restricting visitation. As a result, many parks have begun providing transit systems. However, a recent study has shown that the long-term financial sustainability of these systems is challenging, primarily due to their ongoing dependence on capital funding (Booz Allen Hamilton and the Volpe National Transportation Systems Center, 2011). For example, the fuel, maintenance, and replacement costs of the vehicles alone make these transit systems difficult to sustain. Therefore, there is growing interest in providing transportation to and within parks via bicycling and walking, which are sustainable and more cost-effective. Moreover, by encouraging multi-modal transportation systems (e.g. transit and bicycle or transit and walking), it is possible to minimize the size of a transit system, while still ensuring adequate mobility standards to visitors. The facilities needed to accommodate bicycling and walking are significantly smaller than the roads and parking lots needed for automobiles. In addition, walking and bicycling bring in other benefits, such as supporting national initiatives for better health, improving air quality, and engaging visitors more directly with the natural surroundings. As summarized in the NPS's Green Parks Plan (GPP), 'encouraging visitors to step out of their vehicle – either to use alternative transportation or to bike or walk – positively impacts their experiences while reducing their environmental impact' (NPS, 2013). Yet this is not at all an easy task due to both well-established personal attitudes (i.e. people tend to see the car as the only transport mode) (Bamberg *et al.*, 2003; White *et al.*, 2011) and lack of adequate infrastructures and facilities (i.e. few national parks have alternative transportation systems and bicycling path networks) (Lumsdon, 2000).

Based on experiences in the USA and the Netherlands, this chapter presents and discusses four key elements that may encourage bicycling and walking to and within natural areas: infrastructure, connectivity, convenience and promotion. The consideration of such different contexts in terms of both level of bicycling/walking infrastructures (the Netherlands being universally known for these activities) and characteristics of natural areas (the Netherlands being far less wild than the USA) favours the identification of elements that are widely applicable to any context. First, the chapter provides a contextual overview of areas in the USA and the Netherlands where bicycling and walking are currently encouraged through a variety of measures. Then the four elements are presented and discussed in detail, and conclusions drawn based on the presented information. The chapter originates from a recent work by Villwock-Witte *et al.* (2012). However, that document is intended to outline components of a good bicycle and pedestrian plan, not elements that encourage bicycling and walking, as is the focus here.

Natural areas providing bicycling and walking opportunities in the USA and the Netherlands

Like other countries around the world, the USA and the Netherlands host various natural areas that offer facilities (e.g. bicycle paths) and initiatives (e.g. car traffic restrictions) to encourage bicycling and walking. Natural areas in the two countries present rather different characteristics though. Parks in the USA may have extremely large sizes (i.e. up to hundreds of thousand hectares) and are often in remote locations, far from urban areas. On the contrary, parks in the Netherlands have moderate sizes (i.e. hundreds to thousands of hectares) and are located in moderately to densely populated areas, therefore being easily connected to urban areas via bicycle paths or hiking trails. Below is a description of the areas in the two countries that were considered for the analyses presented in this chapter.

United States

Acadia NP, located in south-eastern Maine, is number nine on the list of the top ten most visited national parks in the USA with approximately 2.3 million visitors in 2013 (NPS Stats, 2014). The scenic beauty of Acadia NP and the carriage roads are the primary draws of the park. The Island Explorer, the park's shuttle service, has several routes that connect the local communities to the park and its sites. Many of the routes are paralleled by walking paths, which allow a user to travel using multiple modes.

Grand Canyon NP, located in north-western Arizona, is number two on the list of top ten most visited national parks in the USA with approximately 4.5 million visitors in 2013 (NPS Stats, 2014). The park draws both national and international visitors to see the breath-taking geology exposed by the canyon's depth. During the peak summer season, the park is packed with visitors on the South Rim, most of them travelling in individual motorised vehicles, which cause congestion both within the park and in the gateway city of Tusayan. In 1997, a group of volunteers planned and developed a Greenway trail system along the South Rim (Olson, 2012). While the implemented trail is only a portion of a larger vision, it provides opportunities for bicycling and walking along the South Rim.

San Antonio Missions National Historical Park (NHP), located near San Antonio, Texas, has approximately 1.7 million annual visitors (Sherwood and Murphy, 2014). The park preserves four Spanish frontier missions and several other cultural and natural sites located along an eight-mile stretch of the San Antonio River. Following on recent collaborations between the park and the city of San Antonio, a visitor to the area may utilize the bike share system to go from downtown to the sites within the park (Sherwood and Murphy, 2014).

Saratoga NHP, located in east-central New York state, had approximately 54,000 recreational visitors in 2013 (NPS Stats, 2014). It preserves the site of one of the 15 most decisive battles in world history, where the Americans defeated the British and subsequently received French support for the American battle for independence (National Park Foundation, 2004). The park initially only offered an auto-touring road; however, as visitors expressed interest in touring the park by bicycle, the park widened the road to make an extra-wide bicycle lane to accommodate bicyclists.

Sequoia and Kings Canyon NP, located in central-eastern California, had approximately 1.5 million visitors in 2013 (NPS Stats, 2014). The park, which is well-known for the giant sequoia trees, has one shuttle system that connects the park and the gateway community of Visalia, and a second one that operates within the park.

Valley Forge NHP, located near Philadelphia, Pennsylvania, had approximately two million recreational visitors in 2013 (NPS Stats, 2014). The park, which marks the site of George Washington's Continental Army's third winter encampment (National Park Foundation, 2004), has an auto-touring route. However, roughly following the touring route is a paved pathway that is used by both walkers and bicyclists. About two-thirds of the visitors make use of recreational facilities like the pathway.

Yellowstone NP, located in north-western Wyoming and south-western Montana, is number four on the list of top ten most visited national parks in the USA with approximately 3.2 million visitors in 2013 (NPS Stats, 2014). Considered the world's first national park, it is most well known for its thermal features, like the geyser Old Faithful. Some walking paths are found across the park and park roads are popular among touring bicyclists.

White Sands National Monument, located near Alamogordo, New Mexico, had approximately 491,000 visitors in 2013 (NPS Stats, 2014). The park, which is known for the large deposit of gypsum sands, offers a few walking paths and annually hosts the 'Full Moon Bike' ride, letting visitors enjoy car-free roads after dark.

Zion NP, located in south-western Utah, is number seven on the list of top ten most visited national parks in the USA with approximately 2.8 million annual visitors (NPS Stats, 2014). The park, which is known for its cliffs and canyons, including the Narrows (National Park Foundation, 2004), implemented a shuttle system within Zion Canyon starting in 2000 and concurrently restricted private vehicle access to enhance the visitor experience. A shuttle service is also run to connect the gateway community of Springdale and the visitor centre in Zion Canyon (Upchurch, 2014).

The Netherlands

De Hoge Veluwe National Park, located in east-central Netherlands, has about 500,000 annual visitors (Villwock-Witte and Leidekker, 2015). The park, which is the most famous national park in the Netherlands, is criss-crossed by a widespread network of bicycle paths and provides hundreds of bicycles for rent.

Kasteel de Haar, located near Utrecht in the Netherlands, attracts thousands of visitors each year and ranks as one of the 100 most visited museums in Holland (Tromp and Trum, 2013). The present castle, constructed from 1892 to 1912, is surrounded by a 135-acre park offering great opportunities for long walks (Tromp and Trum, 2013).

Mill Network at Kinderdijk-Elshout (called Kinderdijk hereafter), a UNESCO World Heritage Site located near Rotterdam, is an authentic polder landscape encompassing 19 windmills built between 1738 and 1740, and various canals, ditches and storage basins. The site, which attracts approximately 300,000 annual visitors, offers great opportunities for walking and bicycling on the many trails and bicycle paths.

Elements that support bicycling and walking in natural areas

Infrastructure

The first element that supports bicycling and walking to and within natural areas is infrastructure. The presence of bicycling and walking infrastructure creates an environment that communicates to visitors that these modes are available. The term infrastructure in this case broadly refers to facilities allowing the visitor to: bicycle or walk, rent or share a bicycle, and park the bicycle.

Providing space for a person to bicycle or walk will promote the use of these modes. However, the type of space provided will dictate the type of user, as suggested by differences in bicycle spaces at Yellowstone NP, Saratoga NHP and Acadia NP (Figure 20.1). At Yellowstone, only a narrow shoulder is present with bi-directional traffic; at Saratoga, a wider shoulder, which mimics more closely a bicycle lane, is present with uni-directional traffic; at Acadia, the carriage roads are car-free multi-use pathways. Few families ride bicycles through Yellowstone, while Acadia's carriage roads are a mecca for families.

Making bicycles available to visitors can help encourage bicycling within a natural area. Providing bicycle rentals or a bike sharing program are two approaches. The biggest difference between bicycle rentals and bike sharing is that the former requires users to return the bicycle to the place where they rented it, whereas the latter allows users to make one-way trips (Gleason and Miskimins, 2012).

Some examples of parks that offer bicycle rentals include Grand Canyon NP, Acadia NP, Valley Forge NHP, and De Hoge Veluwe NP (Gleason and Miskimins, 2012; Villwock-Witte and Leidekker, 2015). However, these rental programs take different approaches based on site needs and characteristics. Grand Canyon and Valley Forge have concessionaires providing bicycle rentals on-site, whereas Acadia, as a result of its interwoven character with surrounding towns, has bicycle rentals offered through businesses in these towns. Among natural areas offering bike sharing, De Hoge Veluwe NP has the longest tradition. Starting in 1975, the park began offering a complimentary bike share within the park known as 'white bikes' (Villwock and Leidekker, 2015). Park management introduced bike sharing as a subtle way to dissuade visitors from experiencing the park in private vehicles. The fleet of white bikes has grown from the 50 original white bikes to approximately 1,700, clearly demonstrating the popularity of seeing the park using this mode. In the USA, San Antonio Missions NHP is one of three parks (all in urban environments) that can be enjoyed by bike share (Sherwood and Murphy, 2014).

If visitors are traveling to a natural area by bicycle, they will need a place to securely lock their bike, unless the natural area is best seen by bicycle. The bicycle parking can be located near a visitor centre where visitors can obtain information about the site or pay for fees. When a park hosts historical sites, facilities to park the bicycle are needed near each site so that visitors can leave the bicycle there and visit the sites on foot. In order to encourage the use of bicycle as means to reach hiking areas, parking facilities should be installed at trailheads. Kinderdijk is a good example of a natural area that provides designated bike parking. In fact, while most of the World Heritage Site can be seen on the bike, there is also a museum that is not accessible by bicycle. A safe parking facility in this case is provided through sturdy wooden bike racks, secured into the ground (Figure 20.2).

Figure 20.1 Bicycle paths and lanes at Saratoga National Historical Park (top left), Yellowstone National Park (top right) and Acadia National Park (bottom) (sources: National Park Service, top left; Natalie Villwock-Witte, top right and bottom).

Connectivity

Visitors will hardly bicycle or walk to and within natural areas if points of interest, bicycle paths, trails and transportation hubs (e.g. bus stops) are not well connected. Connectivity can be created by making public transportation options available from

Figure 20.2 Bike parking facilities at Kinderdijk (the Netherlands) (source: Natalie Villwock-Witte).

an airport hub or large city to the natural area. A good example of this is Acadia NP, where there are several options for connecting from surrounding airports and major metropolitan cities, like Boston. Once visitors arrive, they can choose to rent bikes from a business in town, or they can make use of the Island Explorer shuttle service to see the park. There are walking trails that parallel and diverge from the

Island Explorer routes, which a visitor can access from the shuttle stops. Zion NP implemented a shuttle system in 2000 (Dunning, 2005), which has resulted in an increase in trail use, including backcountry trails. Furthermore, with the removal of private vehicles from the roadway when the shuttle operates, the road environment is more bicycle-friendly and 'interest in bicycling has grown in popularity' (Dunning, 2005).

Another option to improve connectivity is to create a connection to a larger bike network. A great example of that is offered by the Netherlands, where many natural areas are accessible by the 'cycle junction network': a network of nodes (i.e. junctions, villages) and links (i.e. bicycle paths), which users can easily follow to reach a natural area. In conjunction with train, the network allows easy bicycle access to natural areas from virtually any location across the country.

In the USA, groups like Adventure Cycling are working on initiatives to expand roll-on service for bicycles on Amtrak, which is the primary passenger rail service in the country (Adventure Cycling Association, 2014). Another example of enhanced multi-modal connectivity existed at Valley Forge NHP when a pilot shuttle system was operated on roughly the same route as a pathway for bicycles and pedestrians (Figure 20.3). In the summer, the shuttle was a welcome service to families whose children were tired or to visitors who became overwhelmed by the heat. San Antonio Missions NHP illustrates a model for creating connectivity between downtown San Antonio and the park using a bike share system, which allows a visitor to stay in the city and

Figure 20.3 Example of shuttle bus service paralleling a pathway (source: Natalie Villwock-Witte).

bike to each of the missions within the park, stopping at bike share stations along the way (Sherwood and Murphy, 2014).

In order to fully connect a multi-modal system, managers must carefully consider the challenges associated with integrating different modes. For example, allowing users to roll bicycles onto buses is generally more effective than providing bicycle racks on the front of buses. Picking up a bike and putting it on a rack can be intimidating and requires some level of physical strength, which many visitors may not have. An alternative solution is to provide a shuttle with a bike trailer, as is available at Acadia NP. In addition, connectivity must be considered from the perspective of a family traveling with children. It is preferable to allow families to roll strollers onto the buses, as it may be difficult and time consuming for families to break down strollers before boarding. To facilitate roll-on access for bicycles and strollers, buses without steps are appealing.

Convenience

In order to encourage visitors to walk and bike, natural areas should consider offering preference or incentives to these modes as compared to other modes, in line with 'carrot and stick' approaches (Holding and Kreutner, 1998). This can be done, for example, by providing limited parking for private vehicles or charging for private vehicle parking. Several parks, concerned with the impact that large parking lots have on the sustainability of their resources, have begun to limit parking space within their boundaries. For example, Valley Forge NHP, after conducting a study to identify the underutilization of parking lots, reduced the size of some lots. Similarly, Sequoia and Kings Canyon NP, after realizing the impact of the parking areas on the giant sequoias, removed significant amounts of parking (Rosenberg, 2009), and offered opportunities for visitors to arrive by shuttle services from the gateway communities.

Kasteel De Haar charges visitors arriving by private vehicle an extra fee of €4 per vehicle. When asked about the reasons for such a policy, managers' response was simple: 'There is nowhere to park your vehicle on the premises…it is not so much a policy but more a necessity driven by the possibilities of the park itself… vehicles of park visitors are to be kept…where there is sufficient space' (Scala, 2014). Those who arrive via the cycle junction network on bicycle or use the public transportation system and then walk the remaining distance do not pay an additional fee. Similarly, Kinderdijk has only a small parking lot at the entrance that requires a fee in addition to entrance fees.

Adopting policies that favour bicycling and walking over other travel modes is often difficult and may generate unintended consequences. The case of Zion NP is interesting in this respect. According to Upchurch (2014), the entrance stations to the parking lot that enables access to the shuttle system become severely congested during the peak visitation season. He proposed an increase in the capacity of the entrance stations, primarily because the NPS is concerned with visitor experience. Adding an entrance station or lanes, however, will likely add infrastructure, like pavement, for lanes, and will be more costly, as the park will have to provide additional staff for the added entrance station.

Furthermore, the additional entrance station may induce demand, as arriving by vehicle will become more appealing. While the current entrance fee scheme (US $25 per vehicle, US $12 per person) largely privileges drivers over bicyclists or hikers, improving the infrastructure for walking and biking could possibly change the proportion of visitors arriving by bike or walking, thereby reducing the need to expand both the entrance station and parking lot. If national parks like Zion considered implementing bicycle/pedestrian incentives instead of adding a new entrance station, they would see additional benefits, including reduced infrastructure expansion costs, minimization of costs of additional staff for the entrance booth and increased demand for bicycling businesses in gateway communities.

Promotion

Most visitors to natural areas are preconditioned to use an automobile, because they typically use it on a daily basis. Hence, an increase in the popularity of bicycling and walking as transport modes to and within natural areas can only be achieved if visitors are made aware of the opportunities available for these activities. Some promotion examples include adding information about walking and biking opportunities on websites, incorporating bicycling and walking paths on maps of natural areas, ensuring that maps are available at trailheads for bicycling and walking paths, working with regional partners to ensure that bicycling and walking routes are included on maps, providing maps to bicycle rental shops, and working with transit providers to incorporate bicycling and walking information into their maps, websites and trip planning tools.

One of the most direct ways that a park administration can promote use of bicycling and walking is by providing information about programs and facilities on its website, which is notably the main source of information for tourists planning their visit. Some great examples of US National Wildlife Refuges that have done this can be found in 'Transit and Trail Connections: Assessment of Visitor Access to National Wildlife Refuges' (Volpe Center, 2010). However, to be effective, the information must be clearly visible on the website, not buried under multiple links.

Acadia NP, for example, has developed a 'Car-Free Travel Guide', which informs visitors about how they can travel to the park without a car from the surrounding cities of Bangor, Boston and Portland. The park also has a website that provides detailed information about the Island Explorer shuttle's routes and stops. For bicyclists, the local bike rental shops in Bar Harbor provide handouts that have a map showing how to travel from the harbour to the carriage roads.

However, promotion is not only about information, but also about creating unique opportunities to view a natural area by bicycling or walking. This can be made through special initiatives and programs that allow visitors to experience bicycling and walking under exciting conditions. Such programs, which must rely on adequate infrastructure, may help visitors overcome concerns with unfamiliar travel options (e.g. using a combination of transit and walking). Typically, once such programs are offered, they tend to become very popular. For example, White Sands National Monument began offering a 'Full Moon Bike' ride because one of the managers of the facility wanted visitors to experience the beauty of the white sands under the moonlight: something that usual park hours would not allow. Hence, for this event, park roads are re-opened after dark only to bicyclists. The program is presently so popular that it fills to capacity shortly after registration opens and has encouraged the park administration to launch a 'Full Moon Hike' initiative. As illustrated by programs at White Sands National Monument, closing off existing roadways and re-opening them to walkers and bicyclists is an effective way to promote these alternative modes in natural areas that do not yet have dedicated infrastructures for pedestrians and bicyclists. Other initiatives that may be adopted to encourage cycling and walking are guided tours by bike, as those offered at Valley Forge NHP from early May through late October.

Conclusion

This chapter identified four elements that support biking and walking to and within natural areas: infrastructure, connectivity, convenience and promotion. Based on the experience of natural areas in the USA and the Netherlands, bicycling and walking can be encouraged if public land managers improve bicycling (e.g. bicycle paths) and walking facilities, establish easy and reliable connections between different modes of travel, make bicycling and walking convenient and cost-effective, and promote the use of bicycling and walking facilities. Managers are not expected to implement all of the four elements at once. They could start, for example, by building a piece of infrastructure near the visitor centre, and then developing a program to promote this infrastructure. Based on expected positive feedback, they could expand the infrastructure or create an incentive to use it, such as a periodic discounted or free entrance for visitors who arrive by walking or bicycling. They could also work with local transit agencies to identify potential connections, and finally promote it on the website by providing a direct link for further information. This step-by-step approach is feasible and can significantly increase the share of bicycling and walking in natural areas in the medium term.

Acknowledgements

The author would like to thank Jaime Eidswick, Krista Sherwood and Kevin Witte for collectively providing ideas, input and reviewing early versions of this chapter. She would also like to thank Carla Little for providing technical editing.

References

Adventure Cycling Association (2014) 'Multi-Modal Travel', www.adventurecycling.org/travel-initiatives/multi-modal-travel, accessed 18 June 2014.

Bamberg, S., Ajzen, P. and Schmidt, P. (2003) 'Choice travel mode in the theory of planned behavior: the roles of past behavior, habit, and reasoned action', *Basic and Applied Social Psychology*, vol 25, no 3, pp. 175–187.

Booz Allen Hamilton and the Volpe National Transportation Systems Center (2011) 'National Park Service Alternative Transportation Systems Financial Analysis Phase I and II Findings and Results', National Park Service, Washington, DC.

Daigle, J. and Zimmerman, C. A. (2004) 'The convergence of transportation, information technology, and visitor experience at Acadia National Park', *Journal of Travel Research*, vol 43, no 2, pp. 151–160.

Dunning, A. E. (2005) 'Transit for National Parks and Gateway Communities: Impacts and Guidance', Ph.D. dissertation, Georgia Institute of Technology.

Gleason, R. and Miskimins, L. (2012) 'Exploring Bicycle Options for Federal Lands: Bike Sharing, Rentals and Employee Fleets', Federal Highway Administration, Western Federal Lands Highway Division, Vancouver, WA.

Holding, D. M. and Kreutner, M. (1998) 'Achieving a balance between "carrots" and "sticks" for traffic in national parks: the Bayerischer Wald project', *Transport Policy*, vol 5, no 3, pp. 175–183.

Lumsdon, L. (2000) 'Investigating the needs of the recreational cyclist: the experience of the Peak District National Park', *Town Planning Review*, vol 71, no 4, pp. 477–487.

Minard, A. (2009) 'No More Glaciers in Glacier National Park by 2020?', http://news.nationalgeographic.com/news/2009/03/090302-glaciers-melting.html, accessed 7 March 2014.

National Park Foundation (2004) *The Official Guide to America's National Parks*, 12th Edition, Fodor's, New York.

NPS (2013) '2012 Green Parks Performance Brief, National Park Service', US Department of the Interior, Washington, DC.

NPS (2014) 'Mission', National Park Service, Washington, DC, www.nps.gov/aboutus/mission.htm, accessed 14 August 2014.

NPS Stats (2014) 'National Park Service Visitor Use Statistics', https://irma.nps.gov/stats, accessed 4 November 2014.

Olson, J. (2012) *The Third Mode: Towards a Green Society*, Jeff Olson, Lexington, KY.

Ramp, D., Wilson, V. K. and Croft, D. B. (2006) 'Assessing the impacts of roads in peri-urban reserves: Road-based fatalities and road usage by wildlife in the Royal National Park, New South Wales, Australia', *Biological Conservation*, vol 129, no 3, pp. 348–359.

Rosenberg, S. (2009) 'Out of the Car and into the Park: An Evaluation of the Sequoia Shuttle', National Park Foundation, Washington, DC.

Scala, E. D. (2014) Personal communication, 7 March 2014.

Sherwood, K. and Murphy, J. (2014) 'Expanding a Municipal Bike Share System into an Urban National Park through Community Partnerships: The City of San Antonio and San Antonio Missions National Historical Park', Transportation Research Board, Washington, DC.

Tromp, H. and Trum, R. (2013) *Kasteel de Haar Utrecht*, Komma Creatie, Amersfoort, The Netherlands.

Upchurch, J. (2014) 'Zion National Park: Enhancing Visitor Experience Through Improved Transportation', Transportation Research Board, Washington, DC.

Villwock-Witte, N., Gleason, R. and Shapiro, P. (2012) 'Good practices to encourage bicycling and pedestrians on federal lands: 11 components', *Transportation Research Record*, vol 2307, pp. 80–89.

Villwock-Witte, N. and Leidekker, J. (2015) 'The White Bikes of De Hoge Veluwe National Park: A Case Study for Consideration for U.S. Federal Land Managers', to be published soon in Transportation Research Record.

Volpe Center (2010) 'Transit and Trail Connections: Assessment of Visitor Access to National Wildlife Refuges', John A. Volpe National Transportation Systems Center, Cambridge, MA.

White, D. D., Aquino, J. F., Budruk, M. and Golub, A. (2011) 'Visitors' experiences of traditional and alternative transportation in Yosemite National Park', *Journal of Park and Recreation Administration*, vol 29, no 1, pp. 38–57.

21 Helping gateway communities embrace alternative transportation

Anne Dunning

Introduction

Considerations about protected lands cannot stop at their designated borders. In fact, communities located beyond borders support the lands, and the lands support those communities. In these symbiotic relationships, residents of gateway communities provide a service to land managers when they work on the lands and locally serve visitors by providing lodging, supplies and dining. In turn, residents benefit from the power of protected lands to attract tourism to the local area.

Transportation issues fundamentally transcend the boundaries between protected lands and gateway or inholding communities. Roads link these spaces, and public transit serves people and places in both realms. Planning transportation in and around protected lands requires attention to the many stakeholders who feel the effects of transportation decisions. Proactive communication can generate support for alternative transportation initiatives. The need for communication between protected areas and gateway communities emerges when transportation services cross boundaries. In this case, collaborative partnerships and planning processing involving local stakeholders can contribute to joint financing schemes, land use supportive of transportation initiatives, and strategies for disseminating public information (Dunning, 2005a).

For many public lands, the greatest issues of traffic congestion and related safety occur in internal parking areas and at entrance gates. Coordinating transportation planning in conjunction with gateway communities can help alleviate these internal problems by considering both the resources of a larger area and the full travel needs of visitors. If alternative transportation systems can keep private vehicles from entering public lands, congestion points dissipate. Keeping private vehicles away from entrance gates requires encouraging visitors to choose alternative transportation outside the border in gateway communities. How locals represent and endorse alternative transportation systems to tourists can significantly impact the degree to which visitors use alternative transportation. Hence, gateway communities are critical partners in the planning and management of transportation systems.

This chapter examines strategies for land managers and community leaders to encourage collaboration in planning, supporting and using alternative transportation systems. Many of the documented strategies have covered cases in the United

States although the chapter incorporates international references where available. To date, two key reports in the USA have assessed how public-land managers can best engage and interact with gateway communities:

- 'Partnering for Success: Techniques for Working with Partners to Plan for Alternative Transportation in National Park Service Units' (USDOT Volpe National Transportation Systems Center, 2003);
- 'Innovative Transportation Planning Partnerships to Enhance National Parks and Gateway Communities' (Texas Transportation Institute and Cambridge Systematics, 2009).

These reports, which are freely available on the Internet, offer strong reference material for land managers. The documents provide observations and analyses from case studies for interaction between public lands and gateway communities. This chapter conflates, classifies and elaborates on the findings of these reports, incorporating information from other sources within this structure. The following subsections explore the characteristics of gateway and inholding communities as well as techniques and considerations for engaging local stakeholders. The chapter provides guidance for collaborating with local stakeholders to plan context-sensitive transportation systems with consideration of local resources and accountability for transportation-system performance.

Characteristics of gateway and inholding communities

Gateway communities and inholding communities can be characterized in a number of ways. Oliver (2003) developed the following description, which captures important aspects of why these communities are unique. 'Gateway communities are the towns, cities, and communities that border public lands such as national and state parks, wildlife refuges, forests, historic sites, wilderness areas, national forests, and other public lands. They offer scenic beauty and a high quality of life that attracts millions [...] looking to escape traffic congestion, fast tempo and uniformity of cities and suburbs. Gateway communities provide food, lodging, and business for [visitors of] public lands. They serve as portals to public lands and therefore play an important role in defining the park, forest, or wilderness experience for many visitors [...]. Gateway communities face challenges related to managing growth and development, providing economic prosperity, and preserving their character and sense of place' (Oliver, 2003).

The influence of protected lands extends well beyond their borders (Fredman and Yuan, 2011). Many studies, for example, have documented their economic impacts extending long distances to neighbouring states, provinces and even countries (Headwaters Economics 2014, Land Trust Alliance 2011, and European Commission 2013). Gateway communities have been defined as:

- communities immediately adjacent to a tourist area;
- the last non-tourist places where one commits to entering a tourist area;

- all communities economically dependent on a tourist area;
- all communities economically affected by tourist areas.

Experts can debate the stringency of definitions and qualifications. For instance, a graduate student at Oklahoma State University attempted to quantify the 'gateway-ness' of communities outside national parks and identified per-capita accommodation, food-service sales, and the percentage of housing for seasonal and recreational purpose as the most important contributing economic variables (Lincoln, 2012). Regardless of how definitions are applied, managers of protected lands and community leaders must recognize they affect each other through their decisions and initiatives.

This chapter focuses on communities close enough to protected lands to participate in planning local transportation systems. Some communities act as gateways to the protected lands, offering lodging, food, and supplies visitors need. Other communities exist within the borders of protected lands where private inholdings predated governmental land claims.

Although local transportation generally implies short trips, transfers to/from airports and other regional gateways take place on longer distances. Transportation services need to meet all needs of visitors from the time they enter the region because transportation choices made at the port of entry will affect all subsequent travel choices. For example, if people need to rent a car to move from the airport to the natural or protected area, they will commonly use that rental car throughout their stay when traveling within protected lands and gateway communities. Planning for alternative transportation in and around natural and protected areas must consider how to eliminate visitors' need to rent a private vehicle to reach the area.

Many gateway and inholding communities depend on tourism as a major or primary contributor to their local economies. Where communities have long history with protected lands, that dependence likely emerged over time as local work shifted from rural resource extraction to service for commercial tourism. This transition sometimes has brought resentment as resource-protection policies have infringed on commercial activity, leaving low-wage seasonal service jobs as replacements for traditional employment (Kurtz, 2010). Similarly, farming communities might have found policies protecting wild animals have fostered local livestock predation or crop losses. Even after a community has accepted tourism as its basic industry, an economy dependent on tourism inherently leads to challenges related to the distribution of tourism benefits, use and possible depletion of protected land resources for economic gain, and development of conservation-related economic opportunities. Strategies need to be generated locally to relate specifically to each community's unique situation (Schelhas *et al.*, 2002).

Although most small gateway communities that have embraced tourism depend heavily on nearby protected lands, many communities have grown their own reputations and evolved to become tourist destinations themselves. Under this scenario, protected lands can benefit from tourists attracted to the gateway community just as gateway communities have traditionally benefitted from tourists attracted to the protected lands. This mutual regional contribution raises the stature of local leaders interacting with land managers and creates a more equal footing for negotiations (Gomez, 2005). The other extreme of local identity also exists. Some gateway communities lose a local sense of place and individuality as they evolve to meet global expectations for resort communities (Zonneveld, 2011). Gateway communities also frequently attract an elite seasonal population that contrasts against traditional working-class communities. Seasonally vacant homes, educational divides, and economic stratification from service workers living alongside white-collar residents all can generate local tension.

Experience with transportation systems varies among types of gateway communities. If protected lands exist within proximity of a major population centre, citizens of the respective gateway communities likely have knowledge and experience with a variety of transportation alternatives. Citizens living near protected lands situated in rural areas might never have seen rail transit or ridden a bus. The idea of learning how to modify personal travel behaviour to work within the structure of common-carrier schedules and routes likely intimidates these people. Residents of these remote gateway communities are accustomed to feeling the negative impacts of transportation systems. A survey around Acadia National Park in the north-eastern United States, for example, found traffic to be the most fundamental way local residents felt the park impacted them (More *et al.*, 2008). Demonstrating the importance of this problem, the US Federal Lands Highway Division developed a special planning tool to estimate the problem of negative impacts of traffic surrounding road improvements in protected areas (Hardy *et al.*, 2007).

Stakeholder engagement

Engaging stakeholders in transportation planning can be both difficult and fundamentally important. A study of the planning process at the New River Gorge National River in the mountains of West Virginia (USA) found community resistance and a lack of human capital were two of the primary obstacles to the planning process. This study revealed that land managers' engagement in building relationships was an essential key to encouraging collaboration (Riley, 2009).

Few transportation systems around the world commit sufficient resources to communication with the riding public and community stakeholders (Dunning, 2005a). Incomprehensible transit schedules, incoherent maps, and poorly labelled locations have long served as material for popular jokes. The need for communication and public information are heightened where tourism dominates the local economy because more than half of drivers and other travellers are probably unfamiliar with the area and its transportation options. The local travellers are a constantly changing population, thus communication must be a top priority.

Although land managers provide substantial information for visitors, local residents frequently provide fundamental tourism information, and visitors frequently seek and trust local opinions of how to experience the natural resource. Case studies have shown the importance of community support for generating ridership and visitor use of alternative transportation systems (Dunning, 2005a).

Managers of public lands largely set the tone for how much interaction will occur between land units and gateway communities. A superintendent or general manager committed to community involvement can attract interest and support. In some cases, junior managers can create this interest, but again, they must demonstrate dedication to communication and stakeholder interaction (Dunning, 2005a).

How does a land manager identify the community leaders who should be involved in transportation initiatives? A variety of stakeholders can provide diverse perspectives and support (NPS, 2003). A town mayor is a likely contact, but the mayor might have little knowledge or experience with transportation issues. Business leaders will likely want to be involved to protect the local economy and possibly benefit from transportation initiatives. Engaged stakeholders might bring expertise and resources into planning discussions that inform and support the planning process (NPS, 2003).

Most likely, land managers should reach out to a number of different community representatives. Levels of knowledge, experience, and interest will depend on individuals and the organic leadership culture of each local community. Representatives of the sectors indicated in Table 21.1 frequently provide input that can valuably shape transportation decisions. This table indicates common groups or stakeholder interests, as well as possible titles of representatives.

Table 21.1 Stakeholder groups to engage in transportation initiatives.

Group or sector	*Possible representatives*
Administration of protected land	Superintendence Concessioner liaisons
Local population	Elected representatives (mayors or council people) Town managers
Transportation providers	Transit and shuttle operators Taxi services Bicycle rental services
Business community	Chambers of commerce Tourism agencies Concessioners on the protected land
Local economic development	Town planning departments Planning commissioners
Non-profit organizations	Ecological interest groups Friend groups Natural history associations
General knowledge or expertise	Individuals and organizations with varied titles

Source: adapted from Dunning (2005a)

Beyond community leaders, transportation initiatives need to consider all categories that affect visitor decisions, including the private sector (Texas Transportation Institute and Cambridge Systematics, 2009), as well as common citizens and workers. When Zion National Park, in the United States' south-western desert, started its bus service, no visitor was using the bus in town to get past traffic backed up at the entrance gate. Investigation showed that visitors tended to ask waiters and desk staff in hotels how best to see the park. Park managers worked with the local business community to include private seasonal workers in pre-season training provided to park seasonal workers. After the training, a sizable number of visitors started riding the bus into the park, and traffic congestion at the entrance gate diminished (Dunning, 2005a). Service workers in gateway communities contribute to how visitors experience protected lands, thus they need training and exposure to alternative transportation systems and initiatives.

Relationship dynamics: fostering mutual teaching and learning

Gateway communities and managed protected lands maintain symbiotic relationships of co-dependency and mutual support; however, resentment might tinge interactions. Depending on history among stakeholders, past issues can affect support and negotiations for transportation initiatives. Historic difficulties might have arisen in a number of ways. Residents might resent how quickly or how extensively tourism has come to dominate a local economy (Davis and Morais, 2004). Past collaborative initiatives might have been handled tactlessly or inconsiderately, leaving stakeholders wary about contributing time and effort.

For newer parks, research on Central American gateway communities has shown attitudes toward protected lands depend on community demographics, economic conditions, and actions of land managers. Land managers and local leaders have limited control over the local education level and general regional economic health, yet these factors affect attitudes in some gateway communities. Land managers can most positively affect community attitudes by taking responsibility to help with community problems and by taking community opinions on land management into account (Aguirre, 2006).

Initiatives for alternative transportation as this book describes frequently develop from a top-down approach in which a dominant agency, such as the management of protected lands, decides to act. Without a concerted stakeholder-oriented approach to planning, local leaders in rural areas tend to have a relatively small voice, yet they have important things to say.

Studies of best practices have emphasized the need to listen, communicate, and respect all stakeholders and the missions of stakeholder organizations (Texas Transportation Institute and Cambridge Systematics, 2009). Soliciting participation from stakeholders requires empowering them with knowledge and evidence that their voices will make a difference in the outcome. Public workshops and other initiatives aimed at gathering public input can help lay groundwork for planning efforts and generate support for transportation development. Managers of protected lands should involve themselves with community organizations to learn and to educate stakeholders (USDOT Volpe National Transportation Systems Center, 2003).

Local leaders can raise the volume of their voices by educating themselves on what transportation systems do and what they require. From the start of serious discussion of changes to transportation systems, local leaders should probe into fundamental questions. Why is a change necessary? What kind of outcome is desired? What positive and negative side effects will occur on the protected lands, in the community and in the region? Side effects on society, economy and ecology should all be considered.

Fundamentally, the design of transportation systems works best when the ecological, social and economic contexts of the surrounding area set parameters for implementation. Achieving these parameters requires substantive participation from local leaders and advocates of local resources.

Contributions and participation can be established in formal agreements or developed through informal alliances and interaction. Formality in relationships has benefits and drawbacks. The US National Park Service encourages formal partnership devices (NPS, 2003) such as memoranda of understanding, pledged support, establishment of organizations and development of legal institutions. These establishments create infrastructure on which to base future growth, but in doing that, they set bounds. Everything said by participants goes onto public record. Informal collaborations frequently allow more creative development and consideration of new ideas, but they also lose the credibility of a scrutinized public institution. Frequently, public initiatives take root and grow through informal collaboration until they mature to the point they require formal institutionalization.

Motivations to participate and contribute to transportation planning

Why would a community support initiatives for alternative transportation? While some stakeholders might deem an invitation to participate to be an honour, some stakeholders might view it more as an obligation or request for donated effort. Research has found that, 'by far, personal involvement with transit is the strongest determinant [of transit support]. This includes current use of transit and related behaviour—those who are willing to seek information to learn more about their community's transit services' (Rhindress *et al.*, 2008). If residents of gateway communities watch transportation operations from a distance, they will have a weak sense of a system's characteristics and benefits. To generate community support, transportation proponents should design opportunities and means to attract locals to use the system even if only on a trial basis.

When identifying who should be involved in transportation initiatives, the motivations of those stakeholders must also enter consideration. Leaders and members of a community need to recognize and comprehend how transportation initiatives will affect their world. Improving their understanding will improve their support. Because many rural communities have limited or no experience with coordinated transportation systems, introductions and explanations of concepts and ideas require careful strategy.

Research has identified the following specific strategies that can encourage the public to engage into new civil initiatives (Meyer *et al.*, 2005).

- Develop constituency;
- Nurture cooperation;
- Recruit effective advocates and trusted emissaries;
- Build support among like-minded people;
- Involve the people who are affected;
- Guard against irrational opposition (anticipate, identify and isolate it);
- Communicate;
- Market;
- Plan for short- and long-term joint rewards.

Documented experiences and case studies can contribute to this recruiting (Zimmerman *et al.*, 2004; Dunning, 2005b; Texas Transportation Institute and Cambridge Systematics, 2009).

To sustain ongoing support, offering special off-season access opportunities to locals can help generate goodwill between protected lands and surrounding communities. Local residents typically readily support management for visitor traffic, but they appreciate some special treatment for their dedication to the area. Acknowledging that local residents have special awareness, knowledge, and standing with the land can foster good relationships. Frequently, locals like to find niche opportunities to engage with protected lands in ways the general public might not. The fairness and legality of such arrangements require thorough exploration, but some mechanisms for offering opportunities exist. For instance, Denali National Park in Alaska offers an annual lottery for people to drive on a restricted park road after the tourism season wanes (Dunning, 2005a).

Free passes on transit systems for land-management and concession employees can provide multiple benefits. Most obviously, they add ridership, support, and vibrancy to the system. Beyond the obvious, transportation science suggests the commute trip is a strong target because public transit naturally serves commute trips well: passenger surveys have indicated 59.2 per cent of transit trips in America have served the journey to work (APTA, 2007). Indirectly, when local workers use common-carrier transportation systems for their commute trips, they free capacity on the rest of the transportation system for visitors. Congestion decreases on roadways, in queues, and in parking areas, thus improving visitor experience and travellers' safety in the area. A well-used alternative transportation system also instils confidence for others to try the mode, thus attracting visitor participation. Finally, if locals use and believe in the system, they will recommend it to visitors who ask for local opinion on the best way to get around the area.

The context-sensitive and stakeholder-driven planning process

Specific step-by-step guidance for establishing transportation plans and performance measures for recreational areas appears in the Transportation Planning

Handbook of the Institute of Transportation Engineers (ITE, 2009), as summarized in Figure 21.1. This provides a flowchart of the stakeholder-driven planning process. The chart groups some activities because they develop simultaneously. Everything should be revisited based on specific needs, as indicated by iterative loops in the diagram and consistent with the idea that strong planning processes need to be iterative and flexible.

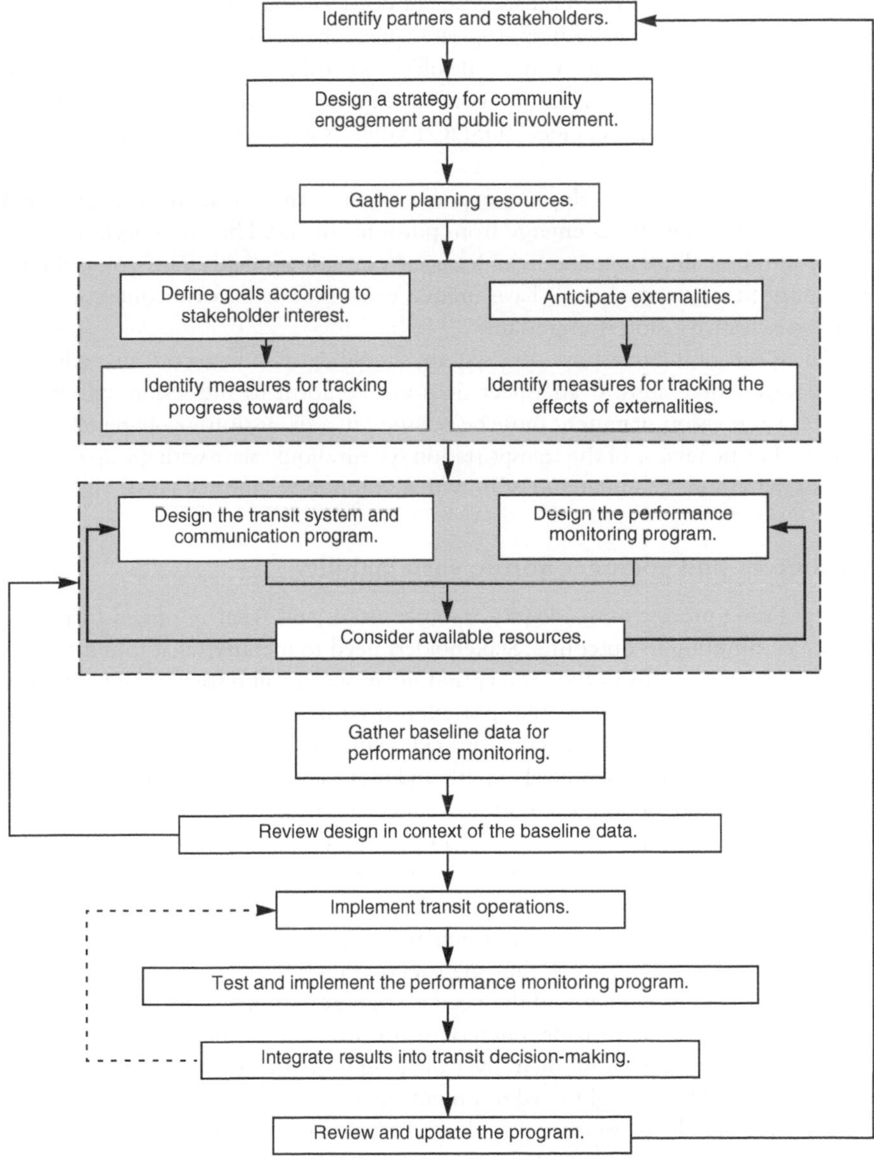

Figure 21.1 Structure of the planning process emphasizing contributions from stakeholders (source: ITE Transportation Planning Handbook, 2009).

Focusing specifically on how to engage stakeholders in such a transportation-planning process, stakeholders fundamentally need to be involved from the start. The community must have its say before the vision is set. Resistance arises when people feel plans are thrust on them, but if all stakeholders contribute from the start, they buy into the idea as it develops.

Experience in the United States has led to general guidance for stakeholder-driven planning. The Texas Transportation Institute and Cambridge Systematics (2009) recommend to 'build on success, but don't be afraid to change' and 'match issues and opportunities with appropriate approaches' (2009). The National Park Service emphasizes the need to be flexible and make the transportation system serve park needs as well as community needs 'in order to leverage existing resources and gain community acceptance' (USDOT Volpe National Transportation Systems Center, 2003).

Stakeholders must develop an explicit vision with well-identified goals and objectives as strong efforts emerge from pursuing an end. The end should be clear in the minds of all participants, which means stating it explicitly. This vision should recognize that protected lands have unique ecological or historic contexts, thus a context-sensitive vision is important.

All aspects of the transportation systems should develop in accordance with the stated vision or visions. If an aspect does not conform to the vision, either that aspect or the vision statement must be revisited in a discussion involving all stakeholders. Future review of the transportation system should start with group's awareness of the vision statement and of how that vision serves mutual needs and goals.

Resources and socioeconomic sustainability

The planning process must identify resource needs and what resources to pursue before committing to outcomes. Stakeholders need to identify what foundation of resources already exists in the locality and situation, and all resources and expertise have to be maximized (Texas Transportation Institute and Cambridge Systematics, 2009). Identification of resources requires knowing the limits of internal knowledge and expertise, such that technical experts and professionals can be sought (USDOT Volpe National Transportation Systems Center, 2003). If financial resources cannot be identified to pay for the planned transportation system for the years to come, participants in the planning process should revisit system design to develop a financially feasible program.

As stated, gateway communities frequently depend on tourism from protected lands for their local economies, thus the land unit itself is the major economic generator, as well as the generator of the need for transportation systems. It often falls to public-land agencies to provide much of the finance needed for initiating transportation systems and operating them over the long term. In some cases, management of the protected lands might need to provide financial incentives or other resources to encourage public participation in the transportation-planning process (USDOT Volpe National Transportation Systems Center, 2003).

Recognizing the many diverse demands for a park's finance, managers should be cautioned not to pay exorbitantly for elegant transportation systems if these systems are not financially sustainable over time. National parks in the United States have encountered difficulties after developing systems with high costs for labour or facilities and finding those costs had to be cut. In such cases, local workers accustomed to high wages find pay cuts difficult, and parks might have financial difficulty fulfilling contracts for services or facilities (Dunning, 2005a).

From the community's standpoint, leaders should hold a strong sense of community assets, capacities and abilities. A New Zealand study found rural communities mostly limit their prospects through apathy and lack of self-esteem (Matthews, 2001), but gateway communities have power. In fact, gateway communities participating in planning transportation systems have developed a record of contributing important support. For example, international clothier L.L. Bean, which is headquartered near Acadia National Park, donated US $2 million to Acadia's Island Explorer shuttle system because the business's headquarters wanted to support the local area. The Zion National Park shuttle system fulfilled its parking needs based on local and state concessions. Cape Cod National Seashore near Boston, Massachusetts, generated capital to buy buses while surrounding communities provided a revenue stream for ongoing operations (Dunning, 2005a). A sense of assets, capacities and capabilities gives communities a strong stance for interacting and negotiating with public-land managers (Matthews, 2001).

Performance monitoring and demonstration

All results from transportation initiatives affecting the public belong in public discussion. These results might be positive, negative, or neutral, and they might relate to a range of subjects, such as system performance, visitor use, ecological impact, financial expenditures, and others. System managers must realize that while local stakeholders might not hold formal knowledge of transportation, they have intelligence to recognize all changes bring good and bad results. These stakeholders merit the belief that they can learn about and objectively evaluate outcomes; they also merit time and effort needed to educate them on how to interpret outcomes.

Most local stakeholders will need training on how to evaluate transportation systems although few people will immediately recognize this need. Most people think they understand transportation: everyone engages local transportation systems every day and has a sense of how well things work from day to day. Despite this base level of familiarity, accurate assessment requires data and a trained eye. People experiencing traffic congestion might not realize when planning has averted a more difficult situation. For instance, Acadia National Park substantially expanded its bus system in 1999. When asked, local leaders declared no improvement in local traffic even though they considered the new bus service successful at attracting passengers. Data both agreed and disagreed with local perception. Roads along a town's highest-demand transit route experienced 1.5 per cent growth in traffic from 1996 to 2001; however, the island that houses Acadia and its gateway and inholding communities saw traffic increase by 18.5–20 per cent on major access routes. Local perception saw difficult traffic in town, but analysis was necessary to juxtapose that bus service helped avert an additional 18.5 per cent increase in traffic in the same space (Dunning, 2005a). Ongoing analysis and education on transportation system impacts can help guide system improvements and maintain stakeholder support.

Beyond transportation impacts, the planning process itself should be evaluated objectively with measures such as those that appear in Table 21.2. Frequently, local participants will declare positive interactions even when better communication techniques should be employed. Especially in small established communities, people frequently hesitate to make formal complaints against people they will continue to encounter. Instead, an honest and non-threatening assessment process must be encouraged. Where local leaders might report congenial relationships, an objective assessment might reveal one-way communication with little opportunity for all stakeholders to participate and contribute fully (Dunning, 2005a). Table 21.2 offers ideas of variables to monitor in community interaction either annually or every five years. Some of the data can come from existing resources, such as meeting rosters indicating who is participating and what interests they represent. Surveys of stakeholder participants can capture perceptions of the quality of interaction, while an external evaluator can help guide evaluation and establishment of standards for future interaction.

Table 21.2 Measures to monitor collaboration among participants in stakeholder consultations.

	Measures	*Reporting*
Input	• Participants (agency count or person count) • Number of interests (count by classification) • Meetings (frequency or count) • Dedicated staff members (count) • Communication policies (description or funding) • Incentives for businesses and organizations supporting transit (description or value)	Annual
Output	• Perception of how well viewpoints are heard (survey scale 1–5) • Perception of progress for goals (survey scale 1–5) • Perception of value of the partnering process (survey scale 1–5)	Quinquennial
Outcome	• Businesses and organizations supporting transit (count and percentage of potential) • Businesses and organizations advertising transit (count and percentage of potential) • Businesses and organizations funding transit (count, percentage of potential, and funding) • Community-generated contributions for transit (source count and amount of funding)	Annual

Source: adapted from Dunning (2005a)

Conclusion

Managing protected lands requires interaction with stakeholders in gateway and inholding communities. Transportation systems require greater concerted involvement due to the nature of mobility decisions affecting all aspects of tourists' visits and the likelihood of systems crossing borders into communities. Land managers and local leaders need to recognize who holds a stake in transportation systems and what motivations they face. This knowledge can lead to transportation visioning and customized systems that serve multiple purposes. The extra effort required for a stakeholder-driven planning process can pay off with local support and encouragement of sound transportation choices among locals and visitors. A robust collaborative process emerges from strong communication, demonstrated mutual respect and openness to teaching and learning from all stakeholders.

References

Aguirre, G. J. A. (2006) 'Linking national parks with its gateway communities for tourism development in Central America: Nindiri, Nicaragua; Bagazit, Costa Rica; and, Portobelo, Panama', *Pasos: Revista de Turismo y Patrimonio Cultural*, vol 4, no 3, pp. 351–371.

APTA (2007) 'A Profile of Public Transportation Passenger Demographics and Travel Characteristics Reported in On-Board Surveys', American Public Transportation Association, Washington, DC.

Davis, J. S. and Morais, D. B. (2004) 'Factions and enclaves: small towns and socially unsustainable tourism development', *Journal of Travel Research*, vol 43, no 3, pp. 3–10.

Dunning, A. (2005a) Transit in Parks: Impacts and Guidance, PhD thesis, Georgia Institute of Technology, United States.

Dunning, A. (2005b) 'Impacts of transit in national parks and gateway communities', *Transportation Research Record*, vol 1931, pp. 129–136.

EC (2013) 'The Economic Benefits of the Natura 2000 Network: Synthesis Report', European Commission, http://ec.europa.eu/environment/nature/natura2000/financing/docs/ENV-12-018_LR_Final1.pdf, accessed 10 September 2014.

Fredman, P. and Yuan, M. (2011) 'Primary economic impacts at three spatial levels: The case of Fulufjället National Park, Sweden', *Scandinavian Journal of Hospitality and Tourism*, volume 11, supplement 1, pp. 74–86.

Gomez, V. L. (2005) Tourism and Preservation in Gateway Communities: A Case Study of the Towns Surrounding Mesa Verde National Park, Master's thesis, University of Pennsylvania, United States.

Hardy, M. H., Larkin, J. J., Wunderlich, K. E. and Nedzesky, A. J. (2007) 'Estimating user costs and economic impacts of roadway construction in six federal lands projects', *Transportation Research Record*, vol 1997, pp. 48–55.

Headwaters Economics (2014) 'Protected Lands and Economics: A Summary of Research and Careful Analysis on the Economic Impact of Protected Federal Lands', Headwater Economics, Bozeman, MT, http://headwaterseconomics.org/wphw/wp-content/uploads/Protected_Lands_Economics.pdf, accessed 10 September 2014.

ITE (2015) *Transportation Planning Handbook*, 4th Edition, Institute of Transportation Engineers, Washington, DC.

Kurtz, R. S. (2010) 'Public lands policy and economic trends in gateway communities', *Review of Policy Research*, vol 27, no 1, pp. 77–88.

Land Trust Alliance (2011) 'The Economic and Tax-Base Benefits of Land Conservation: What the Research Shows on How Conserving Land Protects the Bottom Line', Land Trust Alliance Fact Sheet, www.landtrustalliance.org/conservation/documents/the-economic-and-tax-base-benefits-of-land-conservation, accessed 6 June 2014.

Lincoln, C. A. (2011) Exploring Links Between Gateway Communities and National Parks: A Quantitative Assessment, Master's thesis, Oklahoma State University, United States.

Matthews, L. (2001) 'Reviving Rural Communities: One small community's first steps to a better future', a Kellogg Rural Leadership Report, http://researcharchive.lincoln.ac.nz/bitstream/10182/5943/1/matthews_2001.pdf, accessed 20 May 2014.

Meyer, M. D., Campbell, S., Leach, D. and Coogan, M. (2005) 'Collaboration: Key to success in transportation', *Transportation Research Record*, vol 1924, pp. 153–162.

More, T. A., Urdaneta, B. and Stevens, T. H. (2008) 'Shifting national park policies and local people: A case study of Acadia National Park', *Journal of Park and Recreation Administration*, vol 26, no 4, pp. 105–125.

Rhindress, M., Lynch, F., Bregman, S., Reichman, R. E., Coopersmith, N. J. and Dunning, J. A. (2008) 'Understanding How to Motivate Communities to Support and Ride Public Transportation', Transit Cooperative Research Program Report 122, Transportation Research Board, Washington, DC.

Riley, C. J. A. (2009) Contributing and Constraining Factors to Collaborative Land Use Planning: Consequences of Proposed Housing Development 'in and around' the New River Gorge National River, Master's thesis, West Virginia University, United States.

Oliver, N. D. (2011) Planning Growth – Preserving Character, Master's thesis, University of Tennessee, United States.

Texas Transportation Institute and Cambridge Systematics (2009) 'Innovative Transportation Planning Partnerships to Enhance National Parks and Gateway Communities', a report prepared as part of NCHRP project 08–36 of the National Cooperative Highway Research Program, Transportation Research Board, Washington, DC.

USDOT Volpe National Transportation Systems Center (2003) 'Partnering for Success: Techniques for Working with Partners to Plan for Alternative Transportation in National Park Service Units', National Park Service, Alternative Transportation Program, Washington, DC, http://ntl.bts.gov/lib/48000/48200/48216/NPS_Partnering_for_Success_2003.pdf, accessed 12 March 2014.

Zimmerman, C., Daigle, J. and Pol, J. (2004) 'Tourism business and intelligent transportation systems: Acadia National Park, Maine', *Transportation Research Record*, vol 1895, pp. 182–187.

Zonneveld, R. (2011) 'Lost in transitions: staging global tourism in local small towns', Doctoral dissertation submitted to Lincoln University, New Zealand.

Part V

Conclusion

22 A glimpse into future research on sustainable transportation in natural settings

Francesco Orsi

The United Nations World Tourism Organization tells us that international tourist arrivals in 2013 have been 1,087 million (UNWTO, 2014) and that they are expected to increase at a 3.3 per cent rate until 2030 when they will reach 1.8 billion (UNWTO, 2011). While no such data are available for domestic tourism at the global scale, it is estimated that domestic tourism trips outnumber international trips by more than five times to one (Peeters and Dubois, 2010). Given these figures and the growing interest in nature-based tourism, natural areas worldwide are expected to see dramatically increasing visitor flows in the years to come. This poses enormous challenges about how to minimize the negative impacts of massive visitation on natural resources and visitors' recreational experience. Transportation will increasingly play a key role in addressing such challenges as both a source of disturbance to be limited and a tool to manage visitor flows more efficiently. While this book has provided a wide overview of the current knowledge and practical experience of sustainable transportation in natural settings, the field is relatively young and much is yet to be explored. The increased ability to elicit visitors' preferences, the advent of new transportation and communication technologies, and the rise of new countries as popular tourist destinations offer wide room for innovative research. In particular, investigation on the following four major areas unveils promising opportunities to further reduce the negative impact of transportation on natural resources and recreational experiences, and to take greater advantage of transportation's potential as a management tool in natural contexts:

- sustainable transportation to natural areas;
- sustainable transportation through ICT and ITS;
- sustainable transportation in developing countries;
- sustainable transportation and ecosystem services.

Sustainable transportation to natural areas

The great emphasis placed on making transportation sustainable within natural areas is rarely associated with serious discussions about ensuring sustainable transportation to natural areas. This gap is reflected in the many natural areas that, while offering highly efficient alternative transportation systems within their boundaries, leave visitors no option but the private vehicle for being reached from the closest national transportation hub. The consequences of this paradox in terms of sustainability are rather obvious. On the one hand, visitors, being forced to use their vehicle to reach a natural area, will also be encouraged to use it for moving within the area. On the other hand, people without a private vehicle will be given no (or a very remote) possibility to visit the area. Consistent with what said in Chapter 2, however, the sustainability of a transportation system should be assessed considering its large-scale and long-term effects. In this respect, an alternative transportation system that guarantees mobility within a natural area but offers no connection with cities and the larger transportation network generates unsustainable patterns.

Research is needed to better understand the environmental and socio-economic effects of transportation to/from natural areas and the conditions under which such transportation may actually benefit natural resources and human communities. The topic is of particular importance for natural contexts where people live and work: here sustainable transportation should not simply mean the minimization of environmental impacts, but the concrete possibility for local communities to move easily between villages as well as between villages and cities. This line of research, which should necessarily look at the region instead of the single natural area, could explore issues pertaining to regional transportation networks and policies. Being informed by detailed information on preferences and needs of both visitors and residents, the research should aim to design (or improve) transportation networks that favour accessibility via public transit for visitors and residents, and to define policies and public-private partnerships that ensure the economic sustainability of transportation services. This whole field of research could greatly benefit from the integration of outdoor recreation management and transportation studies in peri-urban contexts.

Sustainable transportation through ICT and ITS

Transportation management has taken systematic advantage of Information and Communication Technology (ICT) infrastructures for the collection and communication of data about the flow of people and vehicles. Today traditional data collection performed by operators through sensors or cameras also relies on a widespread diffusion of personal devices (e.g. smartphones) providing detailed information about the position of users in space and time. Moreover, these same devices enable users to get real-time information about transport conditions from both the transportation authority and other users. This possibility of multiple way communication is at the core of so-called intelligent transportation systems (ITS), which are contributing to the solution of various transportation issues such as traffic congestion or excessive wait times at bus stops. While Chapter 5 has shown the potential of ITS for increasing the attractiveness of alternative transportation in protected areas, much still has to be done in this field.

Research may improve our knowledge about how ITS can contribute to the sustainability of an overall transportation system in natural settings. In other words, we should learn to design and use ITS to assure not only that ridership of alternative transportation is maximized, but also that visitor flows are balanced both spatially and temporally, all visitor categories are given adequate transport opportunities and the quality of life of local communities is enhanced. This can be achieved by focusing on three fields of investigation: technology, motivations and outcomes. Research on technology is aimed at designing and testing devices (e.g. monitors at bus stops) and computer programs (e.g. smartphone applications) that may provide visitors and residents with real-time information about public transit, traffic conditions, crowding at popular sites, etc. Research on motivations is aimed at understanding visitors' and residents' attitudes towards the availability of such technologies and related information, and at estimating the impacts of these on an individual's behaviour. Research on outcomes is aimed at predicting the overall effects of the adoption of the above-mentioned technologies given the previously analysed attitudes. This part can rely on complex simulation models to detect all the possible consequences (e.g. diversion of visitor flows, increasing or decreasing queues at given times) of an ITS and eventually get to an optimal design of the overall transportation system.

Sustainable transportation in developing countries

As emphasized by this book, the design and adoption of sustainable transportation systems in natural and protected areas are still a distinctive trait of affluent countries with well-established tourism industries. Developing countries generally lack the economic resources, tourism infrastructures and governance apparatus that make the implementation of such systems actually possible. Nevertheless, the attraction of exotic destinations, the progressive expansion of international travel ranges and the rise, in many developing countries, of a new middle class that can engage in tourism activities will increase tourism-related issues in natural areas of developing countries and call for sustainable transportation measures. The study of transportation for nature-based tourism in developing countries and the analysis of its environmental and socio-economic implications open an entirely new field of research, which is key to protecting some of the most important ecosystems in the world and safeguarding local communities' livelihoods.

Research is needed to develop transportation systems that, while enabling access to natural sites, preserve natural resources and sustain local development. Special attention should be paid to three major topics: appropriate technology, design and management, and funding schemes. Lack of technological development and skills is in fact among the major limitations of poor and emerging countries, and calls for the adoption of transportation systems that offer a good compromise between technological simplicity, reliability and adaptability. Research should aim to achieve this compromise through identification of practical solutions and the improvement of existing transportation infrastructures. Research on design and management should be aimed at defining routes and schedules that deliver visitors not just where the greatest natural attractions are, but also where the same visitors can bring the greatest benefits to the local population (e.g. bus stops and harbours near villages). Finally, research should also be devoted to answering a crucial question, namely who pays for such transportation systems? The applicability and effectiveness of various funding schemes, including national park permits or extra fees on accommodation, could be investigated. Clearly, research should account for the specificity of the single study area and solutions should be tailored on local socio-economic conditions and the types of visitor (e.g. mostly national, mostly international).

Sustainable transportation and ecosystem services

There is a common understanding today that sustainable development ultimately depends on our ability to protect ecosystems and the services they provide. In fact, ecosystem services, which are commonly classified in supporting (e.g. nutrient cycling), provisioning (e.g. fresh water), regulating (e.g. flood control) and cultural (e.g. aesthetic value) services (MEA, 2005), sustain the well being of people, including the production of our basic living needs. Transportation is inextricably linked to ecosystem services because these can guarantee the safety of transportation systems by reducing natural hazards, while transportation systems can alter the ability of ecosystems to provide services to human communities. For example, protecting or restoring vegetation upstream of a road reduces the probability of landslides that may damage the road, and a carefully designed road safeguards the vegetation that may help maintain clean water for local communities. A sound planning, including route design and traffic control, along with direct actions on ecosystems, can minimize both the hazards to which the road is exposed and the threats brought by the road to the local population.

Research is needed to better understand how to account for ecosystem services in transportation planning in natural and protected areas so that human well-being is enhanced. Research efforts can be directed towards both areas of new development, where infrastructures have to be built from scratch, and already developed areas, where the modification of existing infrastructures and a renovated management approach might substantially contribute to the sustainability of transportation systems. This line of research could focus on two main topics: the analysis of linkages between transportation and ecosystem services, and integrated design. Research on transportation-ecosystem services linkages is aimed at estimating the consequences of ecosystem management on transportation (e.g. effects of vegetation on road safety) and the impacts of specific transportation-related measures, such as the construction of a road or the management of car traffic, on the provision of ecosystem services. This analysis should embrace the whole complexity of interactions between transportation and ecosystem services, considering, for example, large-scale land use modifications as a consequence of transportation-related measures (e.g. new settlements developing around a new road). Research on integrated design builds on the analysis of transportation-ecosystem services linkages and is aimed at defining methods to plan transportation-related interventions that comprehensively benefit from and minimize negative impacts on ecosystem services.

The four areas described above, together with the knowledge and experiences presented in this book, depict a research path that can let us achieve a greater understanding of the implications of transportation in natural settings and the strategies to handle them properly. The good news today is that sustainable transportation is no longer a fancy term used by some visionary environmentalist, but rather the expectation of many people visiting natural and protected areas worldwide. This gives us moral support and patronage in pursuing forms of mobility that are more respectful of the environment and society. Transportation is a fundamental means to experiencing natural areas as well as a powerful tool for their management. The protection of our natural wonders for the generations to come ultimately depends on how wisely we can use it.

References

MEA (2005) *Ecosystems and Human Well-being: the Assessment Series*, Millennium Ecosystem Assessment, Island Press, Washington, DC.

Peeters, P., Dubois, G. (2010) 'Tourism travel under climate change mitigation constraints', *Journal of Transport Geography*, vol 18, no 3, pp. 447–457.

UNWTO (2011) *Tourism Towards 2030: Global Overview*, United Nations World Tourism Organization, Madrid, Spain.

UNWTO (2014) *Annual Report 2013*, United Nations World Tourism Organization, Madrid, Spain.

Index

4WD vehicles 88–90
Acadia National Park 59, 67, 70, 168, 175, 215; facilities for cycling and walking at 251–58
accessibility 16, 151, 153, 157
access restrictions: for disabled people 83–5; entrance fees 54, 176, 194, 256; lottery 268; parking fees 100, 124, 204, 231; permits 204, 280; quotas 39, 212; road closures 59, 150, 162; road toll/pricing 12, 123, 150, 162
active tourism 141; *see also* slow travel
Adirondack State Park 85
aerial tramways 36; *see also* cableways
aircrafts 88, 198
air pollution 3, 16, 71, 234
Ajzen, I. 135
Ajzen's theory of planned behaviour 131–37
all-terrain-vehicles 85–6
Alpe di Siusi initiative: description of 102; effects of 106–09
AlpenTaxi: main characteristics of 119–20; operation costs of 119; success factors of 122, 209
Alpine Pearls project 3, 101–2
Alternative Transportation Program (ATP) 3, 170–72
alternative transportation systems (ATS); data collection about 172–76; incentives to 151, 163, 208, 247, 257; inventory of 172–76; long-term benefits of 178; as a means to controlling visitor flows and use levels 3; as a means to supporting parks 168; *see also* carrot and stick measures
Americans with Disabilities Act (1990) 84
Areas of Outstanding Natural Beauty 47
atmospheric pollution 18, 28, 30–2, 36; *see also* air pollution
attitudes: assessment of 209; to car use 130–37; statements 128; towards behavioural change 129
automatic vehicle locator (AVL) 224
automobiles *see* cars
Autonomous Region of the Azores (ARA) 182; *see also* Azores Islands
Azores Islands 182–85

Bayerischer Wald National Park 71, 233
best available technology 30, 31, 36, 39
Bhutan: 2020 strategy 201; as a big national park 194; as a car-free country 193; as a roadless country 194; history of trail-to-road transition in 194–200; modes of transport used in 201; multiple routes with trail as levee in 204–05, 212; royalty to enter 194; single route with trailhead as dam in 204, 212
bicycle paths *see* cycling paths and trails
bicycles: freedom of movement offered by 38; management of visitor flows 38; parking facilities for 252–54; rental 252–53; sharing 251, 253, 257; support transportation 38
Black Forest National Park 235
boats and ferries 2, 31, 35–6
Boundary Waters Canoe Area 84
Brecon Beacons National Park 50
Bus Alpin: main characteristics of 117–19; number of passengers on 120; operation costs of 119; success factors of 121, 209
buses: ability to do linear walks with 45, 51, 54; alternatives to 48–9; atmospheric and acoustic emissions of 30, 33; benefits for residents 46; capacity of 30, 33; carbon emissions of 30, 33; contribution to traffic congestion 30, 33; double decker 46; double-length 73; drivers of 50–1, 54, 57; economies of scale 47; evaluation of

attributes of 49–50; fares 30, 33, 46–7, 51, 54; frequency of 49–54, 156, 163; hub and spoke design 50; interior design of 30, 33, 39; level of satisfaction with 49; as the most common form of alternative mobility 33; offering interpretation 45–6, 54; as an opportunity for contact with local people 45; propane-powered 73; proper technology for 30, 33; reasons for using 47–8; ridership of 163–4; routes 30, 33; schedules 30, 33; space onboard 30, 33; stops 30, 33, 243–5, ticketing 51; as a travel opportunity for people without a car 45; as a way of keeping money in the local economy 47; as a way of minimising transport-related impacts 45; ways of knowing about 48; see also alternative transportation systems (ATS)
businesses 12, 14, 22, 30, 35, 73, 75, 77

cableways 31, 36–7; attributes of 153; in the Dolomites 151; risks associated with overly attractive 164
car-free vacations 125
carrot and stick measures 101, 150–1, 229, 239, 257; balance of 164; design of 164
carrying capacity 20, 22, 223–4
cars: as a changing agent in national parks 57; disincentives on 100, 150, 163, 208, 229; flexibility offered by 45; ownership of 124, 141, 228; as a threat to tourism 127; as a threat to wildlife and recreational experience 249; as a way to experience parks 57–8
carbon dioxide 2, 127, 129, 234
chairlifts 36–7, 151; *see also* cableways
Christomannos, T. 1
corridor mentality of cyclists 145
cost-benefit analysis 129
Cradle Mountain National Park 28
crowding 1, 2, 20, 33, 151, 153, 163–5; bicycles and 32, 38; boats and 31, 35–6; buses and 30, 33; cableways and 31, 36–7; trains and 30, 34
cultural change 210
cycle holidays 140; *see also* cycle tourism
cycle touring 140–1; *see* cycle tourism
cycle tourism: benefits of 144; conditions for supporting 144–7; as a conveyor of local economic development 38; demand for 141, 144; as a rapidly growing tourism sector 144; *see also* cycling
Cycle Tourism Club 140
cycling: as an access mode 140, 142; as a niche tourism activity 140; incentives to 257–9; infrastructures and facilities for 252–7; interest in 249, 251, 256; promotion of 258–9; *see also* cycle tourism
cycling paths and trails: benefits to communities near 145; network 250, 252, 256–7; connectivity between trails, transportation hubs and 254–7

day cycling 140; *see also* cycle tourism
day trippers 52, 130, 231
De Hoge Veluwe National Park 252–3
Denali National Park and Preserve 7, 58, 71, 219–23
destination management organisations 234
developing countries 279–80
disabled people: expectations in the way of access 85–6; importance of outdoor experiences for 85; as a potential tourism market 83; travel experiences of 83–4
Doig, D. 194, 196–7, 205
Dolomites 1, 101, 111; Catinaccio-Rosengarten range in the 151–2; visitor preferences towards access modes in the 156–65
dual mission of national parks and protected areas 16, 36, 84, 168, 249
dynamic message signs (DMS) 60–61; visitor response to 66–7; *see also* Intelligent Transportation Systems (ITS)

ecosystem services 280–1
Eifel National Park 232
electronic message signs (EMS) 59; *see also* Intelligent Transportation Systems (ITS)
English National Park Authorities 127
Environment Act (1995) 142
European Outermost Regions (EOR) 182
European Union 3, 83, 92, 181–2

ferries *see* boats and ferries
Fiordland National Park 82, 92
Forest of Bowland 47
funiculars 36; *see also* cableways

Gandhi, I. 196
gateway communities; characteristics of 262–4; definition of 262; dependence on tourism 263; monitoring stakeholder consultations in 272; motivations to engage in transport planning 267–8; perception of mandatory shuttle

buses 74–9; as portals 72; relationship between parks and 70, 72, 79, 261, 266–8; requirements 271; role in supporting park goals 71; socioeconomic sustainability of involving 270–1; stakeholder engagement in 264–6; transportation planning in conjunction with 261; visitor spending in 72; well-being of 79
Glacier National Park 215, 249
global warming 18
Golden Gate National Recreational Area 59
gondola 36, 82, 88, 92, 194; *see also* cableways
Grampians National Park 28
Grand Canyon National Park 59, 71, 174, 176, 250–1, 253
Great Smoky Mountains National Park 58
greenhouse gases 35, 99, 249
Green Parks Plan (GPP) 250
greenways 141
Große Dolomitenstrasse 1

habitat degradation and fragmentation 2, 19, 168
harbours 2, 19, 21, 35, 258
highway advisory radio (HAR) 59; satisfaction with 62; visitor evaluation of 62–3; visitor response to 66–7; *see also* Intelligent Transportation System (ITS)
hiking 151
holiday cycling 140–1; *see also* cycle tourism
human health 16, 18, 20, 87

Iguazu National Park 28
impacts of visitation; on the environment 19–20; on the recreational experience 20–1
indicators and standards of quality 24, 67, 169, 216, 220, 223–4
indigenous communities 193
Intelligent Transportation Systems (ITS): applications of 59–60; influence on travel mode choices 61–7; providing real-time information 59, 67, 168, 278; usefulness for visitors and residents 61–67; visitor response to 66–7; visitor willingness to use 59–63; as a way to increase awareness and use of shuttle services 64, 66–7
Intermodal Surface Transportation Efficiency Act (ISTEA) (1991) 170–1
International Union for Conservation of Nature (IUCN) 193

Jigme Dorji National Park 194, 198–200; *see also* Bhutan

Kasteel de Haar 252, 257
Kinderdijk-Elshout 252–3, 257

Lake District National Park 5, 47; attitudes towards car use in the 132–7; characteristics of the129–30
land use conversion 11, 12, 19
last mile 32, 111, 115, 209, 231
legislation: Americans with Disabilities Act (1990) 84; Environment Act (1995) 142; Intermodal Surface Transportation Efficiency Act (ISTEA) (1991) 170–1; Moving Ahead for Progress in the 21st Century Act (MAP-21) (2012) 169–71; Safe Accountable Flexible Efficient Transportation Equity Act: A Legacy for Users (SAFETEA-LU) (2005–2009) 170; Transportation Equity Act for the 21st Century (TEA-21) (1998) 170–1; United Nations Convention on the Rights of Persons with Disabilities (2006) 84; US Wilderness Act (1964) 193
leisure travel 46, 142
life cycle assessment 129
lifestyle groups 230; *see also* visitor groups
linear walks 45–6 51, 54, 230
local communities 16, 22–3, 251, 265, 278–9; *see also* gateway communities
Loch Lomond and the Trossachs National Park 47
Looe Valley Branch Railway Line 128

management objectives 24, 172, 216–7, 224
Management Plan (MP) 182; budgeting and scheduling of the 187; communication and promotion 188; logical framework of 187; supervision, monitoring and revision of 188
management strategies; acquiring data for designing 100; multiple routes with trail as levee 204–5, 212; simulation of the effects of 156, 159–62; single route with trailhead as dam 204, 212
mandatory shuttle buses; economic assessment of 75–8; effects on local businesses 75–8; effects on parking and traffic 75–9; perception of business owners and employees on 75–9; visitor assessment of 74–5; *see also* shuttle buses

marketing and information 108, 118, 122–3, 175, 231–3, 242, 247; about bus services 47, 51–2
market segmentation 128–9, 131, 231
Matsuo, M. 194, 196–7
mobile devices 278–9; *see also* Intelligent Transportation Systems (ITS)
mobility planning 182
mini metro 36; *see also* cableways
modal share 6, 115
modal shift 100, 116, 150, 163; obstacles to 28
modal split 164
mode choice 61–5, 152–60; simulation of 156, 159–62
modes of transport: in Bhutan 201; bicycles 32, 37–8; boats and ferries 31, 35–6; buses 30, 32–3; cableways 31, 36–7; private motorised 150, 228, 230; soft 189; trains 30, 34–5; in US national parks 172–6; visitor preferences towards different 152–60
monorails 92, 194, 201–2
motorbikes 21, 89, 102; *see also* private motorised vehicles
motorised access 84–5; approval by impaired and not impaired people 88–91; impact on natural and experiential factors 88–90
Mountain Wilderness 119–20, 123
Mount Rainier National Park 57
Moving Ahead for Progress in the 21st Century Act (MAP-21) (2012) 169–71
multi-modal travel 142

Nakao, S. 194, 196, 198–200
national forest 59, 193, 262
National Health and Nutrition Examination Survey (NHANES) 87–8
National Park Achievement Award 79
national parks: aesthetics changed by cars 57; cars as a way to experience 57; cycle tourism in 142; driving for pleasure in 215; dual mission of 84, 168, 249; history of transportation in 57–9, 215–6, 249–50; isolated communities in 194; motorised access to 84–5; planning sustainable transportation in 169; public access to 224; railway access to 57, 216; road network in 170; special qualities of 127; traffic congestion in 168–70; transportation as a management tool in 224; transportation as the primary means of experiencing 224; use of ITS in 59–60
National Park Service (NPS): transit inventory 169, 172–6; transit systems by business model 173; transit systems by mode 172–3; transit systems by funding source 175–6; transit systems by passenger boardings 174–5
natural areas: disabled people's desired access to 83–92; modes of transport in 30–38; obstacles to implementing effective public transportation in 28, 239–41; quietness of 20; trailheads as ways of access into 156–65; transportation as a way to experience 20
Natural England 129
nature-based tourism 1, 84, 212, 215, 229, 277, 279
Nehru, P.J. 196
New Ecological Paradigm 129
New Forest National Park 47
noise 1, 3, 18–9, 22–4; from boats and ferries 31, 35; from buses 30, 33; from cableways 31; of car traffic 46, 71, 73, 115, 216, 239, 249; from trains 30, 34
non-captive visitors 228
North Cascades National Park 57
North Norfolk 47
Northumberland 47
North York Moors National Park 50

Olympic National Park 57
on-demand transportation systems 115–125; principles supporting the implementation of 123–4

park-and-ride shuttles 60–66; *see also* shuttle buses
parking facilities 127, 170, 178, 190, 216, 243, 250
participatory approaches 212; for developing a sustainable mobility plan 185–91; to territorial planning 100
particulate 2, 18, 24, 73, 108
passengers of choice 229
Paul S. Sarbanes Transit in Parks program 169, 170, 176
Peak District National Park 2, 147–9, 230; current cycling opportunities in the 144–5; description of the 142; future cycling opportunities in the 145–7
Pedal Peak Phase II – Moving Up A Gear 147
pedestrian circulation plan 189
policy 6–7, 181, 188, 201, 209, 219, 257
politicians 1, 100, 234–5

portable changeable message signs (PCMS) 59; *see also* Intelligent Transportation Systems (ITS)
private motorised vehicles 150, 209, 228; *see also* cars
protected areas: as the cornerstone of nature-based tourism 2; trailheads as ways of access into 156–65; *see also* national parks
public-private partnerships 23, 278
public transport 3, 21, 23; access to 115; attractiveness of 230; catchment area of 231; conditions for attracting leisure demand in 242–3; financial support to 239; hubs 229, 231; maintenance of 239–40; marketing issues related to 231–3; passes for 101; stakeholder role towards the establishment of 233–5; tariff systems for 246–7; tourists as a chance to increase the demand of 241–2
push and pull measures 100; *see also* carrot and stick measures

Queen Kesang Choedon Wangchuck 196

railroads *see* railways
railways 2, 19, 21, 143; in conjunction with bicycles 140, decommissioned 141, 144; as the first access to national parks 57, 216, 249; as silent infrastructures 34; in Switzerland 115; *see also* trains
real-time information 59, 67, 168, 278–9; *see also* Intelligent Transportation Systems (ITS)
recreational experience 2, 3, 16, 28, 111, 123–4, 127, 193, 249, 277; bicycles and 32, 38; boats/ferries and 31, 35–6; buses and 30, 33; cableways and 31, 37; effects of shuttle use on 58; impacts of visitation on 20; and noise 18; trains and 30, 34
roadkill 19
roadless areas 194
road restrictions *see* access restrictions
roads: approval for the development of 88–9; banned to private vehicles 118; capacity of 219–21; car-free 252; construction in Bhutan 197–200; costs associated with the maintenance of 70; design of 57–8; effect on wildlife 19, 222–3; for driving for pleasure 215; iconic 215; network in national parks 212; as the primary draws of parks 251; private 118–9; public 145
ropeways 201, 205; *see also* cableways
runoff 2, 19
rural areas 45, 53, 127, 234; basic conditions for the provision of public transport in 228–9; vicious and virtuous circle of public transport in 239–41

Safe Accountable Flexible Efficient Transportation Equity Act: A Legacy for Users (SAFETEA-LU) (2005–2009) 170
San Antonio Missions National Historical Park 251, 253, 256
Saratoga National Historical Park 251, 253
scheduled buses 23, 32, 53; benefits of 46–7
Schweiz Mobil 119, 121–2, 125
Sequoia and Kings Canyon National Park 59, 251, 257
Shiretoko National Park 2
Shuswap Lake 2
shuttle buses 29, 32; attributes of 74–5; business owners' and employees' perception of 75–8; effectiveness of 75; effect of shuttle use on visitor experience 65; evaluation of 63; impacts of 71, 75–7
slow travel 37
Smith, N. 82
snow coaches 2, 173
Snowdonia National Park 47
social desirability bias 134; *see also* social norms
soundscapes 18, 71, 74; *see also* noise
South Tyrol 29; characteristics of 101–2; as a green region 100; tourism-related traffic in 102–9
South West New Zealand World Heritage Area 82
stakeholders: as ambassadors of public transport systems 233; cooperation between 233; -driven planning process 265; empowerment through workshops and other initiatives 266; engagement of 100, 264–6; identification of 185; involvement in mobility planning 185–8; role towards the establishment of successful public transport systems 233–4; workshop 186–7
survey instruments: choice experiment 153–4, 164; normative theory 220; qualitative methods 59, 77, 85, 169, 220; questionnaires 47, 61, 87–8, 132, 154, 156, 164; stated preferences 134, 137, 209, 230

sustainability: assessment 12, 15–6; components 16–7, 24; financial 23, 239; indicators and standards 24; requisites 16–24; spatial and temporal dimensions 12–6
sustainable development 12, 100–1, 183, 190, 280
sustainable mobility 5, 11, 39, 100–3, 110–1, 122, 164, 182, 187
Sustainable Mobility Plan (SMP) 182–3; of Ponta Delgada (SMPPD) 183, 185–191
sustainable transportation systems: atmospheric pollution 18; delivery of visitors 16, 20, 30, 33, 35, 37; design of 20–3; in developing countries 279–80; and ecosystem services 280–1; financial sustainability of 23; freedom of movement 21–2; through ICT and ITS 278–9; impacts on flora and fauna 20; impacts on the recreational experience 20–1; land use conversions 19; to natural areas 277–8; noise levels 18–9; as an opportunity to attract visitors 208; public acceptability of 182; quality of life of the local communities 22–3; visual perception 21–2; see also alternative transportation system (ATS) and transportation
Sustainable Urban Transport Plans (SUTP) 181
Swiss Alpine Club 119
Switzerland 34, 115–25
SWOT analysis 187

Taft, W.H. 73
taxi service 5, 50, 119, 265
territorial-tourist governance 111
time-space compression 13
Todd, B.K. 194–6
tourism: cars as a threat to 127; destinations 117; and gateway communities 263; inbound 102–9; and the local economy 266; mass 117; organisations 233; outbound 102–9; purpose 230; regions 116, 209; sustainable 125, 137; in Switzerland 116; traffic analysis 100, 109; type of 185
tourist flows *see* visitor flows
tourist transport system 99–100
traffic: congestion 3, 29–30, 58, 73, 123, 128, 142, 170, 261, 266; emissions from 127–30; inbound 102–7; non-tourism 103–7; outbound 102–7; -related issues 1, 71; tourism-related 102–7
trails 151, 164
trains; atmospheric and acoustic emissions of 30, 34; capacity of 30, 34; fares 30, 35; proper technology 30, 34; role in the history of protected areas 34, 57, 216, 249
trampling 29, 35
transportation; benefits of 1–2, 215; contribution to human pressure on environment 2; demand 14–6, 20, 22, 30, 36; demand-responsive 229 (*see also* on-demand); everyday 228–31, 233, 236; impacts at the destination 13–6; impacts of 2–3, 18–23, 58, 73, 123, 128, 142, 170, 261, 266; impacts on flora and fauna 19–20; impacts on the recreational experience 20; indirect effects of 12; infrastructure 2, 19, 21–2, 29–35, 39, 57, 99, 141, 170, 252–3; integrated 22, 38, 101, 115, 171–2, 188; land use conversions 19; leisure 120, 229, 236; management 151; as a management tool 24, 211, 224; as a means to convey people 20, 34; multi-modal 142, 147, 250, 256–7; network 13–6, 19, 39, 231; noise emissions 18–9; on-demand 115–25; origin-destination 13–6; pollution 18, 127–30; real-time information about 59, 67, 168, 278 (*see also* mobile devices *and* Intelligent Transportation Systems); regional planning 171–2; scales of analysis 13–6; unintended consequences of 6, 39, 164, 210, 218, 257; as a way of experiencing natural areas 1–2, 215; *see also* alternative transportation systems (ATS)
Transportation Equity Act for the 21st Century (TEA-21) (1998) 170–1
Transportation Research Board 172
travel modes *see* modes of transport

United Nations Convention on the Rights of Persons with Disabilities (2006) 84
United Nations World Tourism Organization (UNWTO) 277
US Wilderness Act (1964) 193

Valley Forge National Historical Park 251, 253, 256, 257, 259
Vias Verdes 141; *see also* greenways
visitor behaviour 3, 13, 20–1, 36, 39, 85, 110–1, 128–37, 208; simulation of

156–7, 161–5; willingness to change 128–37, 209
visitor flows 3–4, 23, 35–6, 38, 104–5, 150–1, 164–5, 208, 211–2, 277
visitor groups 4, 20–2, 30–8, 51, 60, 85–93, 100, 116, 121, 128, 133, 159–64; daily spend of 133; LOHAS (Lifestyle of Health and Sustainability) 230; target 123, 230, 233, 236, 241
visual perception 21
visitor preferences 128–37; towards modes of transport 151–2, 153–64
Volpe National Transportation Systems Center of the US Department of Transportation 6, 169, 172, 249

walking: infrastructures 252–3; promotion of 258–9; as a supporter of initiatives for better health 250; as a sustainable and cost-effective activity 249–50
wheelchair 39, 89, 91
White, J.C. 194–6
White Sands National Monument 252, 258–9
Wider Peak District Cycle Strategy 146–7
wilderness 2, 21, 58, 82–93, 133–35, 193, 205, 219, 262; inhabited 194, windshield 198
wildlife 20, 70, 222–4, 249; Dall's sheep 222; grizzly bears 222; refuges 59, 262; roadkill 19
willingness to pay 157
World Health Organization 83
Wulingyuan World Heritage Area 86

Yellowstone National Park 57, 252–3
yield 129, 135–7
Yosemite National Park 57–8, 70, 174, 215, 249

Zion National Park: business owner perception of shuttle services in 75-78; description of 72-3, 252; entrance fee scheme in 258; entrance station to the parking lot in 257-8; shuttle service in 72-4, 174; success of the shuttle service in 174, 256; traffic-related issues in 73; visitor spending in 74